MATHEMATIC
for Mechanical and Production Engineering

HIGHER TECHNICAL EDUCATION COURSES

Series Editor: W. H. Blythe
Director of Engineering, Richmond upon Thames College

This series of books is designed to meet the needs of students studying unit programmes leading to the higher certificate and diploma awards of the Technician Education Council. It is intended to cover the syllabuses of the engineering, construction and science programmes, and to meet the principal behavioural learning objectives of the TEC unit requirements; but the books also cover other closely related syllabuses at this level and, where appropriate, further material is included to achieve this. The texts are designed to assist the lecturer in the process of continuous assessment of student progress, and the student to practise some self-assessment. The authors are all experienced lecturers and the contents are self-contained and based on tried and tested material.

Jenkins: *Structural Mechanics and Analysis Level IV/V*
Morris: *Mathematics for Mechanical and Production Engineering Level IV*
Tyler: *Environmental Science Level IV*
Williams: *Geotechnics Level IV*

MATHEMATICS for Mechanical and Production Engineering Level IV

JACK MORRIS
Senior Lecturer in Mathematics
South East London College

VNR VAN NOSTRAND REINHOLD COMPANY
New York–Cincinnati–Toronto–London–Melbourne

Published by Van Nostrand Reinhold Company Ltd.,
Molly Millars Lane, Wokingham, Berkshire, England

Published in 1981 by Van Nostrand Reinhold Company,
A Division of Litton Educational Publishing Inc.,
135 West 50th Street, New York, NY 10020, USA

Van Nostrand Reinhold Limited,
1410 Birchmount Road, Scarborough, Ontario, M1P 2E7,
Canada

Van Nostrand Reinhold Australia Pty. Limited,
17 Queen Street, Mitcham, Victoria 3132, Australia

Library of Congress Cataloging in Publication Data

Morris, Jack.
Mathematics for mechanical and production
engineering, level IV.

(Higher technical education courses)
Includes index.
1. Engineering mathematics. I. Title. II. Series.
TA330.M6 510′.2462 81-610
ISBN 0-442-30459-5 AACR2
ISBN 0-442-30460-9 (pbk.)

Printed and bound in Great Britain at
The Camelot Press Ltd, Southampton

Preface

This book has been written both as a self-contained Level IV text and also to provide the mathematical background for various engineering programmes of courses validated by the Technician Education Council. All topics have been taken from the Council's Mathematics Bank of Objectives, U78/911, with the additional background material needed by a student with credits in the standard mathematics units.

The book contains rather more topics than could be covered in the time allocated for a full unit. Colleges will doubtless select those areas most suited to the students' programme. The remaining topics will form a useful introduction to higher levels of work.

The topic areas from the Mathematics Bank of Objectives are distributed as follows:

Section A: Algebra
Topic areas BA 17
AB 12, 13
BF 17, 18
BC 16–18
DE 14, 17

Section B: Calculus and its Applications
Topic areas CA 15–19
CB 15, 16, 18
BD 16–18
CC 17, 18

Section C: Differential Equations
Topic areas CD 16–21

Section D: Numerical Methods
Topic areas AB 16–18
BC 18, 19
CA 20
CB 14

Section E: Statistics
Topic areas EB 11–13
EC 14–17, 20, 21
ED 14–17, 19
EE 16
EF 20, 21
EG 16–18

Acknowledgement

This is to acknowledge the considerable help given by Mr. E. J. Williams of Richmond Upon Thames College in correcting my errors and making numerous suggestions to improve the text and layout.

I am also grateful to the Literary Executor of the late Sir Ronald A. Fisher, FRS to Dr. Frank Yates, FRS and to Longman Group Ltd., London, for permission to reprint Tables 22.5 and 22.6 from their book *Statistical Tables for Biological, Agricultural and Medical Research* (6th edition, 1974).

J. M.

Contents

SECTION A
Algebra

PART 1: HYPERBOLIC FUNCTIONS

1.1 The Functions sinh x and cosh x

We introduce two new functions, HYPERBOLIC SINE and HYPERBOLIC COSINE, written sinh x and cosh x respectively. (These are normally pronounced 'shine x' and 'cosh x'.) They are defined in terms of the exponential function:

$$\cosh x = \tfrac{1}{2}(e^x + e^{-x})$$
$$\sinh x = \tfrac{1}{2}(e^x - e^{-x})$$

These have many analogies with the trigonometrical functions cos x and sin x. For example:

$$\begin{aligned} 2\sinh x\cosh x &= 2 \times \tfrac{1}{2}(e^x - e^{-x}) \times \tfrac{1}{2}(e^x + e^{-x}) \\ &= \tfrac{1}{2}(e^{2x} - e^{-2x}) \\ &= \sinh 2x \text{ by definition} \end{aligned}$$

$$\begin{aligned} 2\cosh^2 x - 1 &= 2 \times \tfrac{1}{4}(e^x + e^{-x})^2 - 1 \\ &= \tfrac{1}{2}(e^{2x} + e^{-2x} + 2) - 1 \\ &= \tfrac{1}{2}(e^{2x} + e^{-2x}) \\ &= \cosh 2x \text{ by definition} \end{aligned}$$

These and similar identities are summarised in Table 1.1. All can be obtained using the same method.

Trigonometrical functions are often described as circular functions in contrast to hyperbolic functions.

Table 1.1

Identity involving circular function	Identity involving hyperbolic function
$\cos^2 x + \sin^2 x = 1$	$\cosh^2 x - \sinh^2 x = 1$
$\tan x = \sin x/\cos x$	$\tanh x = \sinh x/\cosh x$
$1 + \tan^2 x = \sec^2 x$	$1 - \tanh^2 x = \text{sech}^2\, x$
$\sin 2x = 2 \sin x \cos x$	$\sinh 2x = 2 \sinh x \cosh x$
$\cos 2x = 2 \cos^2 x - 1$	$\cosh 2x = 2 \cosh^2 x - 1$
$\quad = \cos^2 x - \sin^2 x$	$\quad = \cosh^2 x + \sinh^2 x$
$\quad = 1 - 2 \sin^2 x$	$\quad = 1 + 2 \sinh^2 x$
$\mathrm{d}/\mathrm{d}x\,(\sin x) = \cos x$	$\mathrm{d}/\mathrm{d}x\,(\sinh x) = \cosh x$
$\mathrm{d}/\mathrm{d}x\,(\cos x) = -\sin x$	$\mathrm{d}/\mathrm{d}x\,(\cosh x) = \sinh x$

Example 1.1(a)
Obtain an identity for $\sinh (x + y)$ in terms of $\sinh x$, $\cosh x$, $\sinh y$ and $\cosh y$.

$$\begin{aligned}
\sinh x \cosh y &= \tfrac{1}{2}(\mathrm{e}^x - \mathrm{e}^{-x}) \times \tfrac{1}{2}(\mathrm{e}^y + \mathrm{e}^{-y}) \\
&= \tfrac{1}{4}(\mathrm{e}^{x+y} - \mathrm{e}^{-x+y} + \mathrm{e}^{x-y} - \mathrm{e}^{-x-y}) \\
\cosh x \sinh y &= \tfrac{1}{2}(\mathrm{e}^x + \mathrm{e}^{-x}) \times \tfrac{1}{2}(\mathrm{e}^y - \mathrm{e}^{-y}) \\
&= \tfrac{1}{4}(\mathrm{e}^{x+y} + \mathrm{e}^{-x+y} - \mathrm{e}^{x-y} - \mathrm{e}^{-x-y}) \\
\sinh x \cosh y &+ \cosh x \sinh y \\
&= \tfrac{1}{2}(\mathrm{e}^{x+y} - \mathrm{e}^{-x-y}) \\
&= \sinh (x + y)
\end{aligned}$$

1.2 Hyperbolic Tangent and the Reciprocal Functions

Extending the analogy between circular and hyperbolic functions, we can define hyperbolic tangent, $\tanh x$ (pronounced 'than x' with the 'th' as in 'thick'), as

$$\tan x = \sinh x/\cosh x = (\mathrm{e}^{-x} - \mathrm{e}^{-x})/(\mathrm{e}^x + \mathrm{e}^{-x})$$

The reciprocal functions

$$\text{sech}\, x = \frac{1}{\cosh x} = \frac{2}{\mathrm{e}^x + \mathrm{e}^{-x}}$$

$$\text{cosech}\, x = \frac{1}{\sinh x} = \frac{2}{\mathrm{e}^x - \mathrm{e}^{-x}}$$

$$\coth x = \frac{1}{\tanh x} = \frac{e^x + e^{-x}}{e^x - e^{-x}}$$

are all analogous to their circular counterparts.

Example 1.2(a)
Prove that

$$\frac{\cosh 2x - 1}{\cosh 2x + 1} = \tanh^2 x$$

$$\begin{aligned}\cosh 2x - 1 &= \tfrac{1}{2}(e^{2x} + e^{-2x}) - 1 \\ &= \tfrac{1}{2}(e^{2x} - 2 + e^{-2x}) \\ &= \tfrac{1}{2}(e^x - e^{-x})^2\end{aligned}$$

Similarly

$$\cosh 2x + 1 = \tfrac{1}{2}(e^x + e^{-x})^2$$

$$\begin{aligned}\frac{\cosh 2x - 1}{\cosh 2x + 1} &= \frac{(e^x - e^{-x})^2}{(e^x + e^{-x})^2} \\ &= \left(\frac{e^x - e^{-x}}{e^x + e^{-x}}\right)^2 \\ &= \tanh^2 x\end{aligned}$$

Example 1.2(b)
If $\cosh x = \frac{5}{4}$, find values of all the other hyperbolic functions. Find, in addition, possible values of x correct to two places of decimals.

If

$$\cosh x = \tfrac{5}{4}$$
$$\operatorname{sech} x = \tfrac{5}{4}$$
$$\sinh x = \pm\sqrt{(\cosh^2 x - 1)} = \pm\tfrac{3}{4}$$
$$\operatorname{cosech} x = \pm\tfrac{4}{3}$$
$$\tanh x = \sinh x/\cosh x = \pm\tfrac{3}{5}$$
$$\coth x = \pm\tfrac{5}{3}$$

$$\cosh x + \sinh x = e^x = \tfrac{5}{4} \pm \tfrac{3}{4}$$
$$e^x = 2 \text{ or } e^x = 0.5$$
$$x = \ln 2 \qquad \text{or} \qquad x = \ln 0.5 = -\ln 2$$

Hence $x = \pm \ln 2 = \pm\, 0.69$.

Cosh x and sech x are always positive but all the other hyperbolic functions can take positive or negative values.

The second part of the question shows one way of obtaining x given the value of a hyperbolic function. Example 1.4(b) shows another method.

Notice that ln 0.5 is best evaluated by changing its sign and replacing its argument with its reciprocal.

1.3 Odd and Even Functions

A function $f(x)$, which involves only odd powers of x, has the property that for any value a, $f(a) = -f(-a)$. For example if $f(x) = x^3$, $f(2) = 8$ and $f(-2) = -8$. This is an example of an ODD FUNCTION.

In the same way, functions involving only even powers are such that $f(a) = f(-a)$, so that if $f(x) = x^2$, $f(2.7) = f(-2.7)$. In this case $f(x)$ is an EVEN FUNCTION.

A coefficient or a sign does not affect the nature of a function so that $f(x) = 2x$ is still odd and $f(x) = -3x^2$ is still even. An even function may be

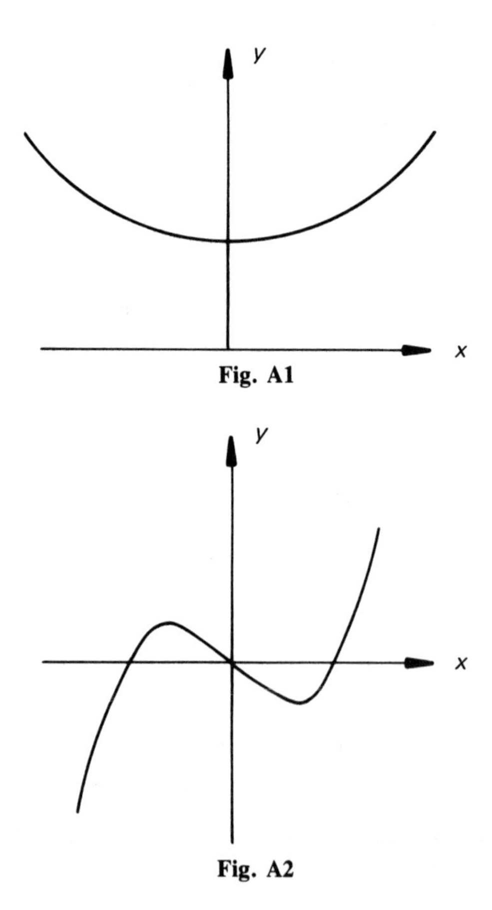

Fig. A1

Fig. A2

the sum of several even terms and, similarly, an odd function may be the sum of several odd terms.

A function such as $\cos x$ or $\cosh x$ is even by the definition above. Similarly $\sin x$ and $\tanh x$ are odd functions. Note that an odd function has a series form containing only odd powers of x and an even function a series containing even powers of x. The graph of an even function is symmetrical about the y-axis, as illustrated in Figure A1. On the other hand, the graph of an odd function is skew symmetric (its reflection in the y-axis is itself reflected in the x-axis), as illustrated in Figure A2.

1.4 Properties of Hyperbolic Functions

Table 1.2 gives the more important properties of the main hyperbolic functions. The values of $\sinh 0$, $\cosh 0$ and $\tanh 0$ can be obtained from their definitions, since $e^x = e^{-x} = 1$ when $x = 0$.

The nature of these functions is obtainable by considering their value when x takes two values differing only in sign. They can also be obtained by considering the series form of the function given in Table 1.2 or from the graphs of $y = \sinh x$, $y = \cosh x$ and $y = \tanh x$ given in Figure A3.

The series forms of $\sinh x$ and $\cosh x$ are obtained using the power series for e^x and e^{-x}, the derivation of which is given in Part 10.1.

$$\begin{aligned} e^x &= 1 + x + x^2/2! + x^3/3! + x^4/4! + \cdots \\ e^{-x} &= 1 - x + x^2/2! - x^3/3! + x^4/4! - \cdots \\ \sinh x &= \tfrac{1}{2}(e^x - e^{-x}) = x + x^3/3! + x^5/5! + \cdots \\ \cosh x &= \tfrac{1}{4}(e^x + e^{-x}) = 1 + x^2/2! + x^4/4! + \cdots \end{aligned}$$

Since $\cosh x \tanh x = \sinh x$, the series form of $\tanh x$ can be confirmed by direct multiplication since

$$\begin{aligned} &(1 + \tfrac{1}{2}x^2 + \tfrac{1}{24}x^4 + \cdots)(x - \tfrac{1}{3}x^3 + \tfrac{2}{15}x^5 + \cdots) \\ &\quad = x + \tfrac{1}{6}x^3 + \tfrac{1}{120}x^5 + \cdots \end{aligned}$$

Table 1.2

	$\cosh x$	$\sinh x$	$\tanh x$
Value of function when $x = 0$	1	0	0
Nature of function	Even	Odd	Odd
Series form	$\sum_{n=0}^{\infty} \frac{x^{2n}}{(2n)!}$	$\sum_{n=0} \frac{x^{2n+1}}{(2n+1)!}$	$x - \frac{1}{3}x^3 + \frac{2}{15}x^5 - \ldots$

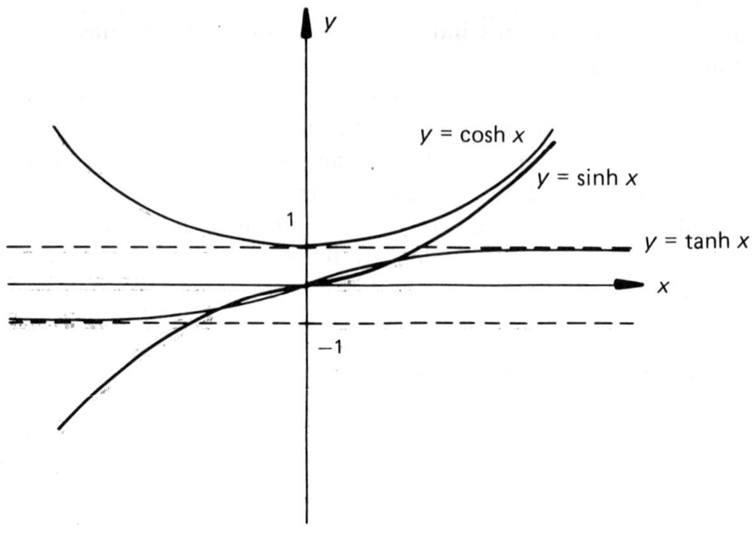

Fig. A3

Example 1.4(a)
Verify that $d/dx\,(\tanh x) = \operatorname{sech}^2 x$ by considering these hyperbolic functions in their exponential form

$$
\begin{aligned}
d/dx\,(\tanh x) &= d/dx\left(\frac{e^x - e^{-x}}{e^x + e^{-x}}\right) \\
&= \frac{(e^x + e^{-x})^2 - (e^x - e^{-x})^2}{(e^x + e^{-x})^2} \\
&= \frac{4}{(e^x + e^{-x})^2} \\
&= \operatorname{sech}^2 x
\end{aligned}
$$

Example 1.4(b)
If $\sinh x = 15/8$, find x correct to two places of decimals.

If $\sinh x = 15/8$

$$
\begin{aligned}
&\tfrac{1}{2}(e^x - e^{-x}) = 15/8 \\
&4(e^x - e^{-x}) = 15 \\
&4e^x - 4e^{-x} = 15 \\
&4e^{2x} - 15e^x - 4 = 0 \\
&(4e^x + 1)(e^x - 4) = 0 \\
&e^x = -0.25 \quad \text{or} \quad e^x = 4
\end{aligned}
$$

Since $e^x > 0$ we can ignore the first solution.

$$e^x = 4$$
$$x = \ln 4 = 1.39 \text{ (corr. 2 pl. dec.)}$$

Unworked Examples 1

Part 1.1

(1) Use the definitions

$$\cosh x = \tfrac{1}{2}(e^x + e^{-x})$$
$$\sinh x = \tfrac{1}{2}(e^x - e^{-x})$$

to obtain
(a) the values of cosh 0 and sinh 0,
(b) the series form of sinh x and cosh x.
(c) the series form of sinh x.

(2) Show, by using their exponential form, that

$$d/dx\,(\sinh x) = \cosh x$$
$$d/dx\,(\cosh x) = \sinh x$$

Deduce two possible solutions of the differential equation $d^2y/dx^2 = y$.

(3) Prove that $\sinh 2x = 2 \sinh x \cosh x$.

(4) Verify that $d/dx\,(\sinh x) = \cosh x$ and that $d/dx\,(\cosh x) = \sinh x$ by considering
(a) the exponential form of sinh x and cosh x,
(b) the series form of sinh x and cosh x.

(5) Show that the minimum value of cosh x is equal to 1.

(6) Evaluate cosh x and sinh x when (a) $x = 1$ (b) $x = -1$. Give answers correct to two places of decimals.

Part 1.2

(7) Prove that as $x \to \infty$, $\tanh x \to 1$.

(8) Show that if $\sec x = \cosh y$ then $\tan x = \sinh y$ and $\sin x = \tanh y$.

(9) Prove that $\tanh 2A = 2 \tanh A/(1 + \tanh^2 A)$.

(10) Show that for all values of $x > 0$
(a) $\cosh x > \sinh x$,
(b) $\operatorname{cosech} x > \operatorname{sech} x$,
(c) $\coth x > \operatorname{cosech} x$.

(11) Prove that $\coth 2\theta = \tfrac{1}{2}(\coth \theta + \tanh \theta)$.

(12) State whether the following functions are odd or even: (a) $\tanh x$ (b) $\coth x$ (c) $\sinh 2x$ (d) $x \cosh x$ (e) $\sinh x + \tanh x$.

(13) Both $2x$ and $\sinh x$ are odd functions. Use this and the fact that one solution of the equation $2x = \sinh x$ is approximately 2.178 to find another solution.

(14) Prove that $\cosh x + \sinh x = e^x$ and $\cosh x - \sinh x = e^{-x}$. Deduce that $(\cosh x \pm \sinh x)^n = \cosh nx \pm \sinh nx$.

(15) Evaluate the integral

$$\int \frac{dx}{(1 + x^2)^{3/2}}$$

by using the substitution (a) $x = \tan \theta$ (b) $x = \sinh u$. Draw a conclusion

(16) If $x = 0.4$, show that $\sinh x = 0.41$, $\cosh x = 1.08$ and $\tanh x = 0.38$, all values being corrected to two places of decimals.

(17) Solve the equation $\sinh x = 0.4$.

(18) Prove that $d/dx\ (\tanh x) = \operatorname{sech}^2 x$ by putting $\tanh x = \sinh x/\cosh x$.

1.5 Osborn's Rule

Reference to Table 1.1 indicates a connection between identities involving circular functions and those involving hyperbolic function. In each case the circular function is replaced by its hyperbolic equivalent and, where the relationship involves the square of an odd function, there is a change of sign. This relationship, which can be applied to any trigonometrical identity, is called OSBORN'S RULE.

Example 1.5(a)

Prove that $\cos 2\theta = (1 - \tan^2 \theta)/(1 + \tan^2 \theta)$ and use Osborn's rule to obtain the corresponding identity involving hyperbolic functions

$$\begin{aligned}(1 - \tan^2 \theta)/(1 + \tan^2 \theta) &= (1 - \tan^2 \theta)/\sec^2 \theta \\ &= \cos^2 \theta - \sin^2 \theta \\ &= \cos 2\theta\end{aligned}$$

Using Osborn's rule, replacing θ by x,

$$\cosh 2x = (1 + \tanh^2 x)/(1 - \tanh^2 x)$$

Example 1.5(b)

It can be shown that $\sin 3A = 3 \sin A - 4 \sin^3 A$. Write down a corresponding identity involving hyperbolic functions and verify it algebraically.

Since $\sin^3 A$ 'includes' as a factor the square of an odd function ($\sin^2 A$), Osborn's rule gives

$$\sinh 3A = 3 \sinh A + 4 \sinh^3 A$$

This is readily verified

$$\begin{aligned}
3 \sinh A &= \tfrac{3}{2}(e^{A} - e^{-A}) \\
\sinh^3 A &= \tfrac{1}{8}(e^{A} - e^{-A})^3 \\
&= \tfrac{1}{8}(e^{3A} - 3e^{A} + 3e^{-A} - e^{-3A}) \\
4 \sinh^3 A &= \tfrac{1}{2}(e^{3A} - 3e^{A} + 3e^{-A} - e^{-3A}) \\
&= \tfrac{1}{2}(e^{3A} - e^{-3A}) - \tfrac{3}{2}(e^{A} - e^{-A}) \\
3 \sinh A + 4 \sinh^3 A &= \tfrac{1}{2}(e^{3A} - e^{-3A}) \\
&= \sinh 3A
\end{aligned}$$

1.6 Hyperbolic Equations

An equation of the form

$$a \cosh x \pm b \sinh x = c$$

where a and b are constants is an example of a HYPERBOLIC EQUATION. There are two common methods of solution, both of which are illustrated in Example 1.6(a). The first method uses the exponential form of hyperbolic functions to arrive at a quadratic equation in e^x. The second method involves use of the identity $\sinh^2 x = \cosh^2 x - 1$ and is obtained by squaring the rearranged equation. It is important to express the equation in terms of $\cosh x$ because, as this function always takes positive values, the ambiguity of signs will normally be removed.

Example 1.6(a)
Solve the equation $3 \cosh x - 5 \sinh x = 2$ giving the value of x correct to two decimal places. (Only one of the following methods would normally be used.)

Method 1

$$3 \cosh x = \tfrac{3}{2}(e^{x} + e^{-x})$$
$$5 \sinh x = \tfrac{5}{2}(e^{x} - e^{-x})$$
$$3 \cosh x - 5 \sinh x = -e^{x} + 4e^{-x}$$
$$-e^{x} + 4e^{-x} = 2$$
$$e^{2x} + 2e^{x} - 4 = 0$$
$$\begin{aligned}
e^{x} &= (-2 \pm \sqrt{20})/2 \\
&= -1 + \sqrt{5} \\
&= 1.236
\end{aligned}$$
$$x = \ln 1.236 = 0.21 \text{ (corr. 2 dec. pl.)}$$

Method 2

$$3 \cosh x - 5 \sinh x = 2$$
$$3 \cosh x - 2 = 5 \sinh x$$
$$9 \cosh^2 x - 12 \cosh x + 4 = 25 \sinh^2 x$$
$$= 25 \cosh^2 x - 25$$
$$16 \cosh^2 x + 12 \cosh x - 29 = 0$$
$$\cosh x = (12 \pm \sqrt{2000})/32 = 1.022$$
$$x = 0.21 \text{ (from tables)}$$

In method 1, only the positive part of the $\pm$ sign is considered since $e^x > 0$ for all values for x. In method 2, x is obtained from $\cosh x$ using tables. If these are not available, the methods illustrated in Example 1.2(b) or Example 1.4(b) may be used.

Unworked Examples 2

Part 1.5

(1) Verify the identity $1 - \tanh^2 x = \text{sech}^2 x$ by
  (a) using the definitions of $\tanh x$ and $\text{sech}\, x$,
  (b) using Osborn's rule,
  (c) assuming the result $\cosh^2 x - \sinh^2 x = 1$.
(2) Prove that $\text{cosech}\, 2x = \frac{1}{2}\coth x\,(1 - \tanh^2 x)$.
(3) Use Osborn's rule to obtain expansions of $\sinh (x \pm y)$ and $\cosh (x \pm y)$.
(4) Given that $\tan 2x = 2 \tan x/(1 - \tan^2 x)$, obtain a corresponding formula for $\tanh 2x$. Find $\tanh 2x$ if $\tanh x = 0.5$.
(5) Prove that $\cosh 3A = 4 \cosh^3 A - 3 \cosh A$ by using the definition of the hyperbolic cosine function. Show that this result is consistent with Osborn's rule.

Part 1.6

(6) Solve the equation $20 \cosh A + 7 \sinh A = 24$.
(7) Find the solution of $5 \cosh x - 3 \sinh x = 5$ correct to two places of decimals.
(8) Solve the hyperbolic equation

$$2 \cosh x + 2 \sinh x = 5$$

  (a) by squaring and using the identity $\cosh^2 x - \sinh^2 x = 1$
  (b) by using the fact that $\cosh x + \sinh x = e^x$.
(9) Show that if $3 \cosh x - 4 \sinh x = 1$, then $e^x = \sqrt{8} - 1$. Hence find x.
(10) Show that when solving the equation

$$5 \cosh x - 4 \sinh x = 3$$

a quadratic equation with repeated roots is obtained. Hence obtain a solution of the equation.

(11) Solve the hyperbolic equation

$$5 \tanh x + 3 \coth x = 8$$

by multiplying throughout by $\tanh x$. Is $\tanh x = \coth x = 1$ a possible solution?

PART 2: COMPLEX NUMBERS

2.1 Extension of the Number System

When man first used numbers, it was by counting. Everything involved positive integers. As life became more sophisticated and there was a need to measure distances or share property, it was found that the number system needed extending and the idea of fractions (and later decimals) had to be introduced.

Subsequently as civilisation developed, values had to be considered as positive or negative and the idea of directed numbers was introduced.

Let us take this type of extension one stage further. So far we have considered that the square roots of negative numbers 'do not exist'. But why not? If we can appreciate the need for number development dealt with above, there is no reason why we cannot consider the value $\sqrt{-1}$. We will refer to this quantity by the symbol j, so that $j^2 = -1$. All other laws of algebra will be obeyed 'normally' so that any other negative square root can be expressed in terms of j. For example $\sqrt{-4} = \sqrt{-1} \times \sqrt{4} = j2$. (The product of j and a real number is written with the j preceding the number.)

Example 2.1(a)

Solve the quadratic equation $x^2 + 2x + 5 = 0$ expressing the roots in terms of j. Verify that the roots satisfy the equation.

$$\begin{aligned} x &= (-2 \pm \sqrt{(-16)})/2 \\ &= (-2 \pm j4)/2 \\ &= -1 \pm j2 \end{aligned}$$

Check: If $x = -1 \pm j2$,

$$\begin{aligned} x^2 &= (-1 \pm j2)^2 \\ &= 1 \mp j4 + j^2 4 \\ &= 1 \mp j4 - 4 \\ &= -3 \mp j4 \end{aligned}$$

$$\begin{aligned} x^2 + 2x + 5 &= (-3 \mp j4) + (-2 \pm j4) + 5 \\ &= 0 \end{aligned}$$

The symbol $\mp$, read as 'minus or plus', is used here to mean that when $x = -1 + \text{j}2$, $x^2 = -3 - \text{j}4$ and when $x = -1 - \text{j}2$, $x^2 = -3 + \text{j}4$.

There is no further simplication of the expression $-1 + \text{j}2$. The value -1 is called its REAL PART while 2 (not j2) is called its IMAGINARY PART. The whole quantity $-1 + \text{j}2$ is called a COMPLEX NUMBER. A complex number with a zero real part such as j2 is said to be PURELY IMAGINARY

In the solution of Example 2.1(a), two complex numbers $-1 \pm \text{j}2$, with the same real part and with imaginary parts differing only in sign were met. These are called CONJUGATE COMPLEX NUMBERS.

2.2 Equality of Complex Numbers

If two complex numbers $a + \text{j}b$ and $c + \text{j}d$, where a, b, c and d are all real, are equal, we have

$$a + \text{j}b = c + \text{j}d$$

$$a - c = \text{j}(d - b)$$

$$(a - c)^2 = -(d - b)^2$$

Since a, b, c and d are all real, $(a - c)^2 \geqq 0$ and $-(d - b)^2 \leq 0$. These can only be equated if both equal zero, i.e. $(a - c) = (d - b) = 0$, so that $a = c$ and $b = d$. This means that if two complex numbers are equal, we may equate their real and their imaginary parts.

2.3 Arithmetic Operations with Complex Numbers

Some of the operations of arithmetic on complex numbers were illustrated in Example 2.1(a) in the checking. Addition and subtraction follow the normal laws of algebra

$$(a + \text{j}b) + (c + \text{j}d) = (a + c) + \text{j}(b + d)$$

$$(a + \text{j}b) - (c + \text{j}d) = (a - c) + \text{j}(b - d)$$

Multiplication is carried out in a similar way with j^2 replaced by -1

$$\begin{aligned}(a + \text{j}b)(c + \text{j}d) &= ac + \text{j}bc + \text{j}ad + \text{j}^2bd \\ &= (ac - bd) + \text{j}(bc + ad)\end{aligned}$$

Division is effected by 'rationalising' the denominator, multiplying by its conjugate

$$\begin{aligned}\frac{a + \text{j}b}{c + \text{j}d} &= \frac{(a + \text{j}b)(c - \text{j}d)}{(c + \text{j}d)(c - \text{j}d)} = \frac{(ac + bd) + \text{j}(bc - ad)}{c^2 + d^2} \\ &= (ac + bd)/(c^2 + d^2) + \text{j}(bc - ad)/(c^2 + d^2)\end{aligned}$$

Example 2.3(a)
If z_1 represents the complex number 5 + j5 and z_2 represents the complex number 1 + j2, obtain (i) $z_1 + z_2$, (ii) $z_1 - z_2$, (iii) $z_1 z_2$, (iv) z_1/z_2.

(i) $$\begin{aligned} z_1 + z_2 &= (5 + \mathrm{j}5) + (1 + \mathrm{j}2) \\ &= 6 + \mathrm{j}7 \end{aligned}$$

(ii) $$\begin{aligned} z_1 - z_2 &= (5 + \mathrm{j}5) - (1 + \mathrm{j}2) \\ &= 4 + \mathrm{j}3 \end{aligned}$$

(iii) $$\begin{aligned} z_1 z_2 &= (5 + \mathrm{j}5)(1 + \mathrm{j}2) \\ &= 5 + \mathrm{j}15 - 10 \\ &= -5 + \mathrm{j}15 \end{aligned}$$

(iv) $$\begin{aligned} z_1/z_2 &= (5 + \mathrm{j}5)/(1 + \mathrm{j}2) \\ &= \frac{(5 + \mathrm{j}5)(1 - \mathrm{j}2)}{(1 + \mathrm{j}2)(1 - \mathrm{j}2)} \\ &= (15 - \mathrm{j}5)/5 \\ &= 3 - \mathrm{j} \end{aligned}$$

All arithmetic operations on complex numbers, as well as operations involving powers or roots lead to complex results but the real and imaginary parts will not necessarily be integral.

Example 2.3(b)
Obtain $(1 + \mathrm{j})^3$.

$$\begin{aligned} (1 + \mathrm{j})^3 &= 1 + \mathrm{j}3 + \mathrm{j}^2 3 + \mathrm{j}^3 \\ &= 1 + \mathrm{j}3 - 3 - \mathrm{j} \\ &= -2 + \mathrm{j}2 \end{aligned}$$

Here $\mathrm{j}^3 = \mathrm{j}^2\mathrm{j} = -\mathrm{j}$.

Example 2.3(c)
Find $\sqrt{(3 + \mathrm{j}4)}$.

Suppose $\sqrt{(3 + \mathrm{j}4)} = a + \mathrm{j}b$.

$$\begin{aligned} 3 + \mathrm{j}4 &= (a + \mathrm{j}b)^2 \\ &= a^2 - b^2 + \mathrm{j}2ab \end{aligned}$$

Hence

$$\begin{aligned} a^2 - b^2 &= 3 \\ ab &= 2 \end{aligned}$$

By the usual methods of solution, $a = \pm 2$, $b = \pm 1$

$$\sqrt{(3 + j4)} = \pm(2 + j)$$

Unworked Examples 3

Part 2.1

(1) Solve the quadratic equation $x^2 + 4x + 5 = 0$.
(2) What is (a) the real part (b) the imaginary part (c) the conjugate of the complex number $2 - j4$.
(3) Show that the complex roots of $x^2 + x + 1 = 0$ are each the square of the other.

Parts 2.2 and 2.3

(4) Simplify $(3 + j5)(1 - j)$.
(5) Add $4 + j$, $2 + j3$, $7 - j$, $-12 + j$ and $5 - j$.
(6) Find the difference between $4 + j5$ and $2 - j$.
(7) Divide $1 + j7$ by $3 + j$.
(8) Show that the sum and product of conjugate complex numbers are real and the difference is purely imaginary.
(9) Show that $j^{26} = -1$.
(10) If $(a + jb)^2 = 8 - j6$, find values for a and b.
(11) Calculate $1 + j + 2/(1 + j)$.
(12) Find $(z + 2)/(z - 2)$ where $z = 3 + j$.
(13) Obtain $(1 - j2)^4$
  (a) by expanding directly,
  (b) by finding $(1 - j2)^2$ and squaring the result.

2.4 Argand Diagram

Since a complex number consists of two independent parts, real and imaginary, a *representation* of the number can be drawn graphically using a horizontal real axis and a vertical imaginary axis as shown in Figure A4. A graph of this type is called an ARGAND DIAGRAM.

Each *point* represents a number which may be real, imaginary or complex. Thus A represents j3, B represents 4 and C, which would be described as (4,3) on a Cartesian co-ordinate system, represents $4 + j3$. D, the reflection of C in the real axis, represents its conjugate $4 - j3$.

Suppose that P and Q represent $1 + j2$ and $4 + j3$ on the Argand diagram in Figure A5. Their sum, $5 + j5$ is represented by R. If the lines OP and OQ

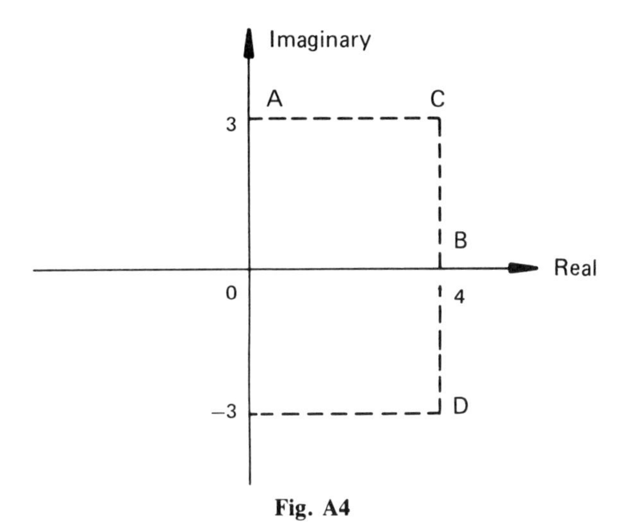

Fig. A4

are drawn, it will be seen that R is the fourth vertex of a parallelogram whose other vertices are O, P and Q. This parallelogram rule can be extended to subtraction. $(1 + j2) - (4 + j3)$ is equivalent to $(1 + j2) + (-4 - j3)$. S represents $-4 - j3$ on the diagram and is easily found because O is the midpoint of QS. T, which completes the parallelogram OPTS represents the difference $(1 + j2) - (4 + j3)$ on the Argand diagram.

These results could have been obtained by using just the triangles which form one-half of the parallelogram in each case. It would then be seen that the laws for addition and subtraction of complex numbers on an Argand

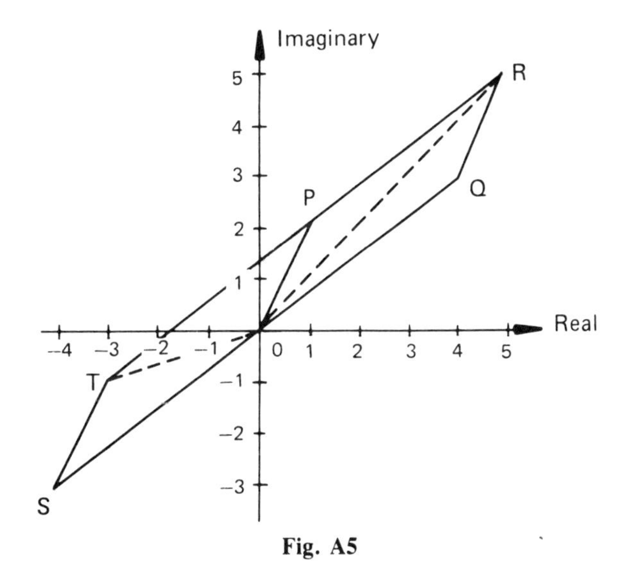

Fig. A5

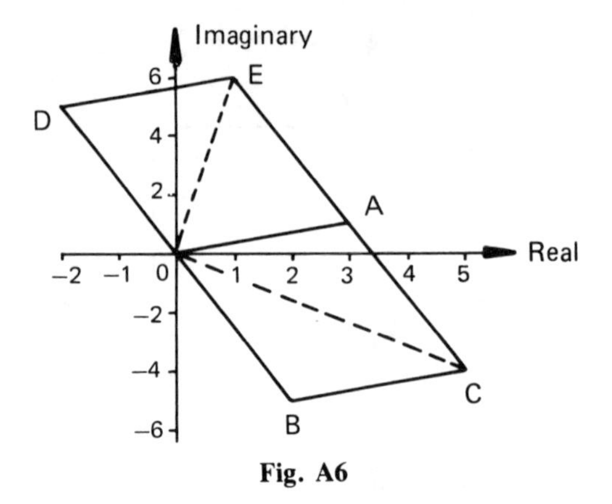

Fig. A6

diagram correspond to the triangle laws of addition and subtraction of vectors, described in Part 7.1. As a result complex numbers are often used to represent planar vectors.

Example 2.4(a)
Draw the complex numbers $3 + j$ and $2 - j5$ on an Argand diagram. Obtain their sum and difference using the diagram and verify that the results are equal to those obtained by calculation.

The diagram is shown in Figure A6. The complex numbers $3 + j$ and $2 - j5$ are represented by A and B, their sum by C and their difference by E. From the diagram, these results are $5 - j4$ and $1 + j6$ respectively.

$$(3 + j) + (2 - j5) = 5 - j4$$
$$(3 + j) - (2 - j5) = 1 + j6$$

Example 2.4(b)
Use an Argand diagram to show that conjugate complex numbers have a real sum and a purely imaginary difference.

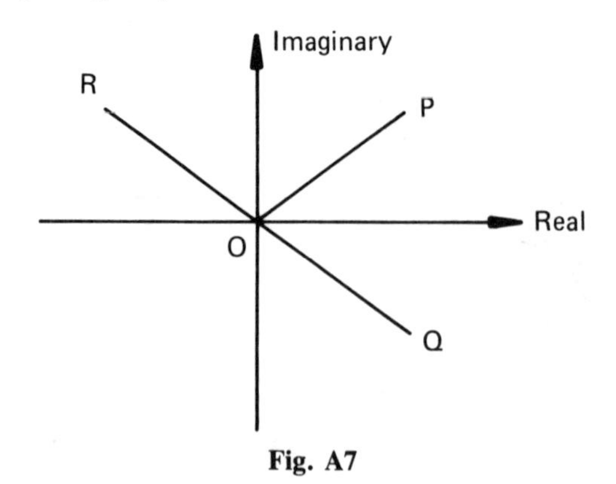

Fig. A7

Let the complex numbers be represented by P and Q on the Argand diagram in Figure A7. OQ must be the reflection of OP in the real axis. If the parallelogram, three of whose vertices are O, P and Q were completed, the fourth vertex would be on the real axis. The sum of conjugate complex numbers is real.

The difference is obtained by finding the fourth vertex of the parallelogram with vertices O, P and R. As this lies on the imaginary axis, the difference is wholly imaginary.

2.5 Polar Representation of Complex Numbers

In Figure A8, the point P would be denoted by (x,y) using a Cartesian system of co-ordinates and by (r,θ) using a polar system. If P were on an Argand diagram, it would represent the complex number $x + jy$. Since $x = r\cos\theta$ and $y = r\sin\theta$, the complex number could be written in the form $r(\cos\theta + j\sin\theta)$, described as the POLAR FORM of a complex number. In this form, r is the distance of P from the origin O and is called the MODULUS of the complex number while θ is the angle that OP makes with the positive direction of the real axis and is known as the ARGUMENT of the complex number.

Of course for a given complex number there are an infinite number of values that can be given to θ (since, for example, $\cos 20° + j\sin 20° = \cos 380° + j\sin 380° = \cos 740° + j\sin 740° = \cdots$, adding 360° to the angle in each case). The value of the angle lying between $-180°$ and 180° (or $-\pi$ and π if we choose to measure the angle in radians) is called the PRINCIPAL VALUE and is the value generally used. Being a distance, the modulus is always measured positive.

Example 2.5(a)

Express in polar form the complex numbers (a) $3 + j4$, (b) $-12 + j5$.

(a) $r = \sqrt{(3^2 + 4^2)} = 5$

$\tan\theta = 4/3 = 1.3333$

$\theta = 53°8'$

$3 + j4 \equiv 5(\cos 53°8' + j\sin 53°8')$

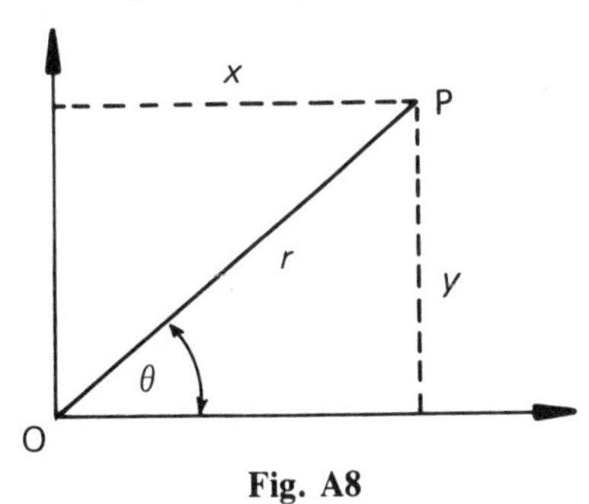

Fig. A8

(b) $r = \sqrt{((-12)^2 + 5^2)} = 13$

$\tan\theta = -5/12 = -0.4167$

$\theta = 157°22'$

$-12 + j5 \equiv 13(\cos 157°22' + j \sin 157°22')$

In both cases the normal formula for converting from Cartesian co-ordinates (x,y) to polar co-ordinates (r,θ), $r = \sqrt{(x^2 + y^2)}$ and $\tan\theta = y/x$ have been used. The principal value of θ has been chosen.

In (b) if $\tan\theta = -0.4167$, $\theta = 157°22'$ or $\theta = -22°38'$. Since the point, representing $12 - j5$ on the Argand diagram lies in the second quadrant, the larger value of θ is chosen.

Example 2.5(b)
State the modulus and argument of the number $z = 4 + j4$.

The modulus of z may be abbreviated to $|z|$ and the argument to arg (z)

$$\begin{aligned} |z| &= \sqrt{(4^2 + 4^2)} \\ &= \sqrt{32} \\ &\approx 5.66 \end{aligned}$$

$$\arg(z) = \tan^{-1} 1 = \pi/4$$

2.6 Multiplication and Division of Complex Numbers

Suppose that z_1 and z_2 are complex numbers which have moduli r_1 and r_2 and arguments θ_1 and θ_2 respectively.

$$\begin{aligned} z_1 z_2 &= r_1(\cos\theta_1 + j \sin\theta_1) \times r_2(\cos\theta_2 + j\sin\theta_2) \\ &= r_1 r_2(\cos\theta_1 + j\sin\theta_1)(\cos\theta_2 + j\sin\theta_2) \\ &= r_1 r_2([\cos\theta_1 \cos\theta_2 - \sin\theta_1 \sin\theta_2] \\ &\quad + j[\cos\theta_1 \sin\theta_2 + \sin\theta_1 \cos\theta_2]) \\ &= r_1 r_2(\cos[\theta_1 + \theta_2] + j\sin[\theta_1 + \theta_2]) \end{aligned}$$

When complex numbers are multiplied, the modulus of the product is equal to the product of the moduli, while the argument of the product is equal to the sum of the arguments. The results can be summarised

$$|z_1 z_2| = |z_1| \cdot |z_2|$$

$$\arg(z_1 z_2) = \arg(z_1) + \arg(z_2)$$

The reverse is true for division of complex numbers. The modulus of the quotient is obtained by dividing moduli and the argument of the quotient by taking the difference between arguments

$$|z_1/z_2| = |z_1|/|z_2|$$

$$\arg(z_1/z_2) = \arg(z_1) - \arg(z_2)$$

When manipulating arguments in these ways it is quite possible for it to fall outside the principal value range. All that is needed is for multiples of 360° to be added or subtracted to bring it into the appropriate range (although it may be more convenient to deal with, say, cos 240° than with the equivalently valued cos −120°).

In polar form, it is simple to extract square roots of complex numbers. The square root of the modulus is extracted and the argument is halved. This is illustrated in Example 2.6(c).

Example 2.6(a)

Obtain $2 - j2\surd 3$, $\surd(3 + j)$, their product and quotient all in polar form. Show that the results are consistent with results obtained using the formulae given in Part 2.6.

$$(2 - j2\surd 3)(\surd 3 + j) = 4\surd 3 - j4$$

$$(2 - j2\surd 3)/(\surd 3 + j) = \frac{(2 - j2\surd 3)(\surd 3 - j)}{(\surd 3 + j)(\surd 3 - j)}$$

$$= (-j8)/4$$

$$= -j2$$

$$|2 - j2\surd 3| = \surd(2^2 + (-2\surd 3)^2) = 4$$

$$\arg (2 - j2\surd 3) = \tan^{-1} (-\surd 3) = -\pi/3$$

$$|\surd 3 + j| = \surd((\surd 3)^2) + 1^2)) = 2$$

$$\arg (\surd 3 + j) = \tan^{-1} (1/\surd 3) = \pi/6$$

$$|4\surd 3 - j4| = \surd((4\surd 3)^2 + (-4)^2) = 8$$

$$\arg (4\surd 3 - j4) = \tan^{-1} (-1\surd 3) = -\pi/6$$

$$|-j2| = \surd(0^2 + (-2)^2) = 2$$

$$\arg (-j2) = \tan^{-1} \infty = -\pi/2$$

These results are consistent with the laws of multiplication and division of complex numbers in polar form.

Example 2.6(b)

Taking equal scales on the real and imaginary axes, plot the points $2 + j4$ and $1 + j$ on an Argand diagram. Obtain the product and quotient of these complex numbers from the diagram, verifying the results algebraically.

If P_1 represents $1 + j$ and P_2 represents $2 + j4$ in Figure A9, $|P_1| \approx 1.4$, $|P_2| \approx 4.5$, $\arg (P_1) = 45°$ and $\arg (P_2) \approx 63°$. For their product we require B such that

$$|B| = 1.4 \times 4.5 = 6.3$$

$$\arg (B) = 63° + 45° = 108°$$

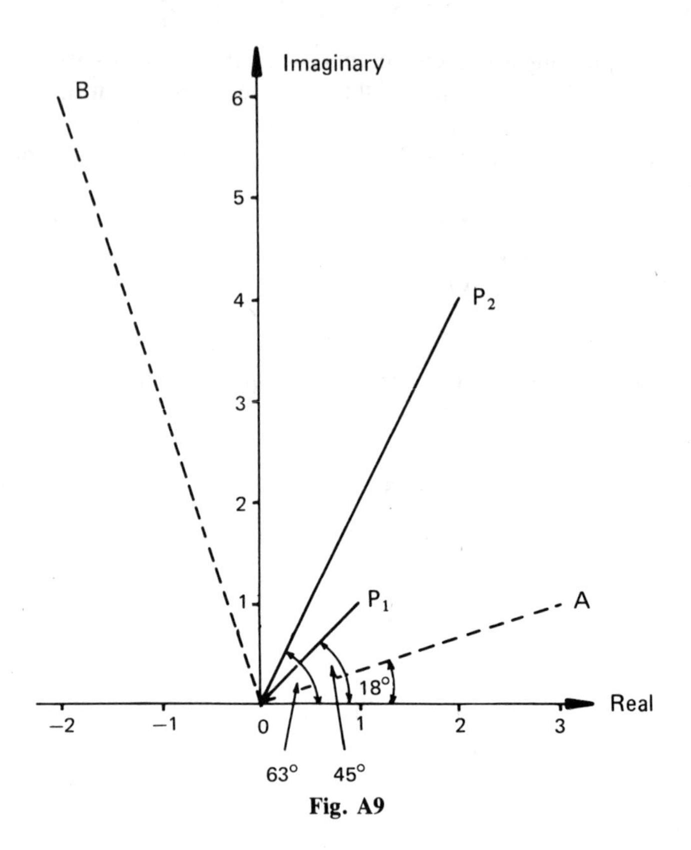

Fig. A9

and for their quotient we require A such that

$$|A| = 4.5/1.4 = 3.2$$
$$\arg(A) = 63° - 45° = 18°$$

A and B are shown in Figure A9 and represent $3 + j$ and $-2 + j6$ respectively. Algebraically,

$$(2 + j4)/(1 + j) = \tfrac{1}{2}(2 + j4)(1 - j)$$
$$= 3 + j$$
$$(2 + j4)(1 + j) = -2 + j6$$

Example 2.6(c)
Verify the result of Example 2.3(c) using the polar form of a complex number

Let

$$z = 3 + j4$$
$$|z| = \sqrt{(3^2 + 4^2)} = 5$$
$$\arg(z) = \tan^{-1} 4/3 = 53°8'$$
$$z = 5(\cos 53°8' + j \sin 53°8')$$

$$\sqrt{z} = \sqrt{5}(\cos 26°34' + \text{j} \sin 26°34')$$
$$= 2 + \text{j}$$

Taking $\sqrt{5}\,(\cos \frac{1}{2}(360° + 53°8') + \text{j} \sin \frac{1}{2}(360° + 53°8'))$ gives $-(2 + \text{j})$.

Unworked Examples 4

Part 2.4

(1) Draw the complex numbers $z_1 = 2 - \text{j}$ and $z_2 = 5 + \text{j}3$ on an Argand diagram. Use the diagram to find (a) $z_1 + z_2$, (b) $z_1 - z_2$, (c) $z_2 - z_1$.

(2) Find $(4 + \text{j}) + (5 - \text{j}3) - (2 + \text{j}2)$ using an Argand diagram. Verify the result algebraically.

(3) Plot the points representing the numbers $\pm\text{j}$ and $\sqrt{3}$ on an Argand diagram. What figure is obtained if the points are joined?

(4) Points A, B and C form an isoscles right-angled triangle with $\angle B = 90°$. If A represents $-2 + \text{j}$ and C represents $5 + \text{j}2$ on an Argand diagram, find the possible complex numbers represented by B. Find also the length of AB.

Part 2.5

(5) Express in polar form (a) $1 - \text{j}$, (b) $8 + \text{j}6$.

(6) Convert $2(\cos 37° + \text{j} \sin 37°)$ into the form $a + \text{j}b$ with a and b each correct to three places of decimals.

(7) Show that if z_1 and z_2 are conjugate complex numbers $|z_1| = |z_2|$ and $\arg(z_1) = -\arg(z_2)$.

Part 2.6

(8) Show that, if $z_1 = 2 - \text{j}$, $z_1{}^2 = 3 - \text{j}4$. Find the moduli of z_1 and of $z_1{}^2$ and verify that one is the square of the other. Is the argument of $z_1{}^2$ twice that of z_1?

(9) Use the result of Question 7 to deduce that the product of conjugate complex numbers is real.

(10) Find $\sqrt{(7 + \text{j}24)}$.

(11) Find the product of $\cos \pi/4 + \text{j} \sin \pi/4$, $2(\cos \pi/3 + \text{j} \sin \pi/3)$ and $5(\cos \pi/12 + \text{j} \sin \pi/12)$.

(12) Express the complex numbers $\cos 150° + \text{j} \sin 150°$ and $\cos 45° + \text{j} \sin 45°$ in the form $x + \text{j}y$. Verify that their produuct and quotient can be obtained in either form using appropriate rules.

(13) Draw the complex numbers $1 + \text{j}$ and $1 + \text{j}2$ on an Argand diagram. Use the diagram to obtain their product.

2.7 Exponential Form of a Complex Number

The series for $\cos\theta$ and $\sin\theta$, with θ measured in radians are developed in Part 10.1. We will, at this stage, quote the results.

$$\cos\theta = 1 - \theta^2/2! + \theta^4/4! - \theta^6/6! + \cdots$$
$$\sin\theta = \theta - \theta^3/3! + \theta^5/5! - \theta^7/7! + \cdots$$

so that $\cos\theta + \mathrm{j}\sin\theta$ can be taken as

$$1 + \mathrm{j}\theta - \theta^2/2! - \mathrm{j}\theta^3/3! + \theta^4/4! + \theta^5/5! - \cdots$$

i.e.

$$1 + \mathrm{j}\theta + \mathrm{j}^2\theta^2/2! + \mathrm{j}^3\theta^3/3! + \mathrm{j}^4\theta^4/4! + \cdots$$

which is the expanded form of $e^{\mathrm{j}\theta}$. Hence

$$r(\cos\theta + \mathrm{j}\sin\theta) = re^{\mathrm{j}\theta}$$

The latter form of a complex number is called its EXPONENTIAL FORM. Taking the exponential form with modulus 1, we have

$$e^{\mathrm{j}\theta} = \cos\theta + \mathrm{j}\sin\theta$$

Replacing θ by $-\theta$, $\cos\theta$ is unchanged but $\sin\theta$ becomes $-\sin\theta$, so that

$$e^{-\mathrm{j}\theta} = \cos\theta - \mathrm{j}\sin\theta$$

This gives consistent results with the division process outlined since

$$\frac{1}{\cos\theta + \mathrm{j}\sin\theta} = \frac{\cos\theta - \mathrm{j}\sin\theta}{\cos^2\theta + \sin^2\theta}$$
$$= \cos\theta - \mathrm{j}\sin\theta$$

The consistency can also be seen when differentiating w.r.t. θ

$$\frac{d}{d\theta}(\cos\theta + \mathrm{j}\sin\theta) = -\sin\theta + \mathrm{j}\cos\theta$$
$$= \mathrm{j}(\cos\theta + \mathrm{j}\sin\theta)$$

which is equivalent to

$$\frac{d}{d\theta}(e^{\mathrm{j}\theta}) = \mathrm{j}e^{\mathrm{j}\theta}$$

Example 2.7(a)
Express $e^{1+\mathrm{j}\pi/3}$ in the form $x + \mathrm{j}y$.

$$e^{1+\mathrm{j}\pi/3} = e.e^{\mathrm{j}\pi/3}$$
$$= e(\cos\pi/3 + \mathrm{j}\sin\pi/3)$$
$$= \tfrac{1}{2}e(1 + \mathrm{j}\sqrt{3})$$
$$\approx 1.36 + \mathrm{j}\,2.35$$

2.8 De Moivre's Theorem

Since $e^{j\theta} = \cos\theta + j\sin\theta$,

$$(e^{j\theta})^n = (\cos\theta + j\sin\theta)^n$$

However,

$$(e^{j\theta})^n = e^{jn\theta} = \cos n\theta + j\sin n\theta$$

Hence

$$(\cos\theta + j\sin\theta)^n = \cos n\theta + j\sin n\theta$$

This result expresses DE MOIVRE'S THEOREM and has important consequences in complex number theory. It applies for all real values of n, positive, negative or fractional.

Example 2.8(a)
Show algebracially that the square of $\sqrt{3} + j$ is equal to $2 + j2\sqrt{3}$. Verify the result using De Moivre's theorem.

$$(\sqrt{3} + j)^2 = 3 + j2\sqrt{3} + j^2 = 2 + j2\sqrt{3}$$
$$|(\sqrt{3} + j)| = \sqrt{(3 + 1^2)} = 2$$
$$\arg(\sqrt{3} + j) = \tan^{-1} 1/\sqrt{3} = \pi/6$$
$$\sqrt{3} + j = 2(\cos\pi/6 + j\sin\pi/6)$$
$$(\sqrt{3} + j)^2 = 4(\cos\pi/6 + j\sin\pi/6)^2$$
$$= 4(\cos\pi/3 + j\sin\pi/3)$$
$$= 2 + j2\sqrt{3}$$

Example 2.8(b)
Use De Moivre's theorem to obtain the *three* cube roots of 1. Plot these roots on an Argand diagram.

In order to use the theorem, 1 must be expressed as a complex number in polar form, i.e. $\cos 0 + j\sin 0$, $\cos 2\pi + j\sin 2\pi$, $\cos 4\pi + j\sin 4\pi$, etc. and the cube root must be expressed in index form

Using the forms above,

$$(\cos 0 + j\sin 0)^{1/3} = \cos 0 + j\sin 0 = 1$$
$$(\cos 2\pi + j\sin 2\pi)^{1/3} = \cos 2\pi/3 + j\sin 2\pi/3$$
$$= \tfrac{1}{2}(-1 + j\sqrt{3})$$
$$(\cos 4\pi + j\sin 4\pi)^{1/3} = \cos 4\pi/3 + j\sin 4\pi/3$$
$$= \tfrac{1}{2}(-1 - j\sqrt{3})$$

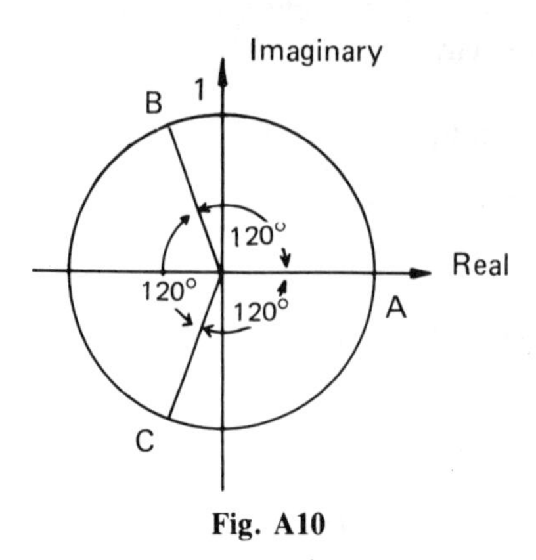

Fig. A10

The three cube roots are 1 and $\frac{1}{2}(-1 \pm j\sqrt{3})$. These are shown as A, B and C on an Argand diagram in Figure A10, where they form equi-angular radii of a unit circle.

Notice that 1 can also be expressed as $\cos 6\pi + j \sin 6\pi$ and other similar forms but no further cube roots of 1 will be obtained.

2.9 Multiple Angle Formulae

De Moivre's theorem can be used to obtain formulae for $\cos n\theta$ in terms of $\cos \theta$, for any integral value of n, and for $\sin n\theta$ in terms of $\sin \theta$ where n is odd. Taking $n = 3$, we have

$$\begin{aligned}\cos 3\theta + j \sin 3\theta &\equiv (\cos \theta + j \sin \theta)^3 \\ &\equiv \cos^3 \theta + 3j \cos^2 \theta \sin \theta + 3j^2 \cos \theta \sin^2 \theta + j^3 \sin^3 \theta \\ &\equiv \cos^3 \theta - 3 \cos \theta \sin^2 \theta + j(3\cos^2 \theta \sin \theta - \sin^3 \theta)\end{aligned}$$

Equating real and imaginary parts of this identity (see Part 2.2) we obtain

$$\begin{aligned}\cos 3\theta &= \cos^3 \theta - 3 \cos \theta \sin^2 \theta \\ &= \cos \theta (\cos^2 \theta - 3 \sin^2 \theta) \\ &= \cos \theta (\cos^2 \theta + 3 \cos^2 \theta - 3) \\ &= 4 \cos^3 \theta - 3 \cos \theta \\ \sin 3\theta &= 3 \cos^2 \theta \sin \theta - \sin^3 \theta \\ &= \sin \theta (3 \cos^2 \theta - \sin^2 \theta) \\ &= \sin \theta (3 - 3 \sin^2 \theta - \sin^2 \theta) \\ &= 3 \sin \theta - 4 \sin^3 \theta\end{aligned}$$

Example 2.9(a)
Using the identity $\cos 2\theta = 2 \cos^2 \theta - 1$, obtain an expression for $\cos 4\theta$ in terms of $\cos \theta$. Verify the result using De Moivre's theorem.

(i) $\cos 4\theta \equiv 2\cos^2 2\theta - 1$
$$\equiv 2(2\cos^2\theta - 1)^2 - 1$$
$$\equiv 2(4\cos^4\theta - 4\cos^2\theta + 1) - 1$$
$$\equiv 8\cos^4\theta - 8\cos^2\theta + 1$$

(ii) $\cos 4\theta + \mathrm{j}\sin 4\theta \equiv (\cos\theta + \mathrm{j}\sin\theta)^4$. Taking only the real part of both sides of the identity,

$$\cos 4\theta = \cos^4\theta + 6\mathrm{j}^2\cos^2\theta\sin^2\theta + \mathrm{j}^4\sin^4\theta$$
$$= \cos^4\theta - 6\cos^2\theta\sin^2\theta + \sin^4\theta$$
$$= \cos^4\theta - 6\cos^2\theta(1 - \cos^2\theta) + (1 - \cos^2\theta)^2$$
$$= 8\cos^4\theta - 8\cos^2\theta + 1$$

Example 2.9(b)
Solve the equation $8c^4 - 8c^2 + 1 = 0$ (a) by treating it as a quadratic equation in c^2, (b) by using the substitution $c = \cos\theta$ and using the result of Example 2.9(a).

(a) $8c^4 - 8c^2 + 1 = 0$
$$c^2 = (8 \pm \sqrt{32})/16$$
$$c^2 = 0.8536 \quad \text{or} \quad c^2 = 0.1464$$
$$c = \pm\sqrt{(0.8536)} = \pm 0.9239$$
$$\text{or} \quad c = \pm\sqrt{(0.1464)} = \pm 0.3826$$

(b) If $c = \cos\theta$, $8\cos^4\theta - 8\cos^2\theta + 1 = 0$,

$\cos 4\theta = 0$

Hence
$$4\theta = \pi/2,\ 3\pi/2,\ 5\pi/2 \text{ or } 7\pi/2$$
$$\theta = \pi/8,\ 3\pi/8,\ 5\pi/8, \text{ or } 7\pi/8$$

The required roots are
$$\cos \pi/8 = 0.9239$$
$$\cos 3\pi/8 = 0.3826$$
$$\cos 5\pi/8 = -0.3826$$
$$\cos 7\pi/8 = -0.9239$$

Other values of θ lead to the same values of c.

Unworked Examples 5

Part 2.7

(i) Express $\mathrm{e}^{(\pi/4)(1+\mathrm{j})}$ in the form $x + \mathrm{j}y$ with x and y each correct to three decimal places.

(2) Find the modulus and argument of $1 + e^{j\pi/3}$.
(3) Prove that (a) $e^{j\pi} = -1$, (b) $j^j = e^{-\pi/2}$.
(Hint: in (b) express $e^{-\pi/2}$ as $(e^{(j\pi/2)j}$.)
(4) Solve the equation $e^x = -1$. Hence deduce that the logarithm of -1 number is imaginary.

Part 2.8

(5) Express $3 + j4$ in polar form. Obtain $(3 + j4)^2$ and $(3 + j4)^3$ (a) algebraically (b) using De Moivre's theorem. Compare the results.
(6) Obtain 5 fifth roots of -1.
(7) Calculate $(\sqrt{3} + j)^{10}$.
(8) Obtain the reciprocal of $\cos\theta + j\sin\theta$ algebraically. Verify this using De Moivre's theorem.

Part 2.9

(9) By comparing the real and imaginary parts of $(\cos\theta + j\sin\theta)^2$, show that $\cos 2\theta = \cos^2\theta - \sin^2\theta$ and that $\sin 2\theta = 2\sin\theta\cos\theta$.
(10) Prove that $\cos 5x = 16\cos^5 x - 20\cos^3 x + 5\cos x$.
(11) Show that $\sin 3A = 3\sin A - 4\sin^3 A$. Use the result to solve the equation

$$8x^3 - 6x - 1 = 0$$

(12) Use De Moivre's theorem to show that
(a) $\cos n\theta = \cos^n\theta - \binom{n}{2}\cos^{n-2}\theta\sin^2\theta + \binom{n}{4}\cos^{n-4}\theta\sin^4\theta + \cdots$
(b) $\sin n\theta = \binom{n}{1}\cos^{n-1}\theta\sin\theta - \binom{n}{3}\cos^{n-3}\theta\sin^3\theta + \cdots$
where $\binom{n}{r}$ means the number of combinations of r items from n.
(13) Show that the roots of the quadratic equation $x^2 - 2x\cos\theta + 1 = 0$ are $\cos\theta \pm j\sin\theta$. Hence deduce that $\cos 2\theta = 2\cos^2\theta - 1$ and that $\sin 2\theta = 2\sin\theta\cos\theta$.

PART 3: THE RELATIONSHIP BETWEEN TRIGONOMETRIC AND HYPERBOLIC FUNCTIONS

3.1 Exponential Form of Sines and Cosines

In Part 2.7, the relationships

$$e^{j\theta} = \cos\theta + j\sin\theta$$
$$e^{-j\theta} = \cos\theta - j\sin\theta$$

were encountered. Adding these,

$$2\cos\theta = e^{j\theta} + e^{-j\theta}$$
$$\cos\theta = \tfrac{1}{2}(e^{j\theta} + e^{-j\theta}) \qquad (1)$$

Subtracting,

$$j2\sin\theta = e^{j\theta} - e^{-j\theta}$$
$$\sin\theta = \frac{1}{j2}(e^{j\theta} - e^{-j\theta}) \qquad (2)$$

(1) and (2) are referred to as the exponential form of cosine and sine respectively. Note that $e^{j\theta} + e^{-j\theta}$ is wholly imaginary. These functions may be compared with the hyperbolic functions met in Part 1.1.

Example 3.1(a)
Verify the identity $\cos^2\theta + \sin^2\theta = 1$ holds when the trigonometrical ratios are expressed in exponential form

$$\cos^2\theta = \frac{1}{4}(e^{j\theta} + e^{-j\theta})^2$$
$$= \frac{1}{4}(e^{2j\theta} + 2 + e^{-2j\theta})$$
$$\sin^2\theta = \left(\frac{1}{j2}\right)^2(e^{j\theta} - e^{-j\theta})^2$$
$$= -\frac{1}{4}(e^{2j\theta} - 2 + e^{-2j\theta})$$

$$\cos^2\theta + \sin^2\theta = \tfrac{1}{4}\times 2 + (-\tfrac{1}{4}\times -2) = 1$$

3.2 Powers of Trigonometrical Ratios

It was shown in Part 2.9 that $\cos n\theta$ and $\sin n\theta$ can be expressed as polynomials in $\cos\theta$ and $\sin\theta$. Sometimes the reverse problem arises when $\cos^n\theta$ or $\sin^n\theta$ is required in terms of multiple angles. These can be obtained using identities (1) and (2) of Part 3.1.

Thus we might have

$$\cos^4\theta = \tfrac{1}{16}(e^{j\theta} + e^{-j\theta})^4$$
$$= \tfrac{1}{16}(e^{4j\theta} + 4e^{2j\theta} + 6 + 4e^{-2j\theta} + e^{-4j\theta})$$
$$= \tfrac{1}{16}(e^{4j\theta} + e^{-4j\theta} + 4((^{2j\theta} + e^{-2j\theta})) + 6)$$
$$= \tfrac{1}{16}(2\cos 4\theta + 8\cos 2\theta + 6)$$
$$= \tfrac{1}{8}\cos 4\theta + \tfrac{1}{2}\cos 2\theta + \tfrac{3}{8}$$

or

$$\sin^3\theta = \left(\frac{1}{j2}\right)^3 (e^{j\theta} - e^{-j\theta})^3$$

$$= -\frac{1}{j8}(e^{3j\theta} - 3e^{j\theta} + 3e^{-j\theta} - e^{-3j\theta})$$

$$= -\frac{1}{4}\frac{1}{j2}(e^{3j\theta} - e^{-3j\theta} - 3((e^{j\theta} - e^{-j\theta})))$$

$$= -\frac{1}{4}(\sin 3\theta - 3\sin\theta)$$

$$= -\frac{1}{4}\sin 3\theta + \frac{3}{4}\sin\theta$$

Example 3.2(a)
Evaluate $\int_0^{\pi/3} \sin^5 x\, dx$ correct to two places of decimals.

We cannot integrate $\sin^5 x$ directly but if it can be expressed in terms of multiple angles, the integration can be carried out directly.

$$\sin^5 x = \left(\frac{1}{j2}\right)^5 (e^{jx} - e^{-jx})^5$$

$$= \frac{1}{16}\cdot\frac{1}{j2}(e^{5jx} - 5e^{3jx} + 10e^{jx} - 10e^{-jx} + 5e^{-3jx} - e^{-5jx})$$

$$= \frac{1}{16}\cdot\frac{1}{j2}((e^{5jx} - e^{-5jx}) - 5(e^{3jx} - e^{-3jx}) + 10(e^{jx} - e^{-jx}))$$

$$= \frac{1}{16}(\sin 5x - 5\sin 3x + 10\sin x)$$

$$\int_0^{\pi/3} \sin^5 x dx = \frac{1}{16}\int_0^{\pi/3} (\sin 5x - 5\sin 3x + 10\sin x)dx$$

$$= \frac{1}{16}\left[-\frac{1}{5}\cos 5x + \frac{5}{3}\cos 3x - 10\cos x\right]_0^{\pi/3}$$

$$= \frac{1}{16}\left(-\frac{1}{10} - \frac{5}{3} - 5\right) - \frac{1}{16}\left(-\frac{1}{5} + \frac{5}{3} - 10\right)$$

$$= 0.11 \text{ (corr. 2 pl. dec.)}$$

Alternative methods of solution are considered in Part 9.

3.3 Connecting Hyperbolic and Trigonometrical Functions

Comparing the identities

$$\cos\theta = \frac{1}{2}(e^{j\theta} + e^{-j\theta})$$

$$\sin\theta = \frac{1}{j2}(e^{j\theta} - e^{-j\theta})$$

which the hyperbolic functions

$$\cosh x = \tfrac{1}{2}(e^{x} + e^{-x})$$
$$\sinh x = \tfrac{1}{2}(e^{x} - e^{-x})$$

if we put $x = j\theta$, we have $\cos\theta = \cosh x$ and $j\sin\theta = \sinh x$. These relationships are more usually expressed either as

$$\cosh j\theta = \cos\theta$$
$$\sinh j\theta = j\sin\theta$$

or

$$\cos j\theta = \cosh(-\theta)$$
$$= \cosh\theta$$

$$\sin j\theta = \frac{1}{j}\sinh(-\theta)$$

$$= -\frac{1}{j}\sinh\theta$$

$$= j\sinh\theta$$

Example 3.3(a)
Obtain the real and imaginary parts of $\sin(x + jy)$.

$$\sin(x + jy) = \sin x \cos jy + \cos x \sin jy$$
$$= \sin x \cosh y + j\cos x \sinh y$$

The real part is $\sin x \cosh y$ and the imaginary part is $\cos x \sinh y$.

Example 3.3(b)
Verify that the standard trigonometrical formulae (a) $\sin^2 x + \cos^2 x = 1$ and (b) $\sin 2x = 2\sin x\cos x$ apply even if x is imaginary.

Let $x = j\theta$, with θ real.

(a) $\sin^2 x + \cos^2 x = \sin^2 j\theta + \cos^2 j\theta$

$$= -\sinh^2\theta + \cosh^2\theta$$

$$= 1$$

(b) $$\sin 2x = \sin j2\theta = j \sinh 2\theta$$
$$2 \sin x \cos x = 2 \sin j\theta \cos j\theta$$
$$= j2 \sinh \theta \cosh \theta$$
$$= j \sinh 2\theta$$

Unworked Examples 6

Part 3.1

(1) Verify that $d/dx\,(\sin x) = \cos x$ and $d/dx\,(\cos x) = -\sin x$ by using the exponential form of sines and cosines.
(2) Show that if $x^2 - 2x \cos \theta + 1 = 0$, $\cos \theta = \frac{1}{2}(x + 1/x)$ and hence that $x = e^{j\theta}$ or $e^{-j\theta}$. Verify this by solving $x^2 - 2x \cos \theta + 1 = 0$ as a quadratic equiation in x. Repeat the process with equation $x^2 - 2jx \sin \theta - 1 = 0$.
(3) Obtain $\tan \theta$ in exponential form.
(4) Why are $e^{j\theta}$ and $e^{-j\theta}$ conjugate complex numbers? Deduce that the exponential forms for $\cos \theta$ and for $\sin \theta$ must be real because they involve conjugate complex quantities.

Part 3.2

(5) Expand $\sin^4 \theta$ as the sum of cosines of multiples of θ.
(6) Prove that $16 \sin^5 x = \sin 5x - 5 \sin 3x + 10 \sin x$. Hence solve the trigonometrical equation

$$16 \sin^5 x = \sin 5x$$

giving all values of x in the range 0 to 2π.
(7) Show that $\cos^6 x = \frac{1}{32}(\cos 6x + 6 \cos 4x + 15 \cos 2x + 10)$. Hence evaluate $\int_0^{\pi/2} \cos^6 x \, dx$.
(8) Expand $\cos^5 \theta$ in terms of multiples of θ.
(9) Verify the results

$$2 \cos^2 \alpha = \cos 2\alpha + 1$$
$$8 \cos^4 \alpha = \cos 4\alpha + 4 \cos 2\alpha + 3$$
$$32 \cos^6 \alpha = \cos 6\alpha + 6 \cos 4\alpha + 15 \cos 2\alpha + 10$$

Hence or otherwise deduce that

$$\sqrt{(\tfrac{1}{2}(\cos 6\alpha + 1))} = \cos \alpha\,(4 \cos^2 \alpha - 3)$$

Part 3.3

(10) Deduce that $\tanh j\theta = j \tan \theta$ and that $\tan j\theta = j \tanh \theta$.
(11) Obtain the real and the imaginary parts of $\cos(x + jy)$.
(12) Show that the relationships $\cosh j\theta = \cos \theta$, $\sinh j\theta = j \sin \theta$ account for Osborn's rule. (See Part 1.5)

(13) Show that if $\sinh x = \tan jy$, $\cosh x = \sec jy$ and hence that $\cos jx$ and $\cos jy$ are reciprocals.

(14) Show that if $\sin (A + jB) = \cos (C + jD)$ then $\sin A \cosh B = \cos C \cosh D$ and $\cos A \sinh B = -\sin C \sinh D$. Deduce that $\sin^2 A + \sinh^2 B = \cos^2 C + \sinh^2 D$.

PART 4: DETERMINANTS

4.1 Determinant Solution of Simultaneous Equations in Two Unknowns

Given the simultaneous equations

$$ax + by = c$$
$$dx + ey = f$$

where a, b, c, d, e and f are given constants and x and y are the unknowns. Solving by any standard method, we get

$$x = (ce - bf)/(ae - bd)$$
$$y = (af - cd)/(ae - bd)$$

The expressions in the numerator and denominator of both x and y consist of the difference of two products. This solution can be arranged as three equal fractions using DETERMINANT notation for the denominators:

$$\frac{x}{\begin{vmatrix} c & b \\ f & e \end{vmatrix}} = \frac{y}{\begin{vmatrix} a & c \\ d & f \end{vmatrix}} = \frac{1}{\begin{vmatrix} a & b \\ d & e \end{vmatrix}}$$

where

$$\begin{vmatrix} \mathrm{P} & \mathrm{Q} \\ \mathrm{R} & \mathrm{S} \end{vmatrix}$$

means $\mathrm{P} \times \mathrm{S} - \mathrm{Q} \times \mathrm{R}$. The symbols P, Q, R and S are called ELEMENTS.

This form of solution is called CRAMER'S RULE. In it:

(a) the solution is arranged as a series of equal fractions; the original equations, followed by 1;

(b) the numerators consist of the unknowns, arranged in the order in which they appear in;

(c) the denominators consist of determinants, the elements of which are the coefficients in their correct position, except for the coefficients corresponding to the numerator. These are replaced by the constant terms.

Example 4.1(a)
Evaluate the determinants

(a) $\begin{vmatrix} 2 & 3 \\ 1 & 5 \end{vmatrix}$ (b) $\begin{vmatrix} -3 & 1 \\ 0 & 2 \end{vmatrix}$

(a) $\begin{vmatrix} 2 & 3 \\ 1 & 5 \end{vmatrix} = 2 \times 5 - 1 \times 3 = 7$

(b) $\begin{vmatrix} -3 & 1 \\ 0 & -2 \end{vmatrix} = (-3) \times (-2) - 0 \times 1 = 6$

Example 4.1(b)

If

$$\begin{vmatrix} 3 & x \\ 2 & 1 \end{vmatrix} = 5$$

find x.

$$\begin{vmatrix} 3 & x \\ 2 & 1 \end{vmatrix} = 3 - 2x = 5$$

$x = -1$.

Example 4.1(c)

Use determinants to solve the equations: (a) $3x - 5y = 4$, $x + 4y = 7$; (b) $2l + 5m = 4$, $3l - 2m = 6$.

(a) $$\frac{x}{\begin{vmatrix} 4 & -5 \\ 7 & 4 \end{vmatrix}} = \frac{y}{\begin{vmatrix} 3 & 4 \\ 1 & 7 \end{vmatrix}} = \frac{1}{\begin{vmatrix} 3 & -5 \\ 1 & 4 \end{vmatrix}}$$

$$\frac{x}{51} = \frac{y}{17} = \frac{1}{17}$$

$$x = 3, y = 1$$

(b) $$\frac{l}{\begin{vmatrix} 4 & 5 \\ 6 & -2 \end{vmatrix}} = \frac{m}{\begin{vmatrix} 2 & 4 \\ 3 & 6 \end{vmatrix}} = \frac{1}{\begin{vmatrix} 2 & 5 \\ 3 & -2 \end{vmatrix}}$$

$$\frac{l}{-38} = \frac{m}{0} = \frac{1}{-19}$$

$$l = 2, m = 0$$

4.2 Evaluation of Third and Higher Order Determinants

All of the determinants found in Part 4.1 consisted of four elements arranged in two rows and two columns. These were examples of SECOND ORDER determinants. A third order determinant contains nine elements arranged in three rows and three columns. Similarly a fourth order determinant contains 16 elements and so on.

Cramer's rule applies with any number of simultaneous equations but the order of the determinants involved will be equal to the number of unknowns.

In order to evaluate higher order determinants some terms must be defined. If the row and column in which a particular element appears, are removed, the value of the remaining subdeterminant is called the MINOR of that particular element. The minor of the element -2 in the determinant D:

$$\begin{vmatrix} -2 & 0 & 1 \\ 4 & 7 & 3 \\ 5 & 2 & -4 \end{vmatrix} \quad \text{is} \quad \begin{vmatrix} 7 & 3 \\ 2 & -4 \end{vmatrix} \quad \text{(i.e. } -34\text{)}$$

while the minor of 7 is

$$\begin{vmatrix} -2 & 1 \\ 5 & -4 \end{vmatrix} \quad \text{(i.e. 3)}$$

There is a minor for each element. If each *position* in the determinant is given a + or − sign, chessboard style as shown in Figure A11, the COFACTOR of an element is the value of its minor multiplied by $+1$ or by -1 as appropriate. The top left-hand position is always designated +, so that the determinant D, the elements -2 and 7, already considered, are both occupying + positions and their cofactors are equal to their minors. On the other hand, the element 0 occupies a − position and the value of its cofactor is $-(4 \times (-4) - 5 \times 3)$, i.e. 31.

The value of a third or higher order determinant is equal to the sum of the products of each element and its cofactor taken along *any* row or column. It is said to be EXPANDED along a row or down a column.

Example 4.2(a)
Obtain the cofactors of the elements of the determinant:

$$\begin{vmatrix} 4 & -2 & 3 \\ 4 & -3 & -4 \\ 5 & 0 & -5 \end{vmatrix}$$

The results are fully laid out in Table 4.1.

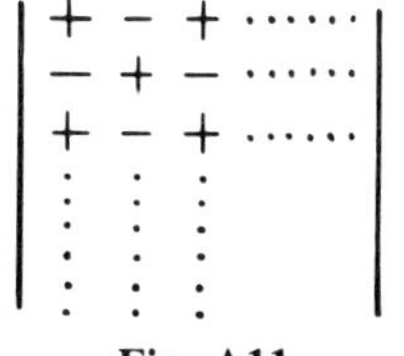

Fig. A11

Table 4.1

Position of element: Row	Position of element: Column	Value of element	Value of its minor	Positional value	Value of cofactor
1	1	4	$(-3) \times (-5) - 0 \times (-4) = 15$	+	15
1	2	−2	$4 \times (-5) - 5 \times (-4) = 0$	−	0
1	3	3	$4 \times 0 - 5 \times (-3) = 15$	+	15
2	1	4	$(-2) \times (-5) - 0 \times 3 = 10$	−	−10
2	2	−3	$4 \times (-5) - 5 \times 3 = -35$	+	−35
2	3	−4	$4 \times 0 - 5 \times (-2) = 10$	−	−10
3	1	5	$(-2) \times (-4) - (-3) \times 3 = 17$	+	17
3	2	0	$4 \times (-4) - 4 \times 3 = -28$	−	28
3	3	−5	$4 \times (-3) - 4 \times (-2) = -4$	+	−4

Example 4.2(b)
Using Table 4.1, show that the value of the determinant is equal to 105 and that this value is obtained irrespective of the row or column chosen for its expansion.

The expansion will be considered along each row and column in turn. Using the notation R 1 for expansion along the first row and C 2 for expansion down the second column, etc.

R 1: Value $= 4 \times 15 + (-2) \times 0 + 3 \times 15 = 105$

R 2: Value $= 4 \times (-10) + (-3) \times (-35) + (-4) \times (-10) = 105$

R 3: Value $= 5 \times 17 + 0 \times 28 + (-5) \times (-4) = 105$

C 1: Value $= 4 \times 15 + 4 \times (-10) + 5 \times 17 = 105$

C 2: Value $= -2 \times 0 + (-3) \times (-35) + 0 \times 28 = 105$

C 3: Value $= 3 \times 15 + (-4) \times (-10) + (-5) \times (-4) = 105$

Normally, of course, the expansion would be obtained without the production of a table and only along one row or column. In this example either R 3 or C 2 would be chosen because they both contain a zero element.

Example 4.2(c)
Solve the equations

$$3x + 4y + 2z = 4$$
$$x - 5y + 3z = -1$$
$$2x + 3y + z = 3$$

using Cramer's rule

The solution is

$$\frac{x}{D_1} = \frac{y}{D_2} = \frac{z}{D_3} = \frac{1}{D_4}$$

where

$$D_1 = \begin{vmatrix} 4 & 4 & 2 \\ -1 & -5 & 3 \\ 3 & 3 & 1 \end{vmatrix}$$

$$= 4 \times (-5 -9) - 4 \times (-1 -9) + 2 \times (-3 + 15)$$

$$= 8$$

$$D_2 = \begin{vmatrix} 3 & 4 & 2 \\ 1 & -1 & 3 \\ 2 & 3 & 1 \end{vmatrix}$$

$$= 3 \times (-1 -9) - 4 \times (1 - 6) + 2 \times (3 + 2)$$

$$= 0$$

$$D_3 = \begin{vmatrix} 3 & 4 & 4 \\ 1 & -5 & -1 \\ 2 & 3 & 3 \end{vmatrix}$$

$$= 3 \times (-15 + 3) - 4 \times (3 + 2) + 4 \times (3 + 10)$$

$$= -4$$

$$D_4 = \begin{vmatrix} 3 & 4 & 2 \\ 1 & -5 & 3 \\ 2 & 3 & 1 \end{vmatrix}$$

$$= 3 \times (-5 -9) - 4(1 - 6) + 2 \times (3 + 10)$$

$$= 4$$

$$\frac{x}{8} = \frac{y}{0} = \frac{z}{-4} = \frac{1}{4}$$

$$x = 2, y = 0, z = -1$$

In each case the determinants have been expanded along R 1.

4.3 Simplifying the Expansion Process

The expansion process for determinants can be simplified by applying certain rules:

1. Interchanging the rows and columns does not alter the value of the determinant.

2. Switching two rows or two columns changes the sign of the determinant.
3. If two rows or two columns are proportional the determinant has a zero value.
4. If all the elements of a row or column are multiplied by a factor, the value of the determinant is multiplied by that factor.
5. Any multiple of a row (or column) can be added to any other row (or column) without altering the value of determinant.

These rules can be illustrated using the determinants

$$A = \begin{vmatrix} 1 & 2 & 3 \\ -1 & 4 & 5 \\ 4 & -2 & 3 \end{vmatrix} \qquad B = \begin{vmatrix} 1 & -1 & 4 \\ 2 & 4 & -2 \\ 3 & 5 & 3 \end{vmatrix}$$

$$C = \begin{vmatrix} 1 & 2 & 3 \\ -1 & 4 & 5 \\ 2 & 4 & 6 \end{vmatrix} \qquad D = \begin{vmatrix} 1 & 2 & 3 \\ 4 & -2 & 3 \\ -1 & 4 & 5 \end{vmatrix},$$

$$E = \begin{vmatrix} 1 & 1 & 3 \\ -1 & 2 & 5 \\ 4 & -1 & 3 \end{vmatrix} \qquad F = \begin{vmatrix} 1 & 3 & 3 \\ -1 & 3 & 5 \\ 4 & 2 & 3 \end{vmatrix}$$

and

$$G = \begin{vmatrix} 1 & 0 & 0 \\ -1 & 6 & 8 \\ 4 & -10 & -9 \end{vmatrix}$$

These can be expanded in the usual way and will be found to have values $A = 26$, $B = 26$, $C = 0$, $D = -26$, $E = 13$, $F = 26$ and $G = 26$.

Interchanging the rows and columns of A gives B, the values illustrating Rule 1. This process is called TRANSPOSITION.

A and D are identical except that R 2 and R 3 have been switched, illustrating Rule 2.

C illustrates Rule 3, because R 3 is double R 1.

If the elements of C 2 of A are multiplied by $\frac{1}{2}$, the resulting determinant is E. The values of A and E illustrate Rule 4.

Adding C 1 of A to C 2 gives F, whose value illustrates Rule 5. This is a most useful rule and, if $-2 \times$ C 1 of A is added to C 2 and $-3 \times$ C 1 is added to C 3, G is obtained. This is effectively a second order determinant, whose value is easily obtained and illustrates the most practical method of expanding determinants.

Example 4.3(a)
Evaluate the determinants

(i) $\begin{vmatrix} 1 & 2 & 2 \\ 2 & 3 & 4 \\ 2 & 4 & 4 \end{vmatrix}$ (ii) $\begin{vmatrix} 4 & 2 & 3 \\ 2 & 3 & 4 \\ 3 & 4 & 5 \end{vmatrix}$ (iii) $\begin{vmatrix} 4 & 12 & 2 & 3 \\ 5 & 0 & 5 & 0 \\ -2 & 12 & -1 & 2 \\ 3 & 3 & 1 & -1 \end{vmatrix}$

(i) $\begin{vmatrix} 1 & 2 & 2 \\ 2 & 3 & 4 \\ 2 & 4 & 4 \end{vmatrix} = 0 \quad (R\,1 \propto R\,3)$

(ii) $\begin{vmatrix} 4 & 2 & 3 \\ 2 & 3 & 4 \\ 3 & 4 & 5 \end{vmatrix} \begin{matrix} C\,1 \to C\,1 - 2C\,2 \\ C\,3 \to C\,3 - C\,2 \\ = \end{matrix} \begin{vmatrix} 0 & 2 & 1 \\ -4 & 3 & 1 \\ -5 & 4 & 1 \end{vmatrix} \begin{matrix} C\,2 \to C\,2 - 2\,C3 \\ = \quad = \end{matrix} \begin{vmatrix} 0 & 0 & 1 \\ -4 & 1 & 1 \\ -5 & 2 & 1 \end{vmatrix}$

$= -8 + 5 = -3$

(iii) $\begin{vmatrix} 4 & 12 & 2 & 3 \\ 5 & 0 & 5 & 0 \\ -2 & 12 & -1 & 2 \\ 3 & 3 & 1 & -1 \end{vmatrix} \begin{matrix} R\,2 \\ = 5 \end{matrix} \begin{vmatrix} 4 & 12 & 2 & 3 \\ 1 & 0 & 1 & 0 \\ -2 & 12 & -1 & 2 \\ 3 & 3 & 1 & -1 \end{vmatrix}$

$\begin{matrix} C\,2 \\ = \end{matrix} \; 15 \begin{vmatrix} 4 & 4 & 2 & 3 \\ 1 & 0 & 1 & 0 \\ -2 & 4 & -1 & 2 \\ 3 & 1 & 1 & -1 \end{vmatrix}$

$\begin{matrix} C\,3 \to C\,3 - C\,1 \\ = \end{matrix} \; 15 \begin{vmatrix} 4 & 4 & -2 & 3 \\ 1 & 0 & 0 & 0 \\ -2 & 4 & 1 & 2 \\ 3 & 1 & -2 & 1 \end{vmatrix}$

$\begin{matrix} R\,2 \\ = \end{matrix} \; -15 \begin{vmatrix} 4 & -2 & 3 \\ 4 & 1 & 2 \\ 1 & -2 & -1 \end{vmatrix}$

$\begin{matrix} C\,1 \to C\,1 - 4C\,2 \\ C\,3 \to C\,3 - 2C\,2 \\ = \end{matrix} \; -15 \begin{vmatrix} 12 & -2 & 7 \\ 0 & 1 & 0 \\ 9 & -2 & 3 \end{vmatrix}$

$\begin{matrix} R\,2 \\ = \end{matrix} \; -15 \begin{vmatrix} 12 & 7 \\ 9 & 3 \end{vmatrix}$

$= -15(36 - 63)$

$= 405$

The principle is to simplify as far as possible by extracting common factors and then using Rule 5 to introduce zero elements. Where an element 1 or -1 exists, it is possible to use the rule to obtain zero directly. Where no such element exists, zero will be obtained in two steps.

In (iii), by obtaining a row of elements all except one of which are zero, the order of the determinant is effectively reduced by one, making its evaluation simpler.

Example 4.3(b)
Solve the equation

$$\begin{vmatrix} x & 2 & 3 \\ 8 & x & 6 \\ x & 1 & 1 \end{vmatrix} = 0$$

$$\begin{vmatrix} x & 2 & 3 \\ 8 & x & 6 \\ x & 1 & 1 \end{vmatrix} \text{R 1} = x(x - 6) - 2(8 - 6x) + 3(8 - x^2)$$

$$= -2x^2 + 6x + 8$$

If

$$\begin{vmatrix} x & 2 & 3 \\ 8 & x & 6 \\ x & 1 & 1 \end{vmatrix} = 0$$

then

$$x^2 - 3x - 4 = 0$$
$$(x - 4)(x + 1) = 0$$
$$x = 4 \text{ of } x = -1$$

It is not usually worth attempting to simplify non-numerical determinants.

Example 4.3(c)
Prove that

$$\begin{vmatrix} \cos\theta & 1 & 0 \\ 1 & 2\cos\theta & 1 \\ 0 & 1 & 2\cos\theta \end{vmatrix} = \cos 3\theta$$

$$\begin{vmatrix} \cos\theta & 1 & 0 \\ 1 & 2\cos\theta & 1 \\ 0 & 1 & 2\cos\theta \end{vmatrix} \text{R 1} = \cos\theta(4\cos^2\theta - 1) - 2\cos\theta$$

$$= 4\cos^3\theta - 3\cos\theta$$

$$= \cos 3\theta \text{ (see Part 2.9).}$$

Unworked Examples 7

Part 4.1

(1) Evaluate the determinants

$$\begin{vmatrix} 1 & 2 \\ 3 & 4 \end{vmatrix} \quad \text{and} \quad \begin{vmatrix} 1 & 1 \\ 1 & 1 \end{vmatrix}$$

(2) Find x if

$$\begin{vmatrix} 2 & x \\ 3 & 4 \end{vmatrix} + 1 = 0$$

(3) Solve the equations

$$\begin{aligned} x + 4y &= 1 \\ 2x - y &= 11 \end{aligned}$$

(4) Find a and b if $3a + b = 2a + 4 = 3b - 3a$ by rearranging the equations suitably and using Cramer's method.

(5) Prove that

$$\begin{vmatrix} a & b \\ b & a \end{vmatrix} = (a + b)(a - b)$$

Part 4.2

(6) Evaluate the determinant

$$\begin{vmatrix} 3 & 4 & 1 \\ 2 & -1 & 3 \\ 4 & 0 & 3 \end{vmatrix}$$

(7) Find the value of

$$\begin{vmatrix} 3 + a & 2 & 2a \\ 2 & 4 + a & a \\ a - 1 & 1 & a - 3 \end{vmatrix}$$

when $a = -1$.

(8) Use determinants to solve the equations

$$\begin{aligned} 2p + 3q + 4r &= 9 \\ p - 2q + 6r &= 5 \\ 7p - q - 3r &= 3 \end{aligned}$$

(9) Find the value of the determinant

$$\begin{vmatrix} 1 & 2 & 3 \\ 2 & 1 & 2 \\ 3 & 1 & 1 \end{vmatrix}$$

Replace each element by its cofactor and find the value of the resulting determinant.

(10) If the solution of the equations

$$3x - 5y + z = 0$$
$$2x + 4y + 3z = 0$$
$$2x - 7y + 4z = 0$$

were written in the determinant form

$$\frac{x}{D_1} = \frac{y}{D_2} = \frac{z}{D_3} = \frac{1}{D_4}$$

explain why D_1, D_2 and D_3 would all have a zero value, without evaluating the determinants. If $D_4 \neq 0$, deduce that $x = y = z = 0$.

Part 4.3

(11) For all values of a, find the value of

$$\begin{vmatrix} 1 & a & a^2 \\ a & a^2 & a^3 \\ a^2 & a & 1 \end{vmatrix}$$

(12) Evaluate

$$\begin{vmatrix} 54 & -23 \\ 87 & -38 \end{vmatrix}$$

(13) Show that

$$\begin{vmatrix} p & 0 & 0 & 0 \\ 0 & q & 0 & 0 \\ 0 & 0 & r & 0 \\ 0 & 0 & 0 & s \end{vmatrix} = pqrs$$

(14) Evaluate

$$\begin{vmatrix} 1 & 1 & 1 & 1 \\ 0 & 1 & 1 & 1 \\ 0 & 0 & 1 & 1 \\ 0 & 0 & 0 & 1 \end{vmatrix}$$

(15) Evaluate

$$\begin{vmatrix} 3 & 3 & 1 & 3 \\ 4 & 3 & 1 & 3 \\ 6 & 7 & 2 & 7 \\ 3 & 3 & 1 & 4 \end{vmatrix}$$

(16) Solve the equations

(a)
$$\begin{aligned} 3x - 5y + 2z &= 2 \\ 4x - y + 3z &= 9 \\ 6x - 3y + 2z &= 7 \end{aligned}$$

(b)
$$\begin{aligned} 4l + m - 8n &= 3 \\ 2l + 6m - 5n &= -5 \\ 3l + 4m - 2n &= 3 \end{aligned}$$

(17) Find x if

$$\begin{vmatrix} x & 1 & x \\ -1 & x & 3 \\ 2 & 1 & 4 \end{vmatrix} = 8$$

(18) Show that

$$\begin{vmatrix} a+b & a & a \\ a & a+b & a \\ a & a & a+b \end{vmatrix} = b^2(3a + b)$$

Verify that the result is true by putting $a = 1, b = 3$ in both the determinant and its algebraic value.

PART 5: LINEAR DEPENDENCY

5.1 Linear Dependence of Equations

Suppose that we use Cramer's method to solve the set of equations

$$5x + 3y + z = 2 \quad (1)$$

$$x - y + 2z = 3 \quad (2)$$

$$7x + y + 5z = 8 \quad (3)$$

Here $x/D_1 = y/D_2 = z/D_3 = 1/D_4$ and

$$D_1 = \begin{vmatrix} 2 & 3 & 1 \\ 3 & -1 & 2 \\ 8 & 1 & 5 \end{vmatrix} \begin{matrix} \text{C } 1 \rightarrow \text{C } 1 - 2\text{C } 3 \\ \text{C } 2 \rightarrow \text{C } 2 - 3\text{C } 3 \\ = \end{matrix} \begin{vmatrix} 0 & 0 & 1 \\ -1 & -7 & 2 \\ -2 & -14 & 5 \end{vmatrix} = 0 \ (\text{C } 1 \propto \text{C } 2)$$

$$D_2 = \begin{vmatrix} 5 & 2 & 1 \\ 1 & 3 & 2 \\ 7 & 8 & 5 \end{vmatrix} \begin{matrix} \text{C 1} \rightarrow \text{C 1} - 5\text{C 3} \\ \text{C 2} \rightarrow \text{C 2} - 2\text{C 3} \\ = \end{matrix} \begin{vmatrix} 0 & 0 & 1 \\ -9 & -1 & 2 \\ -18 & -2 & 5 \end{vmatrix} = 0 \; (\text{C 1} \propto \text{C 2})$$

$$D_3 = \begin{vmatrix} 5 & 3 & 2 \\ 1 & -1 & 3 \\ 7 & 1 & 8 \end{vmatrix} \begin{matrix} \text{C 2} \rightarrow \text{C 2} + \text{C 1} \\ \text{C 3} \rightarrow \text{C 3} - 3\text{C 1} \\ = \end{matrix} \begin{vmatrix} 5 & 8 & -13 \\ 1 & 0 & 0 \\ 7 & 8 & -13 \end{vmatrix} = 0 \; (\text{C 2} \propto \text{C 3})$$

$$D_4 = \begin{vmatrix} 5 & 3 & 1 \\ 1 & -1 & 2 \\ 7 & 1 & 5 \end{vmatrix} \begin{matrix} \text{C 1} \rightarrow \text{C 1} - 5\text{C 3} \\ \text{C 2} \rightarrow \text{C 2} - 3\text{C 3} \\ = \end{matrix} \begin{vmatrix} 0 & 0 & 1 \\ -9 & -7 & 2 \\ -18 & -14 & 5 \end{vmatrix} = 0 \; (\text{C 1} \propto \text{C 2})$$

The method appears to break down! The reason for this is found in the original equations because (1) + 2 × (2) = (3) and so, although there appear to be three equations, one of these is a combination of the other two. The set of equations is said to be INDETERMINATE while the equations themselves are described as LINEARLY DEPENDENT. Only linearly independent equations of this type are determinate.

If Equation (3) had been

$$7x + y + 5z = 9$$

there would have been an algebraic contradiction between this and (1) + 2 × (2), because $7x + y + 5z$ cannot be equal to both 8 and 9.

A mathematically impossible set of equations such as this is said to be INCONSISTENT. Changing the constant term from 8 to 9 would alter the values of determinants D_1, D_2 and D_3 and make them non-zero, leaving D_4 zero.

On the other hand, it is quite possible to have all or any of D_1, D_2 or D_3 equal to zero and D_4 non-zero, when one or more of the variables is equal to zero (see Example 4.2 (c) in which $y = 0$).

The nature of the equations can therefore be determined by examining D_1, D_2, D_3 and D_4 to see whether they are zero or non-zero. The possibilities are

Table 5.1

Value D_1, D_2 and D_3	Value of D_4	Nature of equations
Non-zero	Non-zero	Determinate. All of the variables non-zero
Some or all zero	Non-zero	Determinate. Some of variables zero
Zero	Zero	Indeterminate
Non-zero	Zero	Inconsistent

summarised in Table 5.1. When the final determinant is non-zero, we refer to a UNIQUE solution of the equations.

Example 5.1(a)
Verify that the equations

$$2x - 5y + 3z = 5$$
$$x + 4y - z = 1$$
$$4x + 3y + z = 7$$

are consistent but indeterminate.

Writing the solution

$$\frac{x}{D_1} = \frac{y}{D_2} = \frac{z}{D_3} = \frac{1}{D_4}$$

$$D_1 = \begin{vmatrix} 5 & -5 & 3 \\ 1 & 4 & -1 \\ 7 & 3 & 1 \end{vmatrix} \begin{matrix} C\,2 \to C\,2 - 4C\,1 \\ C\,3 \to C\,3 + C\,1 \\ = \end{matrix} \begin{vmatrix} 5 & -25 & 8 \\ 1 & 0 & 0 \\ 7 & -25 & 8 \end{vmatrix} = 0 \; (C\,2 \propto C\,3)$$

$$D_2 = \begin{vmatrix} 2 & 5 & 3 \\ 1 & 1 & -1 \\ 4 & 7 & 1 \end{vmatrix} \begin{matrix} C\,2 \to C\,2 - C\,1 \\ C\,3 \to C\,3 + C\,1 \\ = \end{matrix} \begin{vmatrix} 2 & 3 & 5 \\ 1 & 0 & 0 \\ 4 & 3 & 5 \end{vmatrix} = 0 \; (C\,2 \propto C\,3)$$

$$D_3 = \begin{vmatrix} 2 & -5 & 5 \\ 1 & 4 & 1 \\ 4 & 3 & 7 \end{vmatrix} \begin{matrix} C\,2 \to C\,2 - 4C\,1 \\ C\,3 \to C\,3 - C\,1 \\ = \end{matrix} \begin{vmatrix} 2 & -13 & 3 \\ 1 & 0 & 0 \\ 4 & -13 & 3 \end{vmatrix} = 0 \; (C\,2 \propto C\,3)$$

$$D_4 = \begin{vmatrix} 2 & -5 & 3 \\ 1 & 4 & -1 \\ 4 & 3 & 1 \end{vmatrix} \begin{matrix} C\,2 \to C\,2 - 4C\,1 \\ C\,3 \to C\,3 + C\,1 \\ = \end{matrix} \begin{vmatrix} 2 & -13 & 5 \\ 1 & 0 & 0 \\ 4 & -13 & 5 \end{vmatrix} = 0 \; (C\,2 \propto C\,3)$$

Reference to Table 5.1 shows that the values of this combination make the equations consistent but indeterminate.

Example 5.1(b)
Find the value of k which makes the following equations linearly dependent

$$3x + 4y + 2z = 2$$
$$7x + 3y + 4z = 3$$
$$kx - 7y + 4z = 1$$

With the notation adopted in Part 5.1, the equations are linearly dependent if $D_4 = 0$

$$\begin{vmatrix} 3 & 4 & 2 \\ 7 & 3 & 4 \\ k & -7 & 4 \end{vmatrix} = 0$$

Expanding along R 3

$$10k + 7 \times (-2) + 4 \times (-19) = 0$$
$$10k - 90 = 0$$
$$k = 9$$

Example 5.1(c)
If p and q are constants, determine the nature of the equations

$$pa + 5b - 4c = q$$
$$2a + 3b - c = 4$$
$$4a + b - 5c = 2$$

if (i) $p = 5, q = 7$, (ii) $p = 5, q = 6$, (iii) $p = 6, q = 3$. Find a, b and c in any case where a unique solution exists.

With the usual notation we can write

$$\frac{a}{D_1} = \frac{b}{D_2} = \frac{c}{D_3} = \frac{1}{D_4}$$

and expanding the determinants, we obtain

$$D_1 = -14q + 98$$
$$D_2 = -18p + 6q + 48$$
$$D_3 = 2p - 10q + 60$$
$$D_4 = -14p + 70$$

(i) If $p = 5$, $q = 7$; $D_1 = D_2 = D_3 = D_4 = 0$. The equations are indeterminate.
(ii) If $p = 5$, $q = 6$; $D_1 = 14$, $D_2 = -6$, $D_3 = 10$ and $D_4 = 0$. The equations are inconsistent.
(iii) If $p = 6$, $q = 3$; $D_1 = 56$, $D_2 = -42$, $D_3 = 42$, $D_4 = -14$. A unique solution exists.

$$a = D_1/D_4 = -4$$
$$b = D_2/D_4 = 3$$
$$c = D_3/D_4 = -3$$

5.2 Ill-Conditioned Equations

The set of equations

$$x + y = 2$$
$$2x + ky = 3.9$$

where k is a constant to be given, has the determinant solution

$$\frac{x}{2k - 3.9} = \frac{y}{-0.1} = \frac{1}{k - 2}$$

Clearly, if $k = 2$, the equations are inconsistent. If k is close to but different from 2, remarkable results are obtained.

When $k + 1.99$, $x = -8$ and $y = 10$

When $k = 2.01$, $x = 12$ and $y = -10$

In other words altering k very slightly changes the solution of the equations out of all proportion and this is obviously due to the factor $k - 2$, which is very small. Equations where this happens are said to be ILL-CONDITIONED.

In many cases there is not a lot that can be done to help the situation. One possibility is to work with fractional coefficients rather than their approximated decimal equivalents when equations are severely ill-conditioned.

Example 5.2(a)
Verify that the equations

$$28a + 15b = 13$$
$$13a + 7b = 6$$

have the unique solutions $a = 1$, $b = -1$. Divide each equation by 19, correcting all results to two places of decimals. Solve the resulting equations and comment on the conditioning status of the equations.

Since the equations are clearly linearly independent and satisfied by $a = 1$, $b = -1$, this is a unique solution.

Dividing by 19 and correcting gives

$$1.47a + 0.79b = 0.68$$
$$0.68a + 0.37b = 0.32$$

and the solution obtained using determinants is

$$\frac{a}{-0.0012} = \frac{b}{0.008} = \frac{1}{0.0067}$$

i.e. $a = -0.18$, $b = 1.19$.

The marked difference between the solutions of the two sets of equations indicates that they are severely ill-conditioned.

Example 5.2(b)
Show that the equations

$$9x + 5y = 4$$
$$kx + 6y = 5$$

have a solution $x = 1$, $y = -1$ when $k = 11$. If k is increased by 1% find the resulting percentage changes in x and y.

Using determinant notation, the results simplify to

$$\frac{x}{-1} = \frac{y}{45 - 4k} = \frac{1}{54 - 5k}$$

$$x = 1/(5k - 54) \qquad y = (4k - 45)/(5k - 54)$$

when $k = 11$, $x = 1$, $y = -1$. Increasing k by 1%, its value becomes 11.11

$$x = 1/1.55 = 0.65$$
$$y = -0.56/1.55 = -0.36$$

Increasing k by 1% decreases x by 35% and increases y by 64%.

Unworked Examples 8

Part 5.1

(1) Explain the following terms in connection with linear simultaneous equations:
(a) indeterminate equations,
(b) linearly independent equations,
(c) inconsistent equations,
(d) a unique solution.

(2) Show that the equations

$$7a - 3b + 5c = 0$$
$$5a - 2b + 3c = 0$$
$$7a - 4b + 9c = 0$$

form an indeterminate set. Find the ratio $a:b:c$.

(3) Assuming that the equations

$$3x - 5y + 11z = 0$$
$$8x + 7y - 2z = 0$$
$$7x - 4y + 9z = 0$$

are linearly independent, suggest a soultion without a calculation.

(4) Find the value of k which makes the following equations linearly dependent

$$3l + 5m - n = 2$$
$$4l + 7m - 3n = 4$$
$$5l + 7m + kn = -2$$

(5) If

$$\begin{aligned} px + 9y - 3z &= q \\ 3x + 5y + z &= 5 \\ 2x + 3y + 3z &= 8 \end{aligned}$$

(a) show that there is unique solution to the equations when $p = 1$ and $q = -5$, and find it.
(b) show that equation are linearly dependent when $p = 5$ and $q = -1$.

Part 5.2

(6) Verify that the exact solution of the equations

$$\begin{aligned} \tfrac{3}{8}x + \tfrac{1}{3}y &= \tfrac{25}{4} \\ \tfrac{9}{19}x + \tfrac{3}{7}y &= \tfrac{54}{7} \end{aligned}$$

are $x = 38$, $y = -24$. Demonstrate that the equations are ill-conditioned by obtaining all coefficients and constants correct to two places of decimals and obtaining a solution.

(7) Which of the following sets of equations is highly ill-conditioned

(a) $$\begin{aligned} 2x - y &= 2 \\ 2x - 1.01y &= 1 \end{aligned}$$

(b) $$\begin{aligned} 2x - y &= 2 \\ 2x + 1.01y &= 1 \end{aligned}$$

(8) Solve the system of equations

$$\begin{aligned} 3.01x - y &= 3 \\ 3x - ky &= 2 \end{aligned}$$

if k is equal to (i) 1 (ii) 0.99 (iii) 1.01. Give answers to three significant figures.

(9) Solve the equations

$$\begin{aligned} 3x + 5y &= 8 \\ 4x - 3y &= 1 \end{aligned}$$

Are these equations well or ill-conditioned?

PART 6: MATRICES AND THE SOLUTION OF SIMULTANEOUS EQUATIONS

6.1 Matrix Notation

Suppose that P in Figure A12 represents an engineering system with two input values x_1 and y_1 and two output values x_2 and y_2 linked linearly. Written

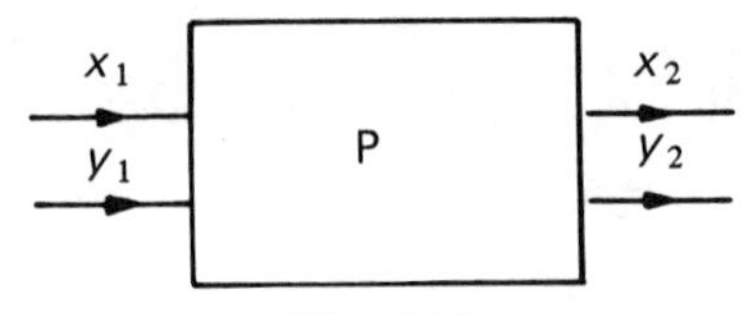

Fig. A12

mathematically, we can say

$$x_2 = ax_1 + by_1$$
$$y_2 = cx_1 + dy_1$$

where a , b, c and d are constants whose values would be determined from system P. Knowing these constants and given values of x_1 and y_1, the values of x_2 and y_2 can be determined directly. On the other hand, given the output values x_2 and y_2, x_1 and y_1 can be determined but only by solving a pair of simultaneous equations.

The sort of equations described occur frequently in many branches of engineering and it is convenient to represent them using the following MATRIX (plural MATRICES) notation.

$$\begin{pmatrix} x_2 \\ y_2 \end{pmatrix} = \begin{pmatrix} a & b \\ c & d \end{pmatrix} \begin{pmatrix} x_1 \\ y_1 \end{pmatrix}$$

In the same way the set of equations

$$x_2 = lx_1 + my_1 + nz_1$$
$$y_2 = px_1 + qy_1 + rz_1$$
$$z_2 = sx_1 + ty_1 + uz_1$$

where l, m, n ... u are constants and would be written

$$\begin{pmatrix} x_2 \\ y_2 \\ z_2 \end{pmatrix} = \begin{pmatrix} l & m & n \\ p & q & r \\ s & t & u \end{pmatrix} \begin{pmatrix} x_1 \\ y_1 \\ z_1 \end{pmatrix}$$

so that in each case a *set* of equations is replaced by a single matrix equation. Matrices may have any number of rows and columns and one with m rows and n columns is described as an $m \times n$ matrix (pronounced "m by n matrix"). However for most applications, the matrix is either SQUARE (i.e. the

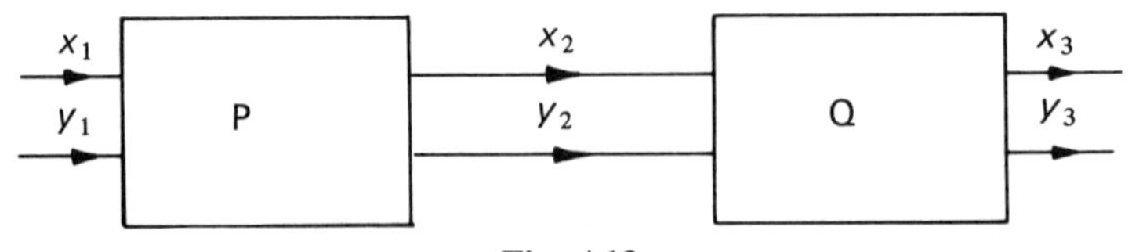

Fig. A13

number of rows is equal to the number of columns) or consists of a single column. Such a matrix is called a COLUMN MATRIX or a COLUMN VECTOR. (This is often just referred as a VECTOR.) As with determinants the individual values are called elements.

If P and Q in Figure A13 represent two linked engineering systems such that the output from P represents the input to Q, then for the first system we have

$$x_2 = ax_1 + by_1$$
$$y_2 = cx_1 + dy_1$$

and for the second we might have

$$x_3 = ex_2 + fy_2$$
$$y_3 = gx_2 + hy_2$$

If P and Q were left unchanged but integrated into one system R, Figure A13 would become the system represented by Figure A14.

Combining the two sets of equations above

$$\begin{aligned} x_3 &= e(ax_1 + by_1) + f(cx_1 + dy_1) \\ &= (ea + fc)x_1 + (eb + fd)y_1 \\ y_3 &= g(ax_1 + by_1) + h(cx_1 + dy_1) \\ &= (ga + hc)x_1 + (gb + hd)y_1 \end{aligned}$$

If all three of these sets of equations are arranged in matrix form we get

$$\begin{pmatrix} x_2 \\ y_2 \end{pmatrix} = \begin{pmatrix} a & b \\ c & d \end{pmatrix} \begin{pmatrix} x_1 \\ y_1 \end{pmatrix} \tag{1}$$

$$\begin{pmatrix} x_3 \\ y_3 \end{pmatrix} = \begin{pmatrix} e & f \\ g & h \end{pmatrix} \begin{pmatrix} x_2 \\ y_2 \end{pmatrix} \tag{2}$$

$$\begin{pmatrix} x_3 \\ y_3 \end{pmatrix} = \begin{pmatrix} ea + fc & eb + fd \\ ga + hc & gb + hd \end{pmatrix} \begin{pmatrix} x_1 \\ y_1 \end{pmatrix} \tag{3}$$

Comparing (i) and (2) with (3) we can establish the result

$$\begin{pmatrix} e & f \\ g & h \end{pmatrix} \begin{pmatrix} a & b \\ c & d \end{pmatrix} = \begin{pmatrix} ea + fc & eb + fd \\ ga + hc & gb + hd \end{pmatrix}$$

This illustrates how two matrices are multiplied together. The elements along a row of the left-hand matrix are each multiplied by the correspondingly

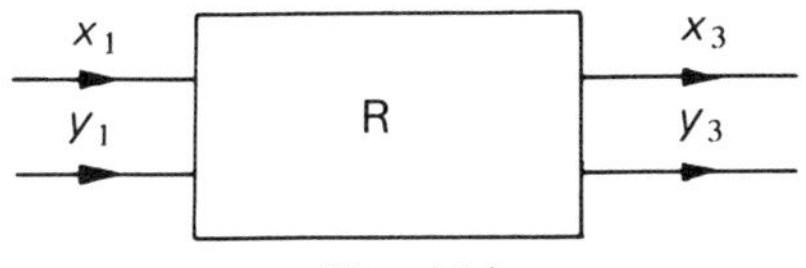

Fig. A14

placed element down a column of the right-hand matrix, the result being the sum of all such products. This is repeated for each row and for every column and means that the number of columns of the left-hand matrix must match the number of rows of the matrix on the right. When this occurs the matrices are said to be CONFORMABLE for multiplication. Only such conformable matrices may be multiplied.

Notice that the rule for multiplying conformable matrices was used in Equations (1), (2), and (3) above. It is convenient to denote matrices by single letters and bold type, i.e. **A**.

Example 6.1(a)
If **A** is the matrix

$$\begin{pmatrix} 2 & 1 & -1 \\ 3 & 4 & 0 \end{pmatrix}$$

and **B** is the matrix

$$\begin{pmatrix} 1 & 3 \\ 2 & 1 \\ -1 & -2 \end{pmatrix}$$

show that **A** and **B** are conformable for multiplication and obtain the products (a) $\mathbf{A} \times \mathbf{B}$ (b) $\mathbf{B} \times \mathbf{A}$.

By the definition of conformability both of the products $\mathbf{A} \times \mathbf{B}$ and $\mathbf{B} \times \mathbf{A}$ may be obtained.

(a)

$$\mathbf{A} \times \mathbf{B} = \begin{pmatrix} 2 & 1 & -1 \\ 3 & 4 & 0 \end{pmatrix} \times \begin{pmatrix} 1 & 3 \\ 2 & 1 \\ -1 & -2 \end{pmatrix}$$

$$= \begin{pmatrix} 5 & 9 \\ 11 & 13 \end{pmatrix}$$

(b)

$$\mathbf{B} \times \mathbf{A} = \begin{pmatrix} 1 & 3 \\ 2 & 1 \\ -1 & -2 \end{pmatrix} \times \begin{pmatrix} 2 & 1 & -1 \\ 3 & 4 & 0 \end{pmatrix}$$

$$= \begin{pmatrix} 11 & 13 & -1 \\ 7 & 6 & -2 \\ -8 & -9 & 1 \end{pmatrix}$$

In (a), the element 5 of the product is obtained as $2 \times 1 + 1 \times 2 + (-1) \times (-1) \times (-1)$. The element 9 is obtained as $2 \times 3 + 1 \times 1 + (-1) \times (-2)$ and so on. The products $\mathbf{A} \times \mathbf{B}$ and $\mathbf{B} \times \mathbf{A}$ (normally written **AB** and **BA**) are not, in general, equal and so it is usual to refer to **AB** as **A** PRE-MULTIPLYING **B** or **B** POST-MULTIPLYING **A**. The conformability of the matrices may be such that only one order of multiplication is possible. However if **A** and **B** are both square matrices of the same size, **AB** and **BA** exist and both will be the same size as **A** or **B**. However, in general, **AB** will still not be equal to **BA**.

Example 6.1(b)
If

$$\mathbf{A} = \begin{pmatrix} 2 & 3 \\ 1 & 5 \end{pmatrix} \quad \text{and} \quad \mathbf{B} = \begin{pmatrix} 4 & 2 \\ -5 & -3 \end{pmatrix}$$

find the matrix products **AB** and **BA**.

$$\begin{aligned} \mathbf{AB} &= \begin{pmatrix} 2 & 3 \\ 1 & 5 \end{pmatrix}\begin{pmatrix} 4 & 2 \\ -5 & -3 \end{pmatrix} \\ &= \begin{pmatrix} -7 & -5 \\ -21 & -13 \end{pmatrix} \\ \mathbf{BA} &= \begin{pmatrix} 4 & 2 \\ -5 & -3 \end{pmatrix}\begin{pmatrix} 2 & 3 \\ 1 & 5 \end{pmatrix} \\ &= \begin{pmatrix} 10 & 22 \\ -13 & -30 \end{pmatrix} \end{aligned}$$

Example 6.1(c)
Using the notation det **A** as the value of the determinant whose elements are those of a square matrix **A**, obtain det **A**, det **B**, det **AB** and det **BA** using the results of Example 6.1(b). Comment on the values obtained.

$$\begin{aligned} \det \mathbf{A} &= 2 \times 5 - 3 \times 1 = 7 \\ \det \mathbf{B} &= 4 \times (-3) - 2 \times (-5) = -2 \\ \det \mathbf{AB} &= (-7) \times (-13) - (-21) \times (-5) \\ &= -14 \\ \det \mathbf{BA} &= 10 \times (-30) - 22 \times (-13) \\ &= -14 \end{aligned}$$

This illustrates the result

$$\det \mathbf{AB} = \det \mathbf{BA} = \det \mathbf{A} \times \det \mathbf{B}$$

Example 6.1(d)
If

$$\mathbf{A} = \begin{pmatrix} 2 & 1 \\ 1 & -1 \end{pmatrix} \quad \mathbf{B} = \begin{pmatrix} 4 & 2 \\ 1 & 1 \end{pmatrix} \quad \mathbf{C} = \begin{pmatrix} 1 & -1 \\ 2 & -1 \end{pmatrix}$$

obtain (**AB**)**C** and **A**(**BC**).

$$\mathbf{AB} = \begin{pmatrix} 2 & 1 \\ 1 & -1 \end{pmatrix}\begin{pmatrix} 4 & 2 \\ 1 & 1 \end{pmatrix}$$

$$= \begin{pmatrix} 9 & 5 \\ 3 & 1 \end{pmatrix}$$

$$(\mathbf{AB})\mathbf{C} = \begin{pmatrix} 9 & 5 \\ 3 & 1 \end{pmatrix}\begin{pmatrix} 1 & -1 \\ 2 & -1 \end{pmatrix}$$

$$= \begin{pmatrix} 19 & -14 \\ 5 & -4 \end{pmatrix}$$

$$\mathbf{BC} = \begin{pmatrix} 4 & 2 \\ 1 & 1 \end{pmatrix}\begin{pmatrix} 1 & -1 \\ 2 & -1 \end{pmatrix}$$

$$= \begin{pmatrix} 8 & -6 \\ 3 & -2 \end{pmatrix}$$

$$\mathbf{A}(\mathbf{BC}) = \begin{pmatrix} 2 & 1 \\ 1 & -1 \end{pmatrix}\begin{pmatrix} 8 & -6 \\ 3 & -2 \end{pmatrix}$$

$$= \begin{pmatrix} 19 & -14 \\ 5 & -4 \end{pmatrix}$$

Both results are identical as long as the order of multiplication of matrices is preserved. The triple product may be written as **ABC** without ambiguity.

6.2 Addition of Matrices

Matrices are conformable for addition if they have equal numbers of rows and equal numbers of columns. To add the matrices, corresponding elements must be totalled so that, for example,

$$\begin{pmatrix} 3 & 1 & -2 \\ 4 & 0 & -6 \end{pmatrix} + \begin{pmatrix} 5 & -1 & 0 \\ -2 & 0 & 2 \end{pmatrix} = \begin{pmatrix} 8 & 0 & -2 \\ 2 & 0 & -4 \end{pmatrix}$$

We can obtain a rule for SCALAR MULTIPLICATION (obtaining the product of a pure number with a matrix) from this.

$$3 \times \begin{pmatrix} 1 & 2 \\ 3 & 4 \end{pmatrix} = \begin{pmatrix} 1 & 2 \\ 3 & 4 \end{pmatrix} + \begin{pmatrix} 1 & 2 \\ 3 & 4 \end{pmatrix} + \begin{pmatrix} 1 & 2 \\ 3 & 4 \end{pmatrix}$$

$$= \begin{pmatrix} 3 & 6 \\ 9 & 12 \end{pmatrix}$$

To multiply a matrix by a scalar quantity k, each element of the matrix is multiplied by k. In the same way a common factor may be extracted from each element of the matrix, so that

$$\begin{pmatrix} 2 & 0 & 4 \\ 2 & 6 & 2 \\ 4 & 4 & -2 \end{pmatrix} = 2\begin{pmatrix} 1 & 0 & 2 \\ 1 & 3 & 1 \\ 2 & 2 & -1 \end{pmatrix}$$

(This process should be contrasted with the extraction of a factor from a determinant, where it need only be common to the elements in a single row or column.)

If a matrix contains fractinal elements, it is more convenient to extract a fractional factor leaving the elements integral so that rather than carry out matrix operations with

$$\begin{pmatrix} 1\frac{1}{2} & \frac{1}{2} & -2 \\ 0 & -\frac{1}{4} & -1\frac{1}{4} \\ 2 & 1 & 2 \end{pmatrix}$$

we would use

$$\frac{1}{4}\begin{pmatrix} 6 & 2 & -8 \\ 0 & -1 & -5 \\ 8 & 4 & 8 \end{pmatrix}$$

Example 6.2(a)

If

$$\mathbf{A} = \begin{pmatrix} 3 & 0 \\ \frac{1}{3} & -\frac{2}{3} \end{pmatrix} \qquad \mathbf{B} = \begin{pmatrix} 1 & 1 \\ -1 & 1 \end{pmatrix}$$

find $3\mathbf{A} + \mathbf{B}$

$$3\mathbf{A} = \begin{pmatrix} 9 & 0 \\ 1 & -2 \end{pmatrix}$$

$$3\mathbf{A} + \mathbf{B} = \begin{pmatrix} 10 & 1 \\ 0 & -1 \end{pmatrix}$$

6.3 Inverse of a Matrix

Let us try pre- and post-multiplying the matrix **A**:

$$\begin{pmatrix} 4 & -1 & 2 \\ 1 & 3 & 4 \\ 2 & 5 & 1 \end{pmatrix}$$

by the matrix

$$\begin{pmatrix} 1 & 0 & 0 \\ 0 & 1 & 0 \\ 0 & 0 & 1 \end{pmatrix}$$

which we will denote by **I**:

$$\begin{pmatrix}1 & 0 & 0\\0 & 1 & 0\\0 & 0 & 1\end{pmatrix}\begin{pmatrix}4 & -1 & 2\\1 & 3 & 4\\2 & 5 & 1\end{pmatrix} = \begin{pmatrix}4 & -1 & 2\\1 & 3 & 4\\2 & 5 & 1\end{pmatrix}$$

$$\begin{pmatrix}4 & -1 & 2\\1 & 3 & 4\\2 & 5 & 1\end{pmatrix}\begin{pmatrix}1 & 0 & 0\\0 & 1 & 0\\0 & 0 & 1\end{pmatrix} = \begin{pmatrix}4 & -1 & 2\\1 & 3 & 4\\2 & 5 & 1\end{pmatrix}$$

Rather surprisingly, perhaps, we find that **A** is unchanged when pre- or post-multiplied by **I**. **I** is obviously a special matrix with properties analogous to the number 1. It is called the UNIT MATRIX or IDENTITY MATRIX and is denoted exclusively by **I**. It is a square matrix of any size and has unit elements down the main diagonal (from top left to bottom right) and zero elements elsewhere. For any conformable matrix **A**

$$\mathbf{AI} = \mathbf{IA} = \mathbf{A}$$

Next consider the products of conformable matrices **P**:

$$\begin{pmatrix}9 & 4 & 6\\5 & 3 & 4\\4 & 2 & 3\end{pmatrix}$$

and **Q**:

$$\begin{pmatrix}1 & 0 & -2\\1 & 3 & -6\\-2 & -2 & 7\end{pmatrix}$$

$$\begin{aligned}\mathbf{PQ} &= \begin{pmatrix}9 & 4 & 6\\5 & 3 & 4\\4 & 2 & 3\end{pmatrix}\begin{pmatrix}1 & 0 & -2\\1 & 3 & -6\\-2 & -2 & 7\end{pmatrix}\\ &= \begin{pmatrix}1 & 0 & 0\\0 & 1 & 0\\0 & 0 & 1\end{pmatrix}\\ &= \mathbf{I}\end{aligned}$$

$$\begin{aligned}\mathbf{QP} &= \begin{pmatrix}1 & 0 & -2\\1 & 3 & -6\\-2 & -2 & 7\end{pmatrix}\begin{pmatrix}9 & 4 & 6\\5 & 3 & 4\\4 & 2 & 3\end{pmatrix}\\ &= \begin{pmatrix}1 & 0 & 0\\0 & 1 & 0\\0 & 0 & 1\end{pmatrix}\\ &= \mathbf{I}\end{aligned}$$

$\mathbf{P}$ and $\mathbf{Q}$ are matrices with properties analogous to reciprocal numbers. Each is called the INVERSE of the other. This result is written $\mathbf{P} = \mathbf{Q}^{-1}$ and $\mathbf{Q} = \mathbf{P}^{-1}$.

It follows that $\mathbf{P}\mathbf{P}^{-1} = \mathbf{P}^{-1}\mathbf{P} = \mathbf{I}$ and that $(\mathbf{P}^{-1})^{-1} = \mathbf{P}$

Example 6.3(a)
Obtain a value for det $\mathbf{I}$. Deduce that det $\mathbf{P}$ and det $\mathbf{P}^{-1}$ are reciprocals.

$$\det \mathbf{I} = \begin{vmatrix} 1 & 0 & 0 & 0 & \cdots \\ 0 & 1 & 0 & 0 & \cdots \\ 0 & 0 & 1 & 0 & \cdots \\ 0 & 0 & 0 & 1 & \cdots \\ \vdots & \vdots & \vdots & \vdots & \end{vmatrix} \quad \text{(of order } n\text{)}$$

$$= \begin{vmatrix} 1 & 0 & 0 & \cdots \\ 0 & 1 & 0 & \cdots \\ 0 & 0 & 1 & \cdots \\ \vdots & \vdots & & \end{vmatrix} \quad \text{(of order } n - 1\text{)}$$

by expanding along R 1. Continuing in this this way $n - 2$ times further

$$\det \mathbf{I} = |1| = 1$$

Since $\mathbf{P}\mathbf{P}^{-1} = \mathbf{I}$,

$$\det \mathbf{P} \det \mathbf{P}^{-1} = \det \mathbf{I} = 1$$

$$\det \mathbf{P}^{-1} = 1/\det \mathbf{P}$$

Example 6.3(b)
Show that

$$\begin{pmatrix} 1 & 6 & 3 \\ -2 & -7 & -5 \\ -1 & -2 & -2 \end{pmatrix}$$

is the inverse of

$$\begin{pmatrix} 4 & 6 & -9 \\ 1 & 1 & -1 \\ -3 & -4 & 5 \end{pmatrix}$$

The simplest way to demonstrate that one matrix is the inverse of the other is to show that their product yields $\mathbf{I}$

$$\begin{pmatrix} 1 & 6 & 3 \\ -2 & -7 & -5 \\ -1 & -2 & -2 \end{pmatrix}\begin{pmatrix} 4 & 6 & -9 \\ 1 & 1 & -1 \\ -3 & -4 & 5 \end{pmatrix} = \begin{pmatrix} 1 & 0 & 0 \\ 0 & 1 & 0 \\ 0 & 0 & 1 \end{pmatrix}$$

6.4 The Inversion Process

Any square matrix $\mathbf{A}$ may be inverted by the following procedure.

(i) Replace each element by its cofactor.

(ii) Transpose the resulting matrix.

(iii) Divide the result of step (ii) by det A.

(See Part 4.3 for the meanings of cofactor and transposition.)

If $\mathbf{A}$ is the matrix

$$\begin{pmatrix} 3 & 1 & 1 \\ 1 & -1 & -2 \\ 3 & 3 & 4 \end{pmatrix}$$

the matrix of cofactors is

$$\begin{pmatrix} 2 & -10 & 6 \\ -1 & 9 & -6 \\ -1 & 7 & -4 \end{pmatrix}$$

Transposing this matrix gives

$$\begin{pmatrix} 2 & -1 & -1 \\ -10 & 9 & 7 \\ 6 & -6 & -4 \end{pmatrix}$$

which is called the ADJOINT of $\mathbf{A}$. It will be found that det $\mathbf{A} = 2$. Hence

$$\mathbf{A}^{-1} = \tfrac{1}{2}\begin{pmatrix} 2 & -1 & -1 \\ -10 & 9 & 7 \\ 6 & -6 & -4 \end{pmatrix}$$

The result can be verified by calculating $\mathbf{AA}^{-1}$ or $\mathbf{A}^{-1}\mathbf{A}$.

Having obtained the matrix of cofactors, it is sensible to carry out a mental check by multiplying each of its elements with the corresponding element of $\mathbf{A}$, along each row and down each column. The result will be equal to det $\mathbf{A}$ in each case. This not only checks the cofactors but also gives the value of det $\mathbf{A}$ needed at step (iii).

If det $\mathbf{A} = 0$, the matrix $\mathbf{A}$ has no inverse. Such a matrix is said to be SINGULAR.

Example 6.4(a)

Obtain the inverse of $\mathbf{A}$:

$$\begin{pmatrix} -1 & -5 & 12 \\ 0 & 1 & -2 \\ 2 & 8 & -19 \end{pmatrix}$$

The matrix of cofactors is

$$\begin{pmatrix} -3 & -4 & -2 \\ 1 & -5 & -2 \\ -2 & -2 & -1 \end{pmatrix}$$

This can be checked giving det $\mathbf{A} = -1$. Hence

$$\mathbf{A}^{-1} = -\begin{pmatrix} -3 & 1 & -2 \\ -4 & -5 & -2 \\ -2 & -2 & -1 \end{pmatrix}$$

$$= \begin{pmatrix} 3 & -1 & 2 \\ 4 & 5 & 2 \\ 2 & 2 & 1 \end{pmatrix}$$

Example 6.4(b)
Show that

$$\begin{pmatrix} 1 & 1 & 1 \\ 1 & 1 & 1 \\ 1 & 1 & 1 \end{pmatrix}$$

is a singular matrix

The corresponding determinant,

$$\begin{vmatrix} 1 & 1 & 1 \\ 1 & 1 & 1 \\ 1 & 1 & 1 \end{vmatrix}$$

equals 0, since rows are proportional. Hence the matrix is singular.

Example 6.4(c)
Show that for conformable matrices $\mathbf{A}$ and $\mathbf{B}$, $(\mathbf{AB})^{-1} = \mathbf{B}^{-1}\mathbf{A}^{-1}$. Verify this if

$$\mathbf{A} = \begin{pmatrix} 2 & 1 \\ 1 & 1 \end{pmatrix} \quad \text{and} \quad \mathbf{B} = \begin{pmatrix} 4 & 2 \\ -1 & -1 \end{pmatrix}$$

Let $\mathbf{AB} = \mathbf{C}$. Pre-multiplying both sides of this matrix equation by $\mathbf{C}^{-1}$ gives

$$\mathbf{C}^{-1}\mathbf{AB} = \mathbf{C}^{-1}\mathbf{C} = \mathbf{I}$$

Post-multiplying now by $\mathbf{B}^{-1}\mathbf{A}^{-1}$,

$$\mathbf{C}^{-1}\mathbf{ABB}^{-1}\mathbf{A}^{-1} = \mathbf{B}^{-1}\mathbf{A}^{-1}$$

The left-hand side can be simplified by taking matrix products in suitable pairs

$$\begin{aligned}\mathbf{C}^{-1}\mathbf{A}(\mathbf{B}\mathbf{B}^{-1})\mathbf{A}^{-1} &= \mathbf{C}^{-1}(\mathbf{A}\mathbf{I})\mathbf{A}^{-1}\\ &= \mathbf{C}^{-1}(\mathbf{A}\mathbf{A}^{-1})\\ &= \mathbf{C}^{-1}\mathbf{I}\\ &= \mathbf{C}^{-1}\\ &= (\mathbf{AB})^{-1}\end{aligned}$$

$$(\mathbf{AB})^{-1} = \mathbf{B}^{-1}\mathbf{A}^{-1}$$

If

$$\mathbf{A} = \begin{pmatrix} 2 & 1 \\ 1 & 1 \end{pmatrix}$$

then

$$\mathbf{A}^{-1} = \begin{pmatrix} 1 & -1 \\ -1 & 2 \end{pmatrix}$$

If

$$\mathbf{B} = \begin{pmatrix} 4 & 2 \\ -1 & -1 \end{pmatrix}$$

then

$$\mathbf{B}^{-1} = \tfrac{1}{2}\begin{pmatrix} 1 & 2 \\ -1 & -4 \end{pmatrix}$$

$$\begin{aligned}\mathbf{AB} &= \begin{pmatrix} 2 & 1 \\ 1 & 1 \end{pmatrix}\begin{pmatrix} 4 & 2 \\ -1 & -1 \end{pmatrix}\\ &= \begin{pmatrix} 7 & 3 \\ 3 & 1 \end{pmatrix}\end{aligned}$$

$$(\mathbf{AB})^{-1} = \tfrac{1}{2}\begin{pmatrix} -1 & 3 \\ 3 & -7 \end{pmatrix}$$

$$\begin{aligned}\mathbf{B}^{-1}\mathbf{A}^{-1} &= \tfrac{1}{2}\begin{pmatrix} 1 & 2 \\ -1 & -4 \end{pmatrix}\begin{pmatrix} 1 & -1 \\ -1 & 2 \end{pmatrix}\\ &= \tfrac{1}{2}\begin{pmatrix} -1 & 3 \\ 3 & -7 \end{pmatrix}\end{aligned}$$

All inverses are carried out as described earlier. The cofactor of, say, the element 2 of **A** is simply 1. There is no need for any further calculation.

Unworked Examples 9

Part 6.1

For Questions 1–9, use the following matrices:

$$\mathbf{A} = \begin{pmatrix} 2 & -1 \\ 4 & 1 \end{pmatrix} \quad \mathbf{B} = \begin{pmatrix} 1 \\ 4 \end{pmatrix} \quad \mathbf{C} = \begin{pmatrix} 3 & 0 \\ 1 & 1 \end{pmatrix} \quad \mathbf{D} = \begin{pmatrix} 1 & 2 & 3 \\ 4 & -1 & -2 \end{pmatrix}$$

(1) Are the following matrix products defined? (i) **BD** (ii) **AC** (iii) **AD** (iv) **DB** (v) **CD**.
(2) Find **AB** and **CB**.
(3) Find det **A** and det **C**.
(4) Calculate **ACB**.

Part 6.2

(5) Find **A** + **C** and 2**A** + 3**C**.
(6) Show that

$$\mathbf{AB} + \mathbf{B} = \begin{pmatrix} -1 \\ 12 \end{pmatrix}$$

(7) Does 3(**A** + **C**) = 3**A** + 3**C**?
(8) Does **A** + **C** = **C** + **A**?
(9) Find **CD** + 3**D**.

Part 6.3

(10) Show that matrices

$$\begin{pmatrix} 2 & -1 & 1 \\ 4 & 4 & 1 \\ -3 & -1 & -1 \end{pmatrix} \quad \text{and} \quad \begin{pmatrix} -3 & -2 & -5 \\ 1 & 1 & 2 \\ 8 & 5 & 12 \end{pmatrix}$$

are the inverses of each other.
(11) Deduce that if matrices **P**, **Q**, **R** are square and **P** is non-singular, if **PQ** = **PR**, then **Q** = **R**.
(12) Show that for any value of θ, the matrices

$$\begin{pmatrix} \cos\theta & -\sin\theta \\ \sin\theta & \cos\theta \end{pmatrix} \quad \text{and} \quad \begin{pmatrix} \cos\theta & \sin\theta \\ -\sin\theta & \cos\theta \end{pmatrix}$$

are each the inverse of the other.

(13) Using the notation $\mathbf{A}^2$ for the matrix product of $\mathbf{A}$ with itself, show that $\det \mathbf{A}^2 = (\det \mathbf{A})^2$. Verify this when

$$\mathbf{A} = \begin{pmatrix} 1 & 2 \\ 3 & 4 \end{pmatrix}$$

Part 6.4

(14) Find the inverse of

$$\begin{pmatrix} a & b \\ c & d \end{pmatrix}$$

(15) Verify that $(\mathbf{AB})^{-1} = \mathbf{B}^{-1}\mathbf{A}^{-1}$ by considering the two matrix products of $\mathbf{AB}$ and $\mathbf{B}^{-1}\mathbf{A}^{-1}$.

(16) Find the value of k which makes

$$\begin{pmatrix} 1 & 3 & 1 \\ 2 & 4 & -2 \\ 4 & k & 0 \end{pmatrix}$$

singular.

(17) Obtain the inverse of

$$\begin{pmatrix} 1 & 5 & -1 \\ 2 & 2 & 1 \\ 5 & 2 & 3 \end{pmatrix}$$

(18) If

$$\mathbf{A} = \begin{pmatrix} 1 & 2 \\ 2 & 4 \end{pmatrix} \quad \text{and} \quad \mathbf{B} = \begin{pmatrix} -2 & 6 \\ 1 & -3 \end{pmatrix}$$

obtain the matrix products $\mathbf{AB}$ and $\mathbf{BA}$. What are the values of $\det \mathbf{AB}$ and $\det \mathbf{BA}$? Deduce that $\mathbf{A}$ and $\mathbf{B}$ are singular.

(19) Show that if

$$\mathbf{A} = \begin{pmatrix} 1 & 2 \\ 3 & 4 \end{pmatrix} \quad \text{and} \quad \mathbf{B} = \begin{pmatrix} -4 & 2 \\ 3 & -1 \end{pmatrix}$$

then $\mathbf{AB} = 2\mathbf{I}$. Deduce $\mathbf{A}^{-1}$.

6.5 Matrix Inversion and the Solution of Simultaneous Equations

As explained in Part 6.1, a set of any number of linear simultaneous equations can be represented by a single matrix equation, so that, for example, the set

$$\begin{aligned} x + 6y + 3z &= 0 \\ -2x - 7y - 5z &= 2 \\ -x - 2y - 2z &= 1 \end{aligned}$$

can be written

$$\begin{pmatrix} 1 & 6 & 3 \\ -2 & -7 & -5 \\ -1 & -2 & -2 \end{pmatrix} \begin{pmatrix} x \\ y \\ z \end{pmatrix} = \begin{pmatrix} 0 \\ 2 \\ 1 \end{pmatrix}$$

This can be denoted by $\mathbf{Au} = \mathbf{b}$, where $\mathbf{A}$ is a square matrix with known elements, $\mathbf{u}$ is a vector with unknown elements and $\mathbf{b}$ is a vector with known elements. If the equations are determinate, the rows of $\mathbf{A}$ must be linearly independent and hence det $\mathbf{A} \neq 0$. This means $\mathbf{A}^{-1}$ exists. Pre-multiplying both sides of the matrix equation $\mathbf{Au} = \mathbf{b}$ by $\mathbf{A}^{-1}$ we obtain

$$\mathbf{u} = \mathbf{A}^{-1}\mathbf{b}$$

from which the elements of $\mathbf{u}$ can be determined. Here $\mathbf{A}$ is the matrix used in Example 6.3(b), so that $\mathbf{A}^{-1}$ is known.

$$\mathbf{u} = \begin{pmatrix} x \\ y \\ z \end{pmatrix} = \begin{pmatrix} 4 & 6 & -9 \\ 1 & 1 & -1 \\ -3 & -4 & 5 \end{pmatrix} \begin{pmatrix} 0 \\ 2 \\ 1 \end{pmatrix}$$

$$= \begin{pmatrix} 3 \\ 1 \\ -3 \end{pmatrix}$$

Hence $x = 3$, $y = 1$, $z = -3$.

Example 6.5(a)

Find the inverse of the matrix

$$\begin{pmatrix} 1 & 2 & 10 \\ 1 & 1 & 5 \\ 4 & 1 & 1 \end{pmatrix}$$

Use the result to solve the simultaneous equations

$$\begin{aligned} l + 2m + 10n &= 0 \\ l + m + 5n &= 2 \\ 4l + m + n &= 10 \end{aligned}$$

By the usual method,

$$\begin{pmatrix} 1 & 2 & 10 \\ 1 & 1 & 5 \\ 4 & 1 & 1 \end{pmatrix}^{-1} = \tfrac{1}{4} \begin{pmatrix} -4 & 8 & 0 \\ 19 & -39 & 5 \\ -3 & 7 & -1 \end{pmatrix}$$

Hence if

$$\begin{pmatrix} 1 & 2 & 10 \\ 1 & 1 & 5 \\ 4 & 1 & 1 \end{pmatrix}\begin{pmatrix} l \\ m \\ n \end{pmatrix} = \begin{pmatrix} 0 \\ 2 \\ 10 \end{pmatrix}$$

$$\begin{pmatrix} l \\ m \\ n \end{pmatrix} = \tfrac{1}{4}\begin{pmatrix} -4 & 8 & 0 \\ 19 & -39 & 5 \\ -3 & 7 & -1 \end{pmatrix}\begin{pmatrix} 0 \\ 2 \\ 10 \end{pmatrix}$$

$$= \tfrac{1}{4}\begin{pmatrix} 16 \\ -28 \\ 4 \end{pmatrix}$$

$$= \begin{pmatrix} 4 \\ -7 \\ 1 \end{pmatrix}$$

$l = 4$, $m = -7$, $n = 1$.

Example 6.5(b)

Verify that

$$\begin{pmatrix} 2 & -1 & 0 & 0 \\ -1 & 2 & -1 & 0 \\ 0 & -1 & 2 & -1 \\ 0 & 0 & -1 & 1 \end{pmatrix},$$

is the inverse of

$$\begin{pmatrix} 1 & 1 & 1 & 1 \\ 1 & 2 & 2 & 2 \\ 1 & 2 & 3 & 3 \\ 1 & 2 & 3 & 4 \end{pmatrix}$$

Hence solve the equations

$$\begin{aligned} a + b + c + d &= 2 \\ a + 2b + 2c + 2d &= 0 \\ a + 2b + 3c + 3d &= -4 \\ a + 2b + 3c + 4d &= -3 \end{aligned}$$

$$\begin{pmatrix} 2 & -1 & 0 & 0 \\ -1 & 2 & -1 & 0 \\ 0 & -1 & 2 & -1 \\ 0 & 0 & -1 & 1 \end{pmatrix}\begin{pmatrix} 1 & 1 & 1 & 1 \\ 1 & 2 & 2 & 2 \\ 1 & 2 & 3 & 3 \\ 1 & 2 & 3 & 4 \end{pmatrix} = \begin{pmatrix} 1 & 0 & 0 & 0 \\ 0 & 1 & 0 & 0 \\ 0 & 0 & 1 & 0 \\ 0 & 0 & 0 & 1 \end{pmatrix}$$

These matrices are inverses. Hence

$$\begin{pmatrix} a \\ b \\ c \\ d \end{pmatrix} = \begin{pmatrix} 2 & -1 & 0 & 0 \\ -1 & 2 & -1 & 0 \\ 0 & -1 & 2 & -1 \\ 0 & 0 & -1 & 1 \end{pmatrix} \begin{pmatrix} 2 \\ 0 \\ -4 \\ -3 \end{pmatrix}$$

$$= \begin{pmatrix} 4 \\ 2 \\ -5 \\ 1 \end{pmatrix}$$

$a = 4$, $b = 2$, $c = -5$, $d = 1$.

6.6 Elimination and the Solution of Simultaneous Equations

So far in Parts 5 and 6 we have mainly considered the solution of two or three simultaneous equations by matrices or determinants. This is because evaluating fourth and higher order determinants is a major undertaking and will be involved at some stage of the solution, whether by determinants or matrices (unless an inverse matrix is given as in Example 6.5(b)).

A set of simultaneous equations, such as those solved in Example 6.5(a), are most efficiently solved by an elimination process. For convenience the equations are repeated and numbered.

$$l + 2m + 10n = 0 \qquad (1)$$

$$l + m + 5n = 2 \qquad (2)$$

$$4l + m + n = 10 \qquad (3)$$

Suppose that l is eliminated by subtracting Equation (2) from Equation (1)

$$\begin{array}{rrrrl} & l + 2m + 10n & = & 0 & \\ - & l + m + 5n & = & 2 & \\ \hline & m + 5n & = & -2 & \qquad (4) \end{array}$$

Since Equation (4) is obtained from the original equations, it gives no further information. In fact any three of Equations (1), (2), (3) or (4) will give the required solution. It would obviously be sensible to use Equation (4) as one of the three equations because it is the simplest. As it is simple, why not obtain a fifth equation in which l has again been eliminated? This could be done by subtracting (3) from $4 \times (2)$.

$$\begin{array}{rrrrl} & 4l + 4m + 20n & = & 8 & \\ - & 4l + m + n & = & 10 & \\ \hline & 3m + 19n & = & -2 & \qquad (5) \end{array}$$

The same elimination process can be applied to Equations (4) and (5). Eliminating m, we have

$$\begin{array}{rl} 3m + 19n = -2 & \text{(Equation (5))} \\ -\quad 3m + 15n = -6 & (3 \times \text{Equation (4)}) \\ \hline 4n = 4 & \\ n = 1 & \end{array} \qquad (6)$$

This is not only a partial solution but also another equation. The simplest set of three equations, out of the six numbered, which will give values of l, m and n, are (1), (4) and (6).

$$l + 2m + 10n = 0 \qquad (1)$$

$$m + 5n = -2 \qquad (4)$$

$$n = 1 \qquad (6)$$

From (6), $n = 1$. Using this value it will be found from (4) that $m = -7$ and substituting the values of m and n in (1), we get $l = 4$. The process by which these equations were obtained can be carried out using matrices as a shorthand notation. The coefficients and constant terms of the original equations are written as the AUGMENTED MATRIX

$$\begin{pmatrix} 1 & 2 & 10 & 0 \\ 1 & 1 & 5 & 2 \\ 4 & 1 & 1 & 10 \end{pmatrix}$$

What happened subsequently was equivalent to 'determinant style' row manipulation, with the addition of rows and the removal of any common factor from a whole row. The process is called one of ROW EQUIVALENCE steps, with the symbol $\sim$ used to link matrices (which are not equal of course).

$$\begin{pmatrix} 1 & 2 & 10 & 0 \\ 1 & 1 & 5 & 2 \\ 4 & 1 & 1 & 10 \end{pmatrix} \sim \begin{pmatrix} 1 & 2 & 10 & 0 \\ 0 & -1 & -5 & 2 \\ 0 & -3 & -19 & 2 \end{pmatrix} \begin{array}{l} \\ (R\,2 \leftarrow R\,2 - R\,1) \\ (R\,3 \leftarrow R\,3 - 4R\,2) \end{array}$$

$$\sim \begin{pmatrix} 1 & 2 & 10 & 0 \\ 0 & 1 & 5 & -2 \\ 0 & 3 & 19 & -2 \end{pmatrix} \quad \text{(signs changed)}$$

$$\sim \begin{pmatrix} 1 & 2 & 10 & 0 \\ 0 & 1 & 5 & -2 \\ 0 & 0 & 4 & 4 \end{pmatrix} \quad (R\,3 \leftarrow R\,3 - 3R\,2)$$

$$\sim \begin{pmatrix} 1 & 2 & 10 & 0 \\ 0 & 1 & 5 & -2 \\ 0 & 0 & 1 & 1 \end{pmatrix} \quad (\text{R } 3 \leftarrow \tfrac{1}{4}\text{R } 3)$$

The final matrix represents the set of Equations (1), (4) and (6) and, with practice, it is possible to write down the solution to the equations directly. This final matrix is said to be in ECHELON FORM.

Example 6.6(a)
Use the elimination process to solve the equations

$$3x + 4y - 2z = 5$$
$$8x + 3y - 3z = 2$$
$$6x + 7y - 5z = 2$$

$$\begin{pmatrix} 3 & 4 & -2 & 5 \\ 8 & 3 & -3 & 2 \\ 6 & 7 & -5 & 2 \end{pmatrix} \sim \begin{pmatrix} 3 & 4 & -2 & 5 \\ -1 & -9 & 3 & -13 \\ 0 & -1 & -1 & -8 \end{pmatrix} \quad \begin{matrix} (\text{R } 2 \leftarrow \text{R } 2 - 3\text{R } 1) \\ (\text{R } 3 \leftarrow \text{R } 3 - 2\text{R } 1) \end{matrix}$$

$$\sim \begin{pmatrix} 3 & 4 & -2 & 5 \\ 1 & 9 & -3 & 13 \\ 0 & 1 & 1 & 8 \end{pmatrix} \quad (\text{signs changed})$$

$$\sim \begin{pmatrix} 0 & -23 & 7 & -34 \\ 1 & 9 & -3 & 13 \\ 0 & 1 & 1 & 8 \end{pmatrix} \quad (\text{R } 1 \leftarrow \text{R } 1 - 3\text{R } 2)$$

$$\sim \begin{pmatrix} 0 & 0 & 30 & 150 \\ 1 & 9 & -3 & 13 \\ 0 & 1 & 1 & 8 \end{pmatrix} \quad (\text{R } 1 \leftarrow \text{R } 1 + 23\text{R } 3)$$

$$\sim \begin{pmatrix} 0 & 0 & 1 & 5 \\ 1 & 9 & -3 & 13 \\ 0 & 1 & 1 & 8 \end{pmatrix}$$

Hence our equations are equivalent to

$$z = 5$$
$$x + 9y - 3z = 13$$
$$y + z = 8$$

so that $z = 5$, $y = 3$, $x = 1$.

Notice that the zero elements need not be laid out in triangular fashion to obtain the required results. With practice some of the steps can be reduced.

Equations (1), (4) and (6) finally obtained in Part 6.6 could have been written as the matrix equation

$$\begin{pmatrix} 1 & 2 & 10 \\ 0 & 1 & 5 \\ 0 & 0 & 1 \end{pmatrix}\begin{pmatrix} l \\ m \\ n \end{pmatrix} = \begin{pmatrix} 0 \\ -2 \\ 1 \end{pmatrix}$$

Such a square matrix, with all elements below the main diagonal equal to zero, is called an UPPER TRIANGULAR MATRIX. If, in addition, the elements down the main diagonal all have the value 1, the word 'unit' is added so that the matrix on the left above would be described as an UPPER UNIT TRIANGULAR MATRIX.

In the same way, a matrix with all its elements above the main diagonal equal to zero is called a LOWER TRIANGULAR MATRIX.

If x and y are non-zero elements, the matrix

$$\begin{pmatrix} x & 0 & 0 & 0 \\ y & x & 0 & 0 \\ y & y & x & 0 \\ y & y & y & x \end{pmatrix}$$

is an example of such a matrix. If the elements x are all equal to 1, it would be described as a unit lower triangular matrix. Upper and lower triangular matrices are denoted by $\mathbf{U}$ and $\mathbf{L}$ respectively. An upper unit matrix can be denoted by $\mathbf{U}_1$.

We will write any matrix $\mathbf{A}$ as the product $\mathbf{A} = \mathbf{LU}_1$. This process is called TRIANGULAR DECOMPOSITION and allows us yet another matrix method for solving simultaneous equations. Consider the solution of the equations

$$\begin{aligned} x + 2y + 2z &= 2 \\ 2x + 5y + 3z &= 9 \\ 3x + 5y + 5z &= 7 \end{aligned}$$

Using the matrix form $\mathbf{Au} = \mathbf{b}$

$$\begin{pmatrix} 1 & 2 & 2 \\ 2 & 5 & 3 \\ 3 & 5 & 5 \end{pmatrix}\begin{pmatrix} x \\ y \\ z \end{pmatrix} = \begin{pmatrix} 2 \\ 9 \\ 7 \end{pmatrix} \qquad (1)$$

we write $\mathbf{A}$ in the form $\mathbf{LU}_1$, inserting as many known elements in $\mathbf{L}$ and $\mathbf{U}_1$ as possible, and using letters for the others

$$\begin{pmatrix} a & 0 & 0 \\ b & d & 0 \\ c & e & f \end{pmatrix} \begin{pmatrix} 1 & g & h \\ 0 & 1 & k \\ 0 & 0 & 1 \end{pmatrix} = \begin{pmatrix} 1 & 2 & 2 \\ 2 & 5 & 3 \\ 3 & 5 & 5 \end{pmatrix}$$
$$\mathbf{L} \qquad\qquad \mathbf{U}_1 \qquad\qquad \mathbf{A}$$

Pre-multiplying $\mathbf{U}_1$ by $\mathbf{L}$, we find

$$a = 1$$
$$b = 2$$
$$c = 3$$

by multiplying the rows of $\mathbf{L}$ by the first column of $\mathbf{U}_1$.

Continuing in the same way using the second column of $\mathbf{U}_1$ we have

$$\begin{aligned} ag &= 2 \\ bg + d &= 5 \\ cg + e &= 5 \end{aligned}$$

Knowing the values of a, b and c allows us to obtain values of g and hence of d and e. Similarly, multiplication using the third column of $\mathbf{U}_1$, gives values of f, h and k.

Normally the matrices are written with the unknown elements left blank and these can be filled in as the calculation proceeds. With practice the whole procedure can be carried out without the need to write down anything further and the final result is of the form

$$\begin{pmatrix} 1 & 0 & 0 \\ 2 & 1 & 0 \\ 3 & -1 & -2 \end{pmatrix} \begin{pmatrix} 1 & 2 & 2 \\ 0 & 1 & -1 \\ 0 & 0 & 1 \end{pmatrix} = \begin{pmatrix} 1 & 2 & 2 \\ 2 & 5 & 3 \\ 3 & 5 & 5 \end{pmatrix}$$

The matrix equation (1) can therefore be written

$$\begin{pmatrix} 1 & 0 & 0 \\ 2 & 1 & 0 \\ 3 & -1 & -2 \end{pmatrix} \begin{pmatrix} 1 & 2 & 2 \\ 0 & 1 & -1 \\ 0 & 0 & 1 \end{pmatrix} \begin{pmatrix} x \\ y \\ z \end{pmatrix} = \begin{pmatrix} 2 \\ 9 \\ 7 \end{pmatrix} \tag{2}$$

Taking the lower triangular matrix we can solve the set of equations

$$\begin{pmatrix} 1 & 0 & 0 \\ 2 & 1 & 0 \\ 3 & -1 & -2 \end{pmatrix} \begin{pmatrix} p \\ q \\ r \end{pmatrix} = \begin{pmatrix} 2 \\ 9 \\ 7 \end{pmatrix}$$

directly using the method of Part 6.6. Here $p = 2$, $q = 5$, $r = -3$. This means that the remainder of the matrix equation (2),

$$\begin{pmatrix} 1 & 2 & 2 \\ 0 & 1 & -1 \\ 0 & 0 & 1 \end{pmatrix} \begin{pmatrix} x \\ y \\ z \end{pmatrix}$$

must be equal to

$$\begin{pmatrix} 2 \\ 5 \\ -3 \end{pmatrix}$$

Hence $z = -3$, $y = 2$, $x = 4$ and these are the solutions of the original equation (1). Simultaneous equations may also be solved by the methods described in Part 19.

Example 6.7(a)
Use triangular decomposition to solve the equations

$$\begin{aligned} x + 4y + 3z &= 2 \\ 3x + 8y + z &= 7 \\ 2x + 6y + 5z &= 3 \end{aligned}$$

$$\begin{pmatrix} 1 & 0 & 0 \\ 3 & -4 & 0 \\ 2 & -2 & 3 \end{pmatrix} \begin{pmatrix} 1 & 4 & 3 \\ 0 & 1 & 2 \\ 0 & 0 & 1 \end{pmatrix} = \begin{pmatrix} 1 & 4 & 3 \\ 3 & 8 & 1 \\ 2 & 6 & 5 \end{pmatrix}$$

Therefore

$$\begin{pmatrix} 1 & 0 & 0 \\ 3 & -4 & 0 \\ 2 & -2 & 3 \end{pmatrix} \begin{pmatrix} 1 & 4 & 3 \\ 0 & 1 & 2 \\ 0 & 0 & 1 \end{pmatrix} \begin{pmatrix} x \\ y \\ z \end{pmatrix} = \begin{pmatrix} 2 \\ 7 \\ 3 \end{pmatrix}$$

But

$$\begin{pmatrix} 1 & 0 & 0 \\ 3 & -4 & 0 \\ 2 & -2 & 3 \end{pmatrix} \begin{pmatrix} 2 \\ -\frac{1}{4} \\ -\frac{1}{2} \end{pmatrix} = \begin{pmatrix} 2 \\ 7 \\ 3 \end{pmatrix}$$

$$\begin{pmatrix} 1 & 4 & 3 \\ 0 & 1 & 2 \\ 0 & 0 & 1 \end{pmatrix} \begin{pmatrix} x \\ y \\ z \end{pmatrix} = \begin{pmatrix} 2 \\ -\frac{1}{4} \\ -\frac{1}{2} \end{pmatrix}$$

$z = -\frac{1}{2}$, $y = \frac{3}{4}$, $x = \frac{1}{2}$.

Unworked Examples 10

Part 6.5

(1) Show that

$$-\tfrac{1}{2}\begin{pmatrix} 5 & -13 & 1 \\ -2 & 4 & 0 \\ -6 & 16 & -2 \end{pmatrix}$$

is the inverse of

$$\begin{pmatrix} 4 & 5 & 2 \\ 2 & 2 & 1 \\ 4 & 1 & 3 \end{pmatrix}$$

Hence solve the equations

$$\begin{aligned} 4l + 5m + 2n &= 10 \\ 2l + 2m + n &= 3 \\ 4l + m + 3n &= -17 \end{aligned}$$

(2) By a matrix inversion method, solve the equations

$$\begin{aligned} a + 3b - 4c &= 8 \\ 5a - b + 2c &= 8 \\ 3a - b + c &= 1 \end{aligned}$$

(3) Deduce that the equations

$$\begin{aligned} x + y + 2z &= 4 \\ 3x + 2y + 3z &= 8 \\ x + 3y + 8z &= 12 \end{aligned}$$

are linearly dependent by attempting a matrix inverse solution method.

(4) Verify that the matrices

$$\begin{pmatrix} 1 & 1 & 1 & 1 \\ 1 & 1 & 1 & 2 \\ 1 & 1 & 2 & 2 \\ 1 & 2 & 2 & 2 \end{pmatrix} \quad \text{and} \quad \begin{pmatrix} 2 & 0 & 0 & -1 \\ 0 & 0 & -1 & 1 \\ 0 & -1 & 1 & 0 \\ -1 & 1 & 0 & 0 \end{pmatrix}$$

are the inverse of each other. Hence solve the simultaneous equations:

$$\begin{aligned} p + q + r + s &= 2 \\ p + q + r + 2s &= 10 \\ p + q + 2r + 2s &= 3 \\ p + 2q + 2r + 2s &= 9 \end{aligned}$$

(5) Solve the equation $\mathbf{Au} = \mathbf{b}$, where $\mathbf{A}$ is the known matrix:

$$\begin{pmatrix} 3 & -5 & 1 \\ 2 & 1 & -2 \\ 3 & -6 & 3 \end{pmatrix}$$

$\mathbf{u}$ is the unknown vector:

$$\begin{pmatrix} x \\ y \\ z \end{pmatrix}$$

and $\mathbf{b}$ is the known vector:

(i) $\begin{pmatrix} 5 \\ 4 \\ 9 \end{pmatrix}$ (ii) $\begin{pmatrix} -2 \\ 11 \\ -9 \end{pmatrix}$ (iii) $\begin{pmatrix} 9 \\ 4 \\ 15 \end{pmatrix}$

Part 6.6

(6) Explain why matrix inversion is not a practical method of solving large simultaneous equations.

(7) Show that the solution of the equations

$$\begin{aligned} a + 2b - 4c + 5d &= 3 \\ 7a + 4b - 2c + 3d &= 13 \\ 9a - 6b - 5c + 6d &= 4 \\ 2a - 2b - 3c + 4d &= 1 \end{aligned}$$

is given by $a = 1$, $b = 1$, $c = 5$, $d = 4$, using an elimination method.

(8) Using row equivalence methods, solve the equations

$$\begin{aligned} 2a - b + 4c &= 17 \\ 3a + 4b + c &= 22 \\ a - 2b + 3c &= 8 \end{aligned}$$

(9) Starting with the matrices $\mathbf{A}$:

$$\begin{pmatrix} 1 & 1 & 3 \\ 5 & 2 & 4 \\ 2 & 1 & 2 \end{pmatrix}$$

and $\mathbf{I}$:

$$\begin{pmatrix} 1 & 0 & 0 \\ 0 & 1 & 0 \\ 0 & 0 & 1 \end{pmatrix}$$

perform row equivalences on $\mathbf{A}$ to reduce it to $\mathbf{I}$ and the idential equivalences to corresponding rows on $\mathbf{I}$. Show that the resulting matrix is $\mathbf{A}^{-1}$.

(10) Write down the solution of the equations

$$\begin{aligned} a &= 4 \\ 3a - 5b &= 2 \\ 2a + b - 3c &= 1 \\ 3a + 4b + 2c + 5d &= 1 \\ a + 2b + 6c + 4d + e &= 2 \end{aligned}$$

(11) By using row equivalences, find the solution of the equation

$$\begin{aligned} 15x - 11y + 17z &= -3 \\ 13x + 14y + 6z &= -6 \\ 7x - 18y + 12z &= 4 \end{aligned}$$

Part 6.7

(12) Obtain the matrix

$$\begin{pmatrix} 2 & 4 & -6 \\ 1 & -2 & 9 \\ -2 & -7 & 4 \end{pmatrix}$$

as the product $\mathbf{LU}_1$ of a lower and upper unit triangular matrix.

(13) Use triangular decomposition to solve the equations

$$\begin{aligned} p + 2q + 4r &= 1 \\ 2p + 6q + 6r &= 1 \\ 3p + 6q + 9r &= 1 \end{aligned}$$

(14) The matrix **A**:

$$\begin{pmatrix} 1 & 4 & 2 \\ 4 & 1 & 5 \\ 3 & 2 & 4 \end{pmatrix}$$

is singular. Verify this by finding det **A**. If **A** is expressed as the product $\mathbf{LU}_1$ what happens to the third column of **L**?

(15) Verify the solution of Question 7 using triangular decomposition.

(16) Compare and contrast the various matrix methods of solving simultaneous equations.

(17) Use triangulation to solve the equations of Question 4.

(18) Solve the equations

$$\begin{aligned} x + 3y + z &= 2 \\ 2x + 4y - 2z &= 1 \\ 5x + 3y + 7z &= 5 \end{aligned}$$

Use any suitable method.

PART 7: VECTORS

7.1 Addition of Vectors

A VECTOR is a physical quantity specified by both magnitude and direction and includes such quantities as displacement, velocity, force and momentum.

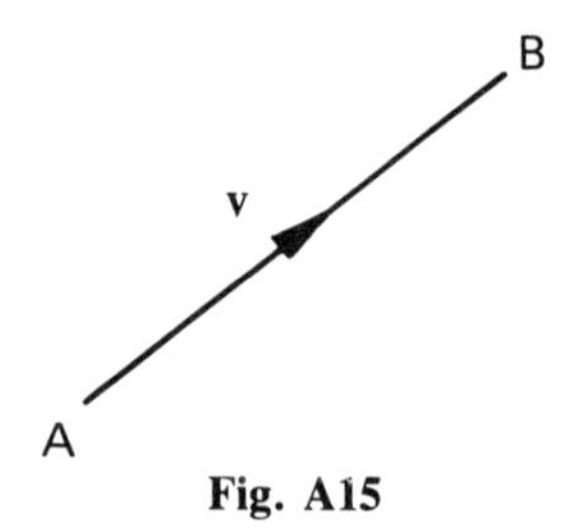

Fig. A15

One in which no direction is implied, such as mass, is called a SCALAR quantity.

A convenient notation for vectors is shown in Figure A15, the arrow indicating the direction and the length indicating magnitude. The magnitude of vector $\mathbf{V}$ is written $|\mathbf{V}|$, bold type being used to distinguish vectors, as with matrices. The vector can also be written $\overrightarrow{AB}$.

Figure A16 indicates a method of vector addition using the triangle law of addition (triangle ABC). If B is the midpoint of CD, BD represents vector $-\mathbf{v}$, and hence a method of obtaining the difference $\mathbf{u} - \mathbf{v}$ is shown.

Given two vectors $\mathbf{u}$ and $\mathbf{v}$ at right angles, as shown in Figure A17, $\mathbf{u} + \mathbf{v} = \mathbf{w}$ but $|\mathbf{w}|^2 = |\mathbf{u}|^2 + |\mathbf{v}|^2$ using Pythagoras' theorem, and $\theta = \tan^{-1} |\mathbf{v}|/|\mathbf{u}|$.

Example 7.1(a)
Find a meaning for the vector $3\mathbf{v}$ given vector $\mathbf{v}$.

Consider three equal vectors $\mathbf{v}$. Being equal they must also be equidirectional and hence vector addition will give the same numerical result as algebraic addition.

$$\mathbf{v} + \mathbf{v} + \mathbf{v} = 3\mathbf{v}$$

The vector $3\mathbf{v}$ is equal in direction to the vector $\mathbf{v}$ but has three times the magnitude.

Example 7.1(b)
Find the sum of vectors of magnitude 10 units in a direction 45° and of magnitude 6 units in a direction −30°, both angles being measured from a fixed reference line.

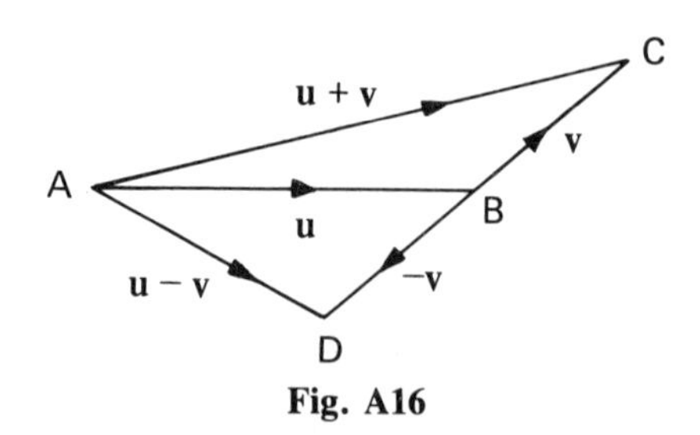

Fig. A16

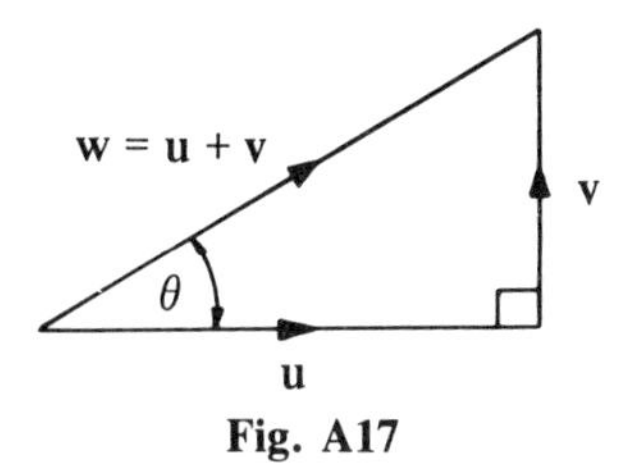

Fig. A17

If, in Figure A18, AB and $\overrightarrow{AD}$ represent the vectors, the angles 45° and 30° must be on alternative sides of the reference line.

Completing the parallelogram ABCD, $\overrightarrow{BC}$ is equal in magnitude and direction to AD and could therefore equally represent the vector of magnitude 6. The sum of the vectors would then be represented by $\overrightarrow{AC}$.

Using the cosine rule

$$AC^2 = 10^2 + 6^2 - 2 \times 10 \times 6 \cos 105°$$
$$= 167.06$$
$$AC = 12.925 \text{ units}$$

This gives the magnitude of the vector sum. Its direction to the reference line α is also required. The angle β can be obtained using the sine rule in triangle ABC

$$\frac{12.925}{\sin 105°} = \frac{6}{\sin \beta}$$

$$\sin \beta = \frac{6 \sin 105°}{12.925} = 0.4484$$

$$\beta = 26°38'$$

$$\text{Direction of vector sum} = 45° - 26°38'$$
$$= 18°22'$$

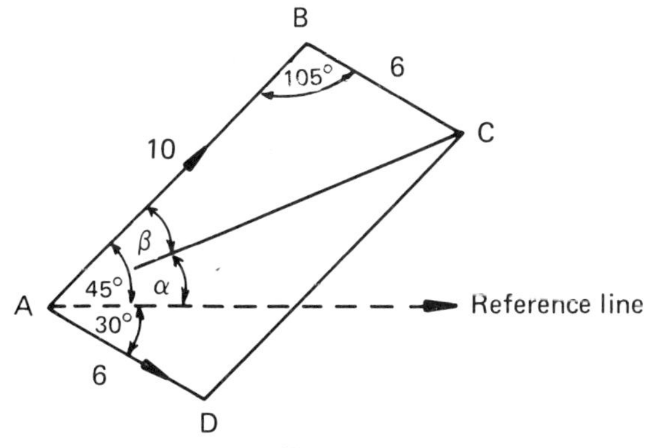

Fig. A18

7.2 Components of a Vector

Figure A17 was used to illustrate the vector addition $\mathbf{u} + \mathbf{v} = \mathbf{w}$. It is sometimes convenient to split a single vector into two COMPONENT VECTORS, acting at right angles. If we wished to do this with **w**, the components might be **u** and **v**, depending on the directions required for the components. If **u** and **v** were chosen as the components of **w**, it will be seen that $\mathbf{u} = \mathbf{w}\cos\theta$ and $\mathbf{v} = \mathbf{w}\sin\theta$.

Example 7.1(b) could have been solved using vector components. Referring to Figure A18 and taking components in the positive direction of the reference line, the net component of AB and AD would be $10\cos 45° + 6\cos 30° = 12.267$ units.

Similarly, the net component in an upward direction would be $10\sin 45° - 6\sin 30°$, which is equal to 4.071 units. These components appear in Figure A19.

The magnitude of the sum of these vectors is $\sqrt{(12.267^2 + 4.071^2)} = 12.295$ units. The direction of the vector is given by α. Here

$$\alpha = \tan^{-1}(4.071/12.267) = 18°22'$$

Given a vector in three dimensions, three components in mutually perpendicular directions are required. It is convenient to relate these to the directions of the co-ordinate axes.

Denoting *unit* vectors along Ox, Oy and Oz by **i**, **j** and **k** respectively, if, in Figure A20, P is the point (x, y, z) in a co-ordinate system, r has components of magnitude x, y and z in the directions parallel to the axes and so we get

$$\mathbf{r} = x\mathbf{i} + y\mathbf{j} + z\mathbf{k}$$

The vector **r** is called a POSITION VECTOR. Since the magnitudes x, y and z are independent, **r** could equally be represented as the column matrix

$$\begin{pmatrix} x \\ y \\ z \end{pmatrix}$$

This is why such matrices were referred to as vectors in Part 6.

Using Pythagoras' theorem in three dimensions $|\mathbf{r}| = \sqrt{(x^2 + y^2 + z^2)}$. If **r** makes angles α, β and γ with the co-ordinate axes

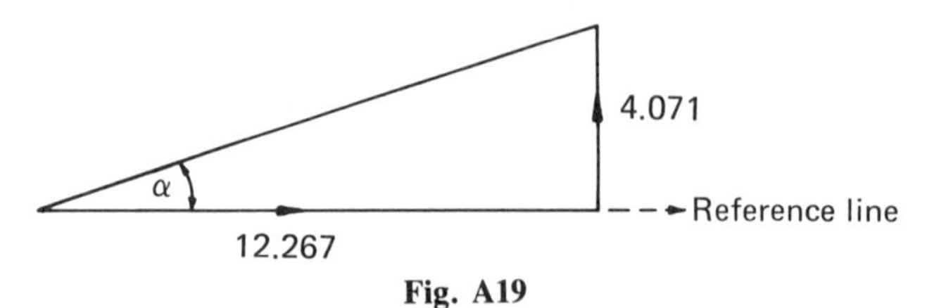

Fig. A19

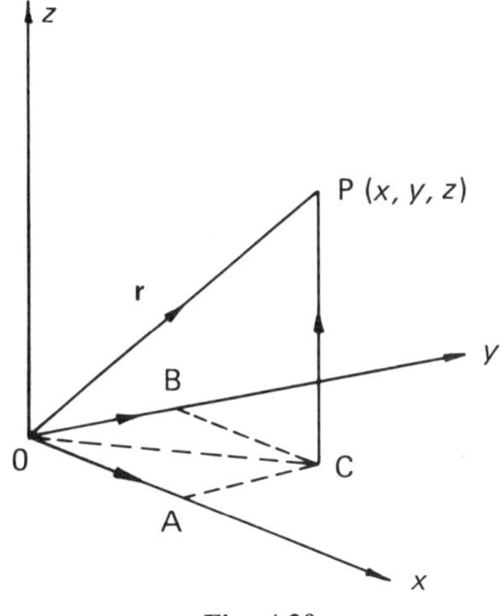

Fig. A20

$$\cos\alpha = x/\sqrt{(x^2 + y^2 + z^2)}$$
$$\cos\beta = y/\sqrt{(x^2 + y^2 + z^2)}$$
$$\cos\gamma = z/\sqrt{(x^2 + y^2 + z^2)}$$

so that $\cos^2\alpha + \cos^2\beta + \cos^2\gamma = 1$. The values of $\cos\alpha$, $\cos\beta$ and $\cos\gamma$ are called the DIRECTION COSINES of **r**.

Example 7.2(a)
Find the magnitude and direction of the sum of the vectors shown in Figure A21 by considering components along and perpendicular to the reference line.

Along the positive direction of the reference line, the net component is $7\cos 30° - 4\cos 20° = 2.303$ units.

In an upward direction, the net component is $7\sin 30° + 4\sin 20° = 4.868$ units. The vector sum $= \sqrt{(2.303^2 + 4.868^2)} = 5.385$ units. The angle made with the positive direction of the reference line is

$$\tan^{-1}(4.868/2.303) = 64°41'$$

Example 7.2(b)
Find the sum of $4\mathbf{i} + 3\mathbf{j} - 2\mathbf{k}$, $7\mathbf{i} - 6\mathbf{j} + \mathbf{k}$ and $-3\mathbf{i} - 4\mathbf{k}$.

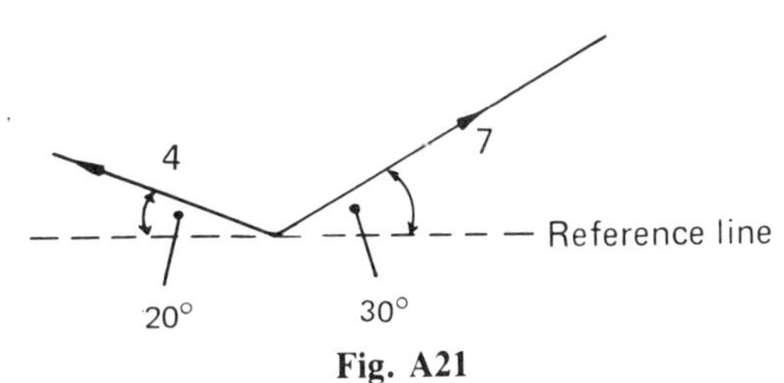

Fig. A21

Since all components of the three vectors are in one of three directions, their sum may be obtained by the addition of equidirectional parts

$$\begin{aligned}(4\mathbf{i} + 3\mathbf{j} - 2\mathbf{k}) + (7\mathbf{i} - 6\mathbf{j} + \mathbf{k}) + (-3\mathbf{i} - 4\mathbf{k})\\ = (4\mathbf{i} + 7\mathbf{i} - 3\mathbf{i}) + (3\mathbf{j} - 6\mathbf{j}) + (-2\mathbf{k} + \mathbf{k} - 4\mathbf{k})\\ = 8\mathbf{i} - 3\mathbf{j} - 5\mathbf{k}\end{aligned}$$

Notice that a three directional vector may have one or even two zero components.

The result could also have been obtained using column matrix notation.

Example 7.2(c)
Find the direction cosines of $2\mathbf{i} + 2\mathbf{j} + \mathbf{k}$

$$|2\mathbf{i} + 2\mathbf{j} + \mathbf{k}| = \sqrt{(2^2 + 2^2 + 1^2)} = 3$$

Calling the angles that this vector makes with the positive direction of the co-ordinate axes α, β and γ

$$\cos\alpha = \tfrac{2}{3}$$
$$\cos\beta = \tfrac{2}{3}$$
$$\cos\gamma = \tfrac{1}{3}$$

Unworked Examples 11

Part 7.1

(1) By drawing a vector diagram show that
 (a) $\mathbf{u} + \mathbf{v} = \mathbf{v} + \mathbf{u}$
 (b) $\mathbf{u} - \mathbf{v} = -(\mathbf{v} - \mathbf{u})$
 (c) $\mathbf{u} + (\mathbf{v} + \mathbf{w}) = (\mathbf{u} + \mathbf{v}) + \mathbf{w}$

(2) Prove that $|\mathbf{u} + \mathbf{v}| \leq |\mathbf{u}| + |\mathbf{v}|$. If $|\mathbf{u} + \mathbf{v}| = |\mathbf{u}| + |\mathbf{v}|$, what can you say about the directions of $\mathbf{u}$ and $\mathbf{v}$?

(3) Show that the sum of two vectors, each equal in magnitude but inclined at 60° opposite sides of a reference line is equal in magnitude to each of the vectors themselves.

(4) Show that if ABC is a triangle, the sum of vectors $\overrightarrow{AB} + \overrightarrow{BC} + \overrightarrow{CA}$ is zero. Extend the result to cover the polygon ABCD . . . YZ by drawing $\overrightarrow{AC}$, $\overrightarrow{AD}$ . . . $\overrightarrow{AY}$.

(5) If the diagonals of a parallelogram represent $\mathbf{u}$ and $\mathbf{v}$, show that the sides represent $\frac{1}{2}(\mathbf{u} + \mathbf{v})$ and $\frac{1}{2}(\mathbf{u} - \mathbf{v})$.

(6) In Figure A22, find the vector sum $\mathbf{v}$ and the angle α.

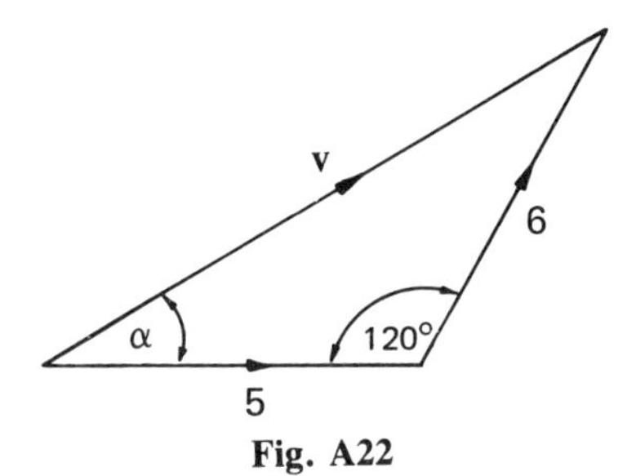

Fig. A22

Part 7.2

(7) Find the sum and difference of $3\mathbf{i} - 4\mathbf{j} - 5\mathbf{k}$ and $2\mathbf{i} - 5\mathbf{j} - 2\mathbf{k}$.

(8) Show that a vector of magnitude $a\sqrt{3}$ which is equally inclined to the co-ordinate axes has components $a\mathbf{i}$, $a\mathbf{j}$ and $a\mathbf{k}$.

(9) Verify the result of Question 6 by considering the sum of perpendicular components.

(10) If a vector has direction cosines $\cos\alpha$, $\cos\beta$ and $\cos\gamma$ show that
(a) $\cos^2\alpha + \cos^2\beta + \cos^2\gamma = 1$
(b) $\sin^2\alpha + \sin^2\beta + \sin^2\gamma = 2$
(c) $\cos\alpha = \pm\sqrt{(\sin\beta + \cos\gamma)(\sin\beta - \cos\gamma)}$

(11) Find the direction cosines of
(a) $2\mathbf{i} - 3\mathbf{j} + 6\mathbf{k}$
(b) $3\mathbf{i} - 4\mathbf{k}$.

(12) If the direction cosines for $\mathbf{v}$ are $\cos\alpha$, $\cos\beta$ and $\cos\gamma$, what can you say about the direction cosines of (a) $3\mathbf{v}$ (b) $-\mathbf{v}$.

(13) If $\mathbf{u} = 2\mathbf{i} + 3\mathbf{j} - 4\mathbf{k}$, $\mathbf{v} = \mathbf{i} - \mathbf{j} - 6\mathbf{k}$ and $\mathbf{w} = 4\mathbf{i} - 3\mathbf{j} - \mathbf{k}$, find
(a) $\mathbf{u} + \mathbf{v} + \mathbf{w}$ (b) $\mathbf{v} - 2\mathbf{w}$ (c) $|\mathbf{u}|$ (d) $|\mathbf{u} - \mathbf{v} + 2\mathbf{w}|$.

7.3 Types of Vector Product

In engineering some physical quantities are defined using the products of two or more vectors. For example, work is done when a force displaces a body and is measured as the product of the force and the component of the displacement in the direction of the force. Although force and displacement are both vectors, work is a scalar quantity.

On the other hand, the moment of a force also involves the product of vector quantities force and a component of displacement. Moments, however, are vector quantities, so that the product of two vectors may itself be a scalar or a vector quantity. Consequently we define two such products, called the SCALAR PRODUCT and the VECTOR PRODUCT according to the nature of the quantity produced.

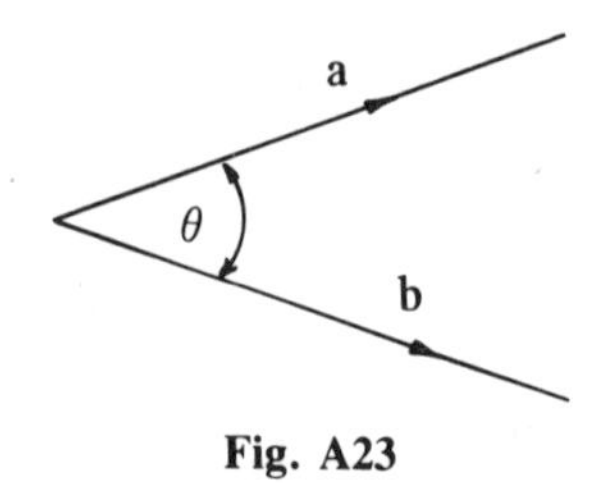

Fig. A23

7.4 Scalar Product

Given vectors **a** and **b** separated by angle θ, as shown in Figure A23, their scalar product written $\mathbf{a} \cdot \mathbf{b}$ is defined by

$$\mathbf{a} \cdot \mathbf{b} = |\mathbf{a}|\,|\mathbf{b}| \cos \theta$$

The notation $\mathbf{a} \cdot \mathbf{b}$ is read as 'a dot b' and the product is sometimes referred to as a 'dot product'.

From the definition it will be seen that $\mathbf{a} \cdot \mathbf{b} = \mathbf{b} \cdot \mathbf{a}$

If **a** and **b** are non-null vectors and $\mathbf{a} \cdot \mathbf{b} = 0$, then $\theta = 90°$. It follows that

$$\mathbf{i} \cdot \mathbf{j} = \mathbf{j} \cdot \mathbf{k} = \mathbf{k} \cdot \mathbf{i} = 0$$

but since **i**, **j** and **k** are *unit* vectors

$$\mathbf{i} \cdot \mathbf{i} = \mathbf{j} \cdot \mathbf{j} = \mathbf{k} \cdot \mathbf{k} = 1$$

It can be shown that the distributive law of multiplication holds for scalar products so that

$$\mathbf{a} \cdot (\mathbf{b} + \mathbf{c}) = \mathbf{a} \cdot \mathbf{b} + \mathbf{a} \cdot \mathbf{c}$$

If $\mathbf{a} = a_1\mathbf{i} + a_2\mathbf{j} + a_3\mathbf{k}$ and $\mathbf{b} = b_1\mathbf{i} + b_2\mathbf{j} + b_3\mathbf{k}$

then

$$\begin{aligned} \mathbf{a} \cdot \mathbf{b} &= (a_1\mathbf{i} + a_2\mathbf{j} + a_3\mathbf{k}) \cdot (b_1\mathbf{i} + b_2\mathbf{j} + b_3\mathbf{k}) \\ &= a_1b_1 + a_2b_2 + a_3b_3 \end{aligned}$$

Example 7.4(a)
Show that $\mathbf{v} \cdot \mathbf{v} = |\mathbf{v}|^2$ for any vector **v**. Deduce that if $\mathbf{v} = a_1\mathbf{i} + a_2\mathbf{j} + a_3\mathbf{k}$, $\mathbf{v} \cdot \mathbf{v} = a_1b_1 + a_2b_2 + a_3b_3$

$$\begin{aligned} \mathbf{v} \cdot \mathbf{v} &= |\mathbf{v}||\mathbf{v}| \cos 0 \\ &= |\mathbf{v}|^2 \end{aligned}$$

If $\mathbf{v} = a_1\mathbf{i} + a_2\mathbf{j} + a_3\mathbf{k}$,

$$\begin{aligned} |\mathbf{v}| &= \sqrt{(a_1^2 + a_2^2 + a_3^2)} \\ \mathbf{v} \cdot \mathbf{v} &= |\mathbf{v}|^2 = a_1^2 + a_2^2 + a_3^2 \end{aligned}$$

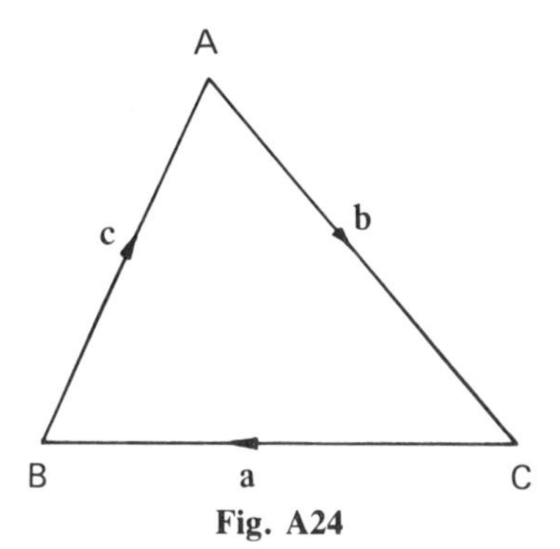

Fig. A24

Since the scalar product involves two equal vectors, the angle between them must be zero.

Example 7.4(b)
In Figure A24 deduce that $\mathbf{a} + \mathbf{b} + \mathbf{c} = 0$. Use this result to verify the cosine rule.

Since a, b and c are represented by the sides of a triangle taken in order,

$$\begin{aligned}
\mathbf{a} + \mathbf{b} + \mathbf{c} &= 0 \\
\mathbf{c} &= -\mathbf{a} - \mathbf{b} \\
\mathbf{c} \cdot \mathbf{c} &= (-\mathbf{a} - \mathbf{b}) \cdot (-\mathbf{a} - \mathbf{b}) \\
&= (\mathbf{a} + \mathbf{b}) \cdot (\mathbf{a} + \mathbf{b}) \\
&= \mathbf{a} \cdot \mathbf{a} + \mathbf{b} \cdot \mathbf{b} + \mathbf{a} \cdot \mathbf{b} + \mathbf{b} \cdot \mathbf{a} \\
&= \mathbf{a} \cdot \mathbf{a} + \mathbf{b} \cdot \mathbf{b} + 2\mathbf{a} \cdot \mathbf{b} \quad (\text{as } \mathbf{a} \cdot \mathbf{b} = \mathbf{b} \cdot \mathbf{a})
\end{aligned}$$

Using the result of Example 7.4(a)

$$\begin{aligned}
|\mathbf{c}|^2 &= |\mathbf{a}|^2 + |\mathbf{b}|^2 + 2|\mathbf{a}| \cdot |\mathbf{b}| \cos (180° - C) \\
&= |\mathbf{a}|^2 + |\mathbf{b}|^2 - 2|\mathbf{a}| \cdot |\mathbf{b}| \cos C
\end{aligned}$$

Using the standard notation for $\triangle$ABC

$$|\mathbf{a}| = a \qquad |\mathbf{b}| = b \qquad |\mathbf{c}| = c$$
$$c^2 = a^2 + b^2 - 2ab \cos C$$

Notice that the angle between **a** and **b** is $180° - C$.

Example 7.4(c)
Find the angle between the vectors $\mathbf{a} = 2\mathbf{i} + \mathbf{j} + \mathbf{k}$ and $\mathbf{b} = \mathbf{i} + \mathbf{j}$.

If the angle is θ,

$$\mathbf{a} \cdot \mathbf{b} = |\mathbf{a}||\mathbf{b}| \cos \theta$$
$$\cos \theta = \mathbf{a} \cdot \mathbf{b}/(|\mathbf{a}||\mathbf{b}|)$$

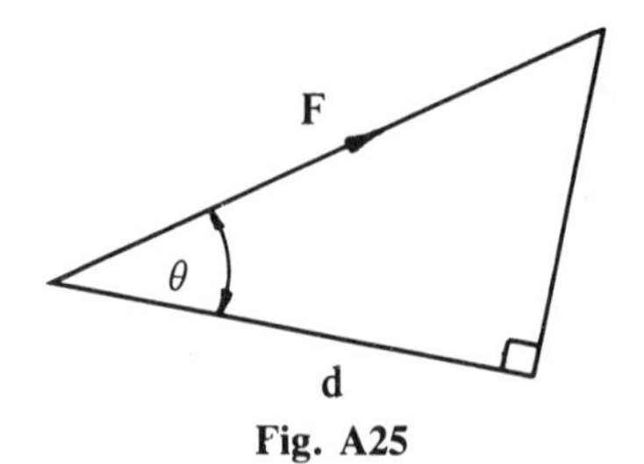

Fig. A25

But

$$\mathbf{a} \cdot \mathbf{b} = (2\mathbf{i} + \mathbf{j} + \mathbf{k}) \cdot (\mathbf{i} + \mathbf{j})$$
$$= 2 + 1 + 0 = 3$$
$$|\mathbf{a}| = \sqrt{(2^2 + 1^2 + 1^2)} = \sqrt{6}$$
$$|\mathbf{b}| = \sqrt{(1^2 + 1^2 + 0)} = \sqrt{2}$$
$$\cos\theta = \frac{3}{\sqrt{6} \times \sqrt{2}} = \frac{3}{\sqrt{12}} = \frac{\sqrt{3}}{2}$$
$$\theta = 30°$$

Example 7.4(d)
Find the work done when a force of magnitude 2 N acting in the direction of the vector $2\mathbf{i} + 3\mathbf{j} + 6\mathbf{k}$ moves a particle from point A $(-1, 2, -8)$ to point B $(3, 8, 4)$, the unit of length being 1 m.

When a force F acts through a distance d such that these are separated by an angle θ (see Figure A25), the work done is

$$Fd\cos\theta = \mathbf{F}\cdot\mathbf{d}$$

The vector $2\mathbf{i} + 3\mathbf{j} + 6\mathbf{k}$ has direction cosines $\frac{2}{7}, \frac{3}{7}, \frac{6}{7}$

$$\mathbf{F} = 2(\tfrac{2}{7}\mathbf{i} + \tfrac{3}{7}\mathbf{j} + \tfrac{6}{7}\mathbf{k}) = \tfrac{4}{7}\mathbf{i} + \tfrac{6}{7}\mathbf{j} + \tfrac{12}{7}\mathbf{k}$$
$$\mathbf{d} = [3 - (-1)]\mathbf{i} + [8 - 2]\mathbf{j} + [4 - (-8)]\mathbf{k}$$
$$= 4\mathbf{i} + 6\mathbf{j} + 12\mathbf{k}$$
$$\text{Work done} = (\tfrac{2}{7}\mathbf{i} + \tfrac{3}{7}\mathbf{j} + \tfrac{6}{7}\mathbf{k})\,(4\mathbf{i} + 6\mathbf{j} + 12\mathbf{k})$$
$$= \tfrac{8}{7} + \tfrac{18}{7} + \tfrac{72}{7}$$
$$= 14 \text{ J}$$

7.5 Vector Product

Given vectors **a** and **b** separated by an angle θ, their vector product, written $\mathbf{a} \times \mathbf{b}$ is defined by

$$\mathbf{a} \times \mathbf{b} = |\mathbf{a}||\mathbf{b}| \sin\theta\, \mathbf{n}$$

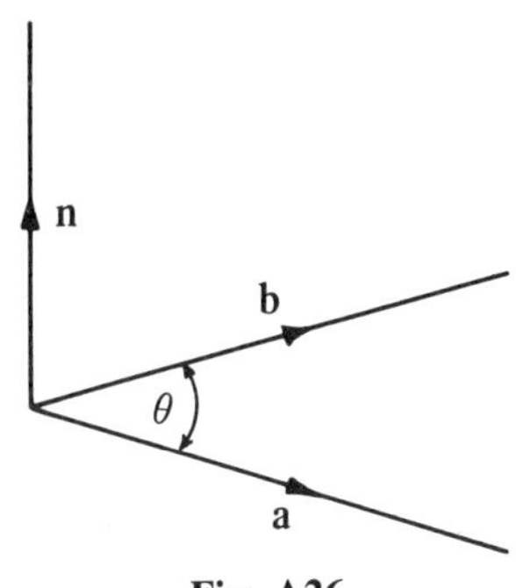

Fig. A26

where **n** is a unit vector, perpendicular to the plane containing **a** and **b**, such that turning a screw from **a** to **b** would move it in the direction **n** (see Figure A26).

The notation $\mathbf{a} \times \mathbf{b}$ is read as 'a cross b' and the product is sometimes referred to as a 'cross product'.

From the definition $\mathbf{b} \times \mathbf{a}$ has the same magnitude as $\mathbf{a} \times \mathbf{b}$ but the opposite sense so that $\mathbf{a} \times \mathbf{b} = -\mathbf{b} \times \mathbf{a}$.

If **a** and **b** are non-null vectors and $\mathbf{a} \times \mathbf{b} = \mathbf{o}$, then $\sin \theta = 0$ and hence **a** and **b** must be parallel or aligned.

From the definition it follows that

$$\begin{aligned} \mathbf{i} \times \mathbf{i} = \mathbf{j} \times \mathbf{j} &= \mathbf{k} \times \mathbf{k} = \mathbf{o} \\ \mathbf{i} \times \mathbf{j} &= \mathbf{k} \\ \mathbf{j} \times \mathbf{k} &= \mathbf{i} \\ \mathbf{k} \times \mathbf{i} &= \mathbf{j} \end{aligned}$$

The distributive law of multiplications holds for vector products as well as scalar products so that $\mathbf{a} \times (\mathbf{b} + \mathbf{c}) = \mathbf{a} \times \mathbf{b} + \mathbf{a} \times \mathbf{c}$. If $\mathbf{a} = a_1\mathbf{i} + a_2\mathbf{j} + a_3\mathbf{k}$ and $\mathbf{b} = b_1\mathbf{i} + b_2\mathbf{j} + b_3\mathbf{k}$,

$$\begin{aligned} \mathbf{a} \times \mathbf{b} &= (a_1\mathbf{i} + a_2\mathbf{j} + a_3\mathbf{k}) \times (b_1\mathbf{i} + b_2\mathbf{j} + b_3\mathbf{k}) \\ &= a_1b_2\mathbf{k} - a_1b_3\mathbf{j} - a_2b_1\mathbf{k} + a_2b_3\mathbf{i} + a_3b_1\mathbf{j} - a_3b_2\mathbf{i} \\ &= (a_2b_3 - a_3b_2)\mathbf{i} - (a_1b_3 - a_3b_1)\mathbf{j} + (a_1b_2 - a_2b_1)\mathbf{k} \\ &= \begin{vmatrix} \mathbf{i} & \mathbf{j} & \mathbf{k} \\ a_1 & a_2 & a_3 \\ b_1 & b_2 & b_3 \end{vmatrix} \end{aligned}$$

using the determinant notation of Part 4.

Example 7.5(a)
Prove that for any vector **a**, $\mathbf{a} \times \mathbf{a}$ is null.

Since the modulus of $\mathbf{a} \times \mathbf{a}$ is $|\mathbf{a}|^2 \sin \theta$ and here $\theta = 0$, $\mathbf{a} \times \mathbf{a}$ must be null.

Example 7.5(b)
Show that if $\mathbf{a} = a_1\mathbf{i} + a_2\mathbf{j} + a_3\mathbf{k}$ and $\mathbf{b} = b_1\mathbf{i} + b_2\mathbf{j} + b_3\mathbf{k}$ and $a_1:a_2:a_3 = b_1:b_2:b_3$ then $\mathbf{a} \times \mathbf{b} = \mathbf{o}$

$$\mathbf{a} \times \mathbf{b} = \begin{vmatrix} \mathbf{i} & \mathbf{j} & \mathbf{k} \\ a_1 & a_2 & a_3 \\ b_1 & b_2 & b_3 \end{vmatrix} = \mathbf{o}$$

since two rows are proportional.

Example 7.5(c)
Find a vector perpendicular to $\mathbf{a} = 3\mathbf{i} + \mathbf{j} - 2\mathbf{k}$ and $\mathbf{b} = 2\mathbf{i} + \mathbf{j} - \mathbf{k}$.

The vector $\mathbf{a} \times \mathbf{b}$ is perpendicular to both $\mathbf{a}$ and $\mathbf{b}$

$$\mathbf{a} \times \mathbf{b} = \begin{vmatrix} \mathbf{i} & \mathbf{j} & \mathbf{k} \\ 3 & 1 & -2 \\ 2 & 1 & -1 \end{vmatrix} = \mathbf{i} - \mathbf{j} + \mathbf{k}$$

($\mathbf{b} \times \mathbf{a}$ will also be perpendicular to $\mathbf{a}$ and $\mathbf{b}$.)

Example 7.5(d)
A rigid body is rotating about a fixed axis in the direction of the unit vector $\mathbf{n}$ with an angular velocity ω radians per second (Figure A27). The angular velocity vector $\boldsymbol{\omega}$ may be considered as a vector with magnitude $|\boldsymbol{\omega}|$ acting in the direction of $\mathbf{n}$. If P is a point such that the position vector $\overrightarrow{OP}$ is $\mathbf{r}$, show that the velocity $\mathbf{v}$ of P is given by $\mathbf{v} = \boldsymbol{\omega} \times \mathbf{r}$

The instantaneous velocity is $|\boldsymbol{\omega}|\overrightarrow{QP}$.

$$|\boldsymbol{\omega}|\overrightarrow{QP} = |\omega||\mathbf{r}| \sin \theta$$

The direction of $\mathbf{v}$ is such that $\mathbf{v} = \boldsymbol{\omega} \times \mathbf{r}$.

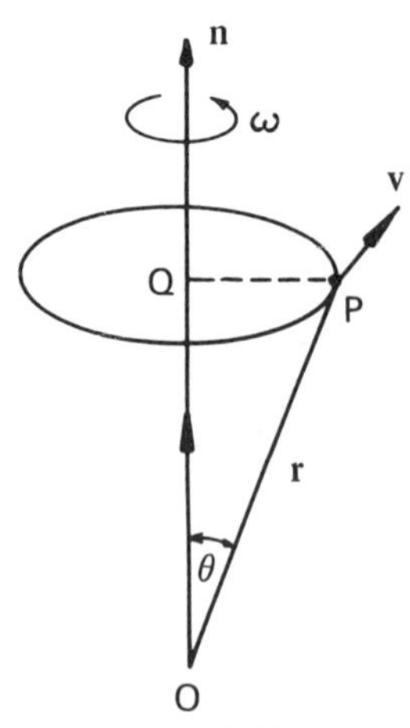

Fig. A27

7.6 Triple Scalar Products

Since $\mathbf{b} \times \mathbf{c}$ is a vector quantity, the triple product $\mathbf{a} \cdot (\mathbf{b} \times \mathbf{c})$ is scalar. It can also be expressed in determinant form. With the usual notation,

$$\mathbf{b} \times \mathbf{c} = \begin{vmatrix} \mathbf{i} & \mathbf{j} & \mathbf{k} \\ b_1 & b_2 & b_3 \\ c_1 & c_2 & c_3 \end{vmatrix}$$

$$= \mathbf{i}\begin{vmatrix} b_2 & b_3 \\ c_2 & c_3 \end{vmatrix} - \mathbf{j}\begin{vmatrix} b_1 & b_3 \\ c_1 & c_3 \end{vmatrix} + \mathbf{k}\begin{vmatrix} b_1 & b_2 \\ c_1 & c_2 \end{vmatrix}$$

$$\mathbf{a} \cdot (\mathbf{b} \times \mathbf{c}) = (a_1\mathbf{i} + a_2\mathbf{j} + a_3\mathbf{k}) \cdot \left(\mathbf{i}\begin{vmatrix} b_2 & b_3 \\ c_2 & c_3 \end{vmatrix} - \mathbf{j}\begin{vmatrix} b_1 & b_3 \\ c_1 & c_3 \end{vmatrix} + \mathbf{k}\begin{vmatrix} b_1 & b_2 \\ c_1 & c_2 \end{vmatrix}\right)$$

$$= a_1\begin{vmatrix} b_1 & b_3 \\ c_2 & c_3 \end{vmatrix} - a_2\begin{vmatrix} b_1 & b_3 \\ c_1 & c_3 \end{vmatrix} + a_3\begin{vmatrix} b_1 & b_2 \\ c_1 & c_2 \end{vmatrix}$$

$$= \begin{vmatrix} a_1 & a_2 & a_3 \\ b_1 & b_2 & b_3 \\ c_1 & c_2 & c_3 \end{vmatrix}$$

In the same way $\mathbf{b} \cdot (\mathbf{c} \times \mathbf{a})$ is equal to

$$\begin{vmatrix} b_1 & b_2 & b_3 \\ c_1 & c_2 & c_3 \\ a_1 & a_2 & a_3 \end{vmatrix}$$

Since this is equivalent to the determinant form of $\mathbf{a} \cdot (\mathbf{b} \times \mathbf{c})$ with *two* pairs of rows interchanged we can say

$$\mathbf{a} \cdot (\mathbf{b} \times \mathbf{c}) = \mathbf{b} \cdot (\mathbf{c} \times \mathbf{a}) = \mathbf{c} \cdot (\mathbf{a} \times \mathbf{b})$$

where the cyclic order $\mathbf{a}$, $\mathbf{b}$, $\mathbf{c}$ is preserved. If this order is changed the result is equal in magnitude but opposite in sign.

Example 7.6(a)
Show that $\mathbf{i} \cdot (\mathbf{j} \times \mathbf{k}) = 1$.

$$\mathbf{i} \cdot (\mathbf{j} \times \mathbf{k}) = \begin{vmatrix} 1 & 0 & 0 \\ 0 & 1 & 0 \\ 0 & 0 & 1 \end{vmatrix}$$

$$= \det \mathbf{I}$$

$$= 1$$

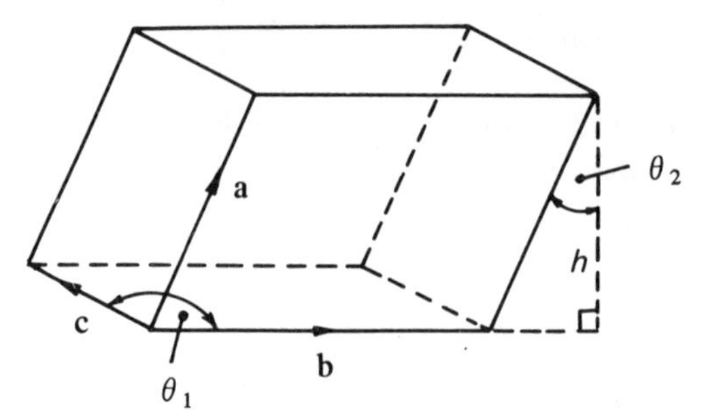

Fig. A28

Example 7.6(b)
Show that $\mathbf{a} \cdot (\mathbf{a} \times \mathbf{b}) = 0$.

$$\mathbf{a} \cdot (\mathbf{a} \times \mathbf{b}) = \begin{vmatrix} a_1 & a_2 & a_3 \\ a_1 & a_2 & a_3 \\ b_1 & b_2 & b_3 \end{vmatrix}$$
$$= 0$$

since two rows of the determinant are equal. If any two of **a**, **b** or **c** are equal the value of $\mathbf{a} \cdot (\mathbf{b} \times \mathbf{c})$ is zero.

Example 7.6(c)
Show that if **a**, **b** and **c** are the vectors shown in Figure A28, $\mathbf{a} \cdot (\mathbf{b} \times \mathbf{c})$ gives the volume of the parallelepiped. Deduce that, if **a**, **b** and **c** are coplanar, $\mathbf{a} \cdot (\mathbf{b} \times \mathbf{c}) = 0$.

The area of the base, a parallelogram, is $|\mathbf{b}||\mathbf{c}| \sin \theta_1 = |\mathbf{b} \times \mathbf{c}|$. The required volume is $h|\mathbf{b} \times \mathbf{c}|$ where h is the vertical height shown. But $h = |\mathbf{a}| \cos \theta_2$.

$$\text{Required volume} = |\mathbf{a}||\mathbf{b} \times \mathbf{c}| \cos \theta_2$$
$$= \mathbf{a} \cdot (\mathbf{b} \times \mathbf{c})$$

If **a**, **b** and **c** are coplanar, the volume vanishes:

$$\mathbf{a} \cdot (\mathbf{b} \times \mathbf{c}) = 0$$

7.7 Triple Vector Product

It is possible to obtain a triple vector product

$$\mathbf{a} \times (\mathbf{b} \times \mathbf{c}) = (a_1\mathbf{i} + a_2\mathbf{j} + a_3\mathbf{k}) \times \begin{vmatrix} \mathbf{i} & \mathbf{j} & \mathbf{k} \\ b_1 & b_2 & b_3 \\ c_1 & c_2 & c_3 \end{vmatrix}$$

The x component of this is

$$\begin{aligned} & a_2(b_1c_2 - b_2c_1) + a_3(b_1c_3 - b_3c_1) \\ &\quad = b_1(a_2c_2 + a_3c_3) - c_1(a_2b_2 + a_3b_3) \\ &\quad = b_1(a_1c_1 + a_2c_2 + a_3c_3) - c_1(a_1b_1 + a_2b_2 + a_3b_3) \\ &\quad = b_1(\mathbf{a} \cdot \mathbf{c}) - c_1(\mathbf{a} \cdot \mathbf{b}) \\ &\quad = (\mathbf{a} \cdot \mathbf{c})b_1 - (\mathbf{a} \cdot \mathbf{b})c_1 \end{aligned}$$

(At line three we added and subtracted a term $a_1b_1c_1$.)

The y and z components can be obtained in the same way and correspond to this. Putting these results together

$$\mathbf{a} \times (\mathbf{b} \times \mathbf{c}) = (\mathbf{a} \cdot \mathbf{c})\mathbf{b} - (\mathbf{a} \cdot \mathbf{b})\mathbf{c}$$

The products $(\mathbf{a} \cdot \mathbf{c})$ and $(\mathbf{a} \cdot \mathbf{b})$ are scalar factors of the vectors $\mathbf{a}$, $\mathbf{b}$ and $\mathbf{c}$.

Example 7.7(a)
Use the result $\mathbf{a} \times (\mathbf{b} \times \mathbf{c}) = (\mathbf{a} \cdot \mathbf{c})\mathbf{b} - (\mathbf{a} \cdot \mathbf{b})\mathbf{c}$ to prove that

$$\mathbf{a} \times (\mathbf{b} \times \mathbf{c}) + \mathbf{b} \times (\mathbf{c} \times \mathbf{a}) + \mathbf{c} \times (\mathbf{a} \times \mathbf{b}) = \mathbf{o}$$

$$\begin{aligned} \mathbf{a} \times (\mathbf{b} \times \mathbf{c}) &= (\mathbf{a} \cdot \mathbf{c})\mathbf{b} - (\mathbf{a} \cdot \mathbf{b})\mathbf{c} \\ \mathbf{b} \times (\mathbf{c} \times \mathbf{a}) &= (\mathbf{b} \cdot \mathbf{a})\mathbf{c} - (\mathbf{b} \cdot \mathbf{c})\mathbf{a} = (\mathbf{a} \cdot \mathbf{b})\mathbf{c} - (\mathbf{b} \cdot \mathbf{c})\mathbf{a} \\ \mathbf{c} \times (\mathbf{a} \times \mathbf{b}) &= (\mathbf{c} \cdot \mathbf{b})\mathbf{a} - (\mathbf{c} \cdot \mathbf{a})\mathbf{b} = (\mathbf{b} \cdot \mathbf{c})\mathbf{a} - (\mathbf{a} \cdot \mathbf{c})\mathbf{b} \\ \mathbf{a} \times (\mathbf{b} \times \mathbf{c}) + \mathbf{b} \times (\mathbf{c} \times \mathbf{a}) &+ \mathbf{c} \times (\mathbf{a} \times \mathbf{b}) = \mathbf{o} \end{aligned}$$

Example 7.7(b)
If $\mathbf{a} = 3\mathbf{i} - 2\mathbf{j} + \mathbf{k}$, $\mathbf{b} = -\mathbf{i} + 4\mathbf{j} - \mathbf{k}$ and $\mathbf{c} = 2\mathbf{i} - \mathbf{j} + \mathbf{k}$ find $\mathbf{a} \times (\mathbf{b} \times \mathbf{c})$ (a) directly (b) using the result $\mathbf{a} \times (\mathbf{b} \times \mathbf{c}) = (\mathbf{a} \cdot \mathbf{c})\mathbf{b} - (\mathbf{a} \cdot \mathbf{b})\mathbf{c}$.

(a) $\mathbf{b} \times \mathbf{c} = \begin{vmatrix} \mathbf{i} & \mathbf{j} & \mathbf{k} \\ -1 & 4 & -1 \\ 2 & -1 & 1 \end{vmatrix} = 3\mathbf{i} - \mathbf{j} - 7\mathbf{k}$

$$\mathbf{a} \times (\mathbf{b} \times \mathbf{c}) = \begin{vmatrix} \mathbf{i} & \mathbf{j} & \mathbf{k} \\ 3 & -2 & 1 \\ 3 & -1 & -7 \end{vmatrix} = 15\mathbf{i} + 24\mathbf{j} + 3\mathbf{k}$$

(b) $\mathbf{a} \cdot \mathbf{c} = 3 \times 2 + (-2) \times (-1) + 1 \times 1 = 9$

$\mathbf{a} \cdot \mathbf{b} = 3 \times (-1) + (-2) \times 4 + 1 \times (-1) = -12$

$9\mathbf{b} + 12\mathbf{c} = 15\mathbf{i} + 24\mathbf{j} + 3\mathbf{k}$

Unworked Examples 12

Parts 7.3 and 7.4

(1) If $\mathbf{a} = 4\mathbf{i} - 2\mathbf{j} + \mathbf{k}$ and $\mathbf{b} = \mathbf{i} - 3\mathbf{j} - \mathbf{k}$ find $\mathbf{a} \cdot \mathbf{b}$ and the angle between $\mathbf{a}$ and $\mathbf{b}$.

(2) Find the work done when a force of magnitude 4 N acting in the direction of the vector $\mathbf{i} - 2\mathbf{j} + 2\mathbf{k}$ moves from point $(-8, 2, 7)$ to point $(4, -7, 7)$ the unit of length being 1 cm.

(3) Prove that, for any scalar quantity k,

$$k\mathbf{a} \cdot \mathbf{b} = \mathbf{a} \cdot k\mathbf{b}$$

(4) Show that, with the usual notation, the angle between vectors $\mathbf{a}$ and $\mathbf{b}$ is

$$\cos^{-1} \frac{a_1a_2 + b_1b_2 + c_1c_2}{\sqrt{(a_1{}^2 + a_2{}^2 + a_3{}^2)}\sqrt{(b_1{}^2 + b_2{}^2 + b_3^2)}}$$

Hence deduce that if l_1, m_1 and n_1 are the direction cosines of $\mathbf{a}$ and l_2, m_2 and n_2 are the direction cosines of $\mathbf{b}$, the above result is equal to $\cos^{-1}(l_1l_2 + m_1m_2 + n_1n_2)$.

(5) If $\mathbf{a}$ and $\mathbf{b}$ are unit vectors in the plane xOy, making angles θ_1 and θ_2 with Ox show that

$$\mathbf{a} = \cos\theta_1\mathbf{i} + \sin\theta_1\mathbf{j}$$
$$\mathbf{b} = \cos\theta_2\mathbf{i} + \sin\theta_2\mathbf{j}$$

Use the method of obtaining the angle between $\mathbf{a}$ and $\mathbf{b}$ to deduce that $\cos(\theta_1 - \theta_2) = \cos\theta_1\cos\theta_2 + \sin\theta_1\sin\theta_2$.

(6) Taking $\mathbf{a} = a_1\mathbf{i} + a_2\mathbf{j} + a_3\mathbf{k}$, $\mathbf{b} = b_1\mathbf{i} + b_2\mathbf{j} + b_3\mathbf{k}$, show that $(\mathbf{a} + \mathbf{b}) \cdot (\mathbf{a} - \mathbf{b}) = \mathbf{a}^2 - \mathbf{b}^2$.

Part 7.5

(7) Using vectors $\mathbf{a}$ and $\mathbf{b}$ of Question 1 find $\mathbf{a} \times \mathbf{b}$.

(8) Show that for any vectors $\mathbf{a}$ and $\mathbf{b}$
(a) $(\mathbf{a} - \mathbf{b}) \times (\mathbf{a} + \mathbf{b}) = 2(\mathbf{a} \times \mathbf{b})$
(b) $|\mathbf{a} \times \mathbf{b}|^2 + (\mathbf{a} \cdot \mathbf{b})^2 = (|\mathbf{a}||\mathbf{b}|)^2$
(c) $(\mathbf{a} + \mathbf{b}) \times \mathbf{b} = \mathbf{a} \times \mathbf{b} = \mathbf{a} \times (\mathbf{a} + \mathbf{b})$.

(9) Find a vector perpendicular to both $\mathbf{i} + 2\mathbf{j} + 7\mathbf{k}$ and $5\mathbf{i} - 6\mathbf{j} + 3\mathbf{k}$.

(10) If, in Figure A29, $\mathbf{F}$ is a force vector and P is a point on its line of action such that relative to a general point O, the position vector $\overrightarrow{OP} = \mathbf{r}$, show that the moment of $\mathbf{F}$ about O is equal to $\mathbf{r} \times \mathbf{F}$. You may assume that the direction of this moment vector is in a direction perpendicular to plane OPQ through P.

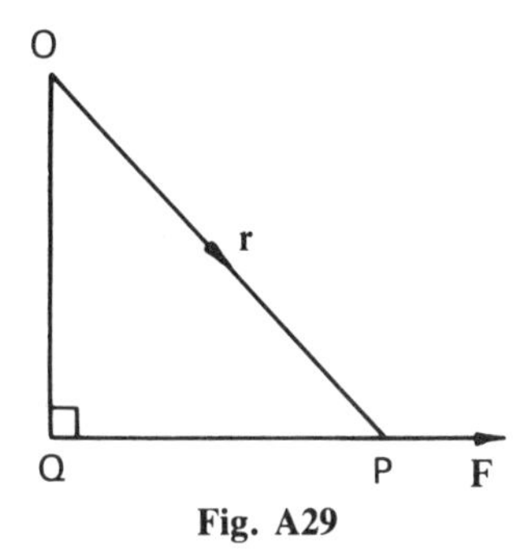

Fig. A29

(11) A rigid body is spinning with angular velocity ω, equal to 6 rad. s^{-1}, about an axis in the direction of the position vector $2\mathbf{i} + 2\mathbf{j} + \mathbf{k}$ which passes through the origin. Find the velocity of the particle at the point (1, 1, 1), if the unit of distance is 1 cm.

Part 7.6

(12) Show that $\mathbf{j} \cdot (\mathbf{j} \times \mathbf{k})$ and $\mathbf{k} \cdot (\mathbf{j} \times \mathbf{k})$ are both null.

(13) If $\mathbf{a} = p\mathbf{i}$, $\mathbf{b} = q\mathbf{j}$, $\mathbf{c} = r\mathbf{k}$ prove that $\mathbf{a} \cdot (\mathbf{b} \times \mathbf{c}) = \mathbf{b} \cdot (\mathbf{c} \times \mathbf{a}) = \mathbf{c} \cdot (\mathbf{a} \times \mathbf{b}) = pqr$.

(14) A tetrahedron has adjacent sides which can be represented by vectors $\mathbf{a}$, $\mathbf{b}$ and $\mathbf{c}$. Prove that its volume is $\frac{1}{6}\mathbf{a} \cdot (\mathbf{b} \times \mathbf{c})$. If the vertices of the tetrahedron are (0, 0, 0), (1, −1, −2), (2, 3, 3) and (3, 4, 5) find possible vectors $\mathbf{a}$, $\mathbf{b}$ and $\mathbf{c}$ and hence obtain the volume of the tetrahedron.

(15) Express $\mathbf{a} \cdot (\mathbf{b} \times \mathbf{c})$ in determinant form. Use the result to explain why $\mathbf{a} \cdot (\mathbf{b} \times \mathbf{c}) = \mathbf{b} \cdot (\mathbf{c} \times \mathbf{a})$ but $\mathbf{a} \cdot (\mathbf{b} \times \mathbf{c}) = -\mathbf{b} \cdot (\mathbf{a} \times \mathbf{c})$.

(16) Using the notation $[\mathbf{a}, \mathbf{b}, \mathbf{c}]$ to represent the triple scalar product $\mathbf{a} \cdot (\mathbf{b} \times \mathbf{c})$ show that $[\mathbf{a}, \mathbf{b}, \mathbf{c} + \mathbf{d}] = [\mathbf{a}, \mathbf{b}, \mathbf{c}] + [\mathbf{a}, \mathbf{b}, \mathbf{d}]$.

(17) Show that
(a) $\mathbf{a} \cdot (\mathbf{b} \times \mathbf{c}) = (\mathbf{a} \times \mathbf{b}) \cdot \mathbf{c}$,
(b) $\mathbf{a} \cdot (\mathbf{a} \times \mathbf{b}) = 0$.

(18) If $\mathbf{a} = 2\mathbf{i} + \mathbf{j} - \mathbf{k}$, $\mathbf{b} = 3\mathbf{i} + 3\mathbf{j} + \mathbf{k}$ and $\mathbf{c} = \mathbf{i} - \mathbf{j} - \mathbf{k}$, evaluate $\mathbf{a} \cdot (\mathbf{b} \times \mathbf{c})$.

(19) If $\mathbf{c} = \mathbf{a} \times \mathbf{b}$, explain why $\mathbf{a} \cdot (\mathbf{b} \times \mathbf{c}) = 0$.

Part 7.7

(20) If $\mathbf{b} = \mathbf{c}$ or if $\mathbf{b} \times \mathbf{c} = \mathbf{a}$, show that $\mathbf{a} \times (\mathbf{b} \times \mathbf{c})$ is null.

(21) If $\mathbf{a} = \mathbf{i} + 2\mathbf{j} + 3\mathbf{k}$, $\mathbf{b} = 2\mathbf{i} + 3\mathbf{j} + \mathbf{k}$ and $\mathbf{c} = 3\mathbf{i} + \mathbf{j} + 2\mathbf{k}$, find
(i) $\mathbf{a} \times (\mathbf{b} \times \mathbf{c})$ (ii) $\mathbf{b} \times (\mathbf{c} \times \mathbf{a})$ (iii) $\mathbf{c} \times (\mathbf{a} \times \mathbf{b})$.

(22) Show that $\mathbf{a} \times (\mathbf{b} \times \mathbf{a}) = (\mathbf{a} \times \mathbf{b}) \times \mathbf{a}$.

(23) Show that $[\mathbf{a} \cdot (\mathbf{b} \times \mathbf{c})]\mathbf{a} = (\mathbf{a} \times \mathbf{b}) \times (\mathbf{a} \times \mathbf{c})$ by choosing any values for the vectors $\mathbf{a}$, $\mathbf{b}$ and $\mathbf{c}$.

SECTION B

Calculus and its Applications

PART 8: FURTHER DIFFERENTIATION

8.1 Explicit and Implicit Functions

Given an equation such as $y = x^2 + 4x - 1$, for any value of x, it is possible to calculate y directly. As a value of x is chosen arbitrarily this unknown is called the INDEPENDENT VARIABLE while y is called the DEPENDENT VARIABLE. The expression $x^2 + 4x - 1$ is called an EXPLICIT function of x. In other cases x and y may not be separable in this way and when this happens, as for example with the equation $y \cos x + x \sin y = xy$ we have an IMPLICIT function.

With this latter equation it is impossible to express y explicitly in terms of x. With an equation such as $x^2 + y^2 = 9$, it is possible to do so but it may be convenient to handle the equation in the implicit form.

8.2 Differentiating Implicit Functions

If we attempt to differentiate an implicit function of x, at some point it will be necessary to differentiate a function of y w.r.t. x. This can be done using the 'function of a function' rule so that

$$\frac{d}{dx}(f(y)) = \frac{d}{dy}(f(y))\frac{dy}{dx}$$

For example,

$$\frac{d}{dx}(y^2) = 2y\frac{dy}{dx}$$

This can be seen differentiating $y = \sqrt{x}$ explicity and the equivalent $y^2 - x = 0$ implicitly. If $y = \sqrt{x}$, $dy/dx = 1/(2\sqrt{x})$.

If $y^2 - x = 0$, $2y\, dy/dx - 1 = 0$. Hence $dy/dx = 1/2y = 1/(2\sqrt{x})$. Normally when an implicit function has been differentiated, dy/dx contains both x and y terms. If it involves products or quotents, they are treated in the usual way, so that

$$\begin{aligned} d/dx\,(xy) &= x\, dy/dx + y \cdot 1 \\ &= x\, dy/dx + y \end{aligned}$$

and

$$\frac{d}{dx}\left(\frac{x}{y}\right) = \frac{y - x\dfrac{dy}{dx}}{y^2}$$

Example 8.2(a)
If $x^2 + y^2 - 4x + 6y + 12 = 0$, obtain dy/dx.

$$\begin{aligned} &x^2 + y^2 - 4x + 6y + 12 = 0 \\ &2x + 2y\, dy/dx - 4 + 6\, dy/dx = 0 \\ &dy/dx\,(2y + 6) = -2x + 4 \\ &dy/dx = -(x - 2)/(y + 3) \end{aligned}$$

Example 8.2(b)
Given $d/dx\,(e^x) = e^x$, show that $d/dx\,(\ln x) = 1/x$

If $y = \ln x$, $x = e^y$. Differentiating both sides w.r.t. x,

$$\begin{aligned} 1 &= e^y\, dy/dx \\ dy/dx &= e^{-y} \\ &= 1/x \end{aligned}$$

Example 8.2(c)
Show that if $y = f(x)$, then $1 = d/dx\,(f(x)) \cdot dx/dy$ and hence that dy/dx and dx/dy are reciprocals.

Differentiating $y = f(x)$ (i) w.r.t. x and (ii) w.r.t. y,

(i) $$\frac{dy}{dx} = \frac{d}{dx}(f(x))$$

(ii) $$\begin{aligned} 1 &= \frac{d}{dy}(f(x)) \\ &= \frac{d}{dx}(f(x)) \cdot \frac{dx}{dy} \end{aligned}$$

Combining these results,

$$1 = \frac{dy}{dx} \cdot \frac{dx}{dy}$$

$$\frac{dx}{dy} = 1 \bigg/ \frac{dy}{dx}$$

Example 8.2(d)
If $e^{2xy} = 3\surd(x^2 + y^2 - 4) + 1$ find possible values of y and of dy/dx when $x = 0$.

When $x = 0$, $e^{2xy} = 1$ and $3\surd(x^2 + y^2 - 4) = 3\surd(y^2 - 4)$. Hence $1 = 3\surd(y^2 - 4) + 1$ so that $y = \pm 2$. Differentiating implicitly w.r.t. x,

$$\begin{aligned} e^{2xy}(2x\,dy/dx + 2y) &= 3(2x + 2y\,dy/dx)/2\surd(x^2 + y^2 - 4) \\ &= (3x + 3y\,dy/dx)/\surd(x^2 + y^2 - 4) \end{aligned}$$

$$\frac{dy}{dx}\left(2xe^{2xy} - \frac{3y}{\surd(x^2 + y^2 - 4)}\right) = \frac{3x}{\surd(x^2 + y^2 - 4)} - 2ye^{2xy}$$

Rearranging,

$$\frac{dy}{dx} = \frac{3x - 2ye^{2xy}\surd(x^2 + y^2 - 4)}{2xe^{2xy}\surd(x^2 + y^2 - 4) - 3y}$$

When $x = 0$, $y = \pm 2$, $dy/dx = 0$.

8.3 Parametric Notation

Figure B1 illustrates how a CYCLOID is generated. This is a curve, shown dotted, described by a point on the circumference of a GENERATING CIRCLE, radius a, rolling along the x-axis. If we consider the point initially at the origin O to have reached the point P_2 when the generating circle has rolled through an angle θ, $OP_1 = P_1P_2 = a\theta$ and the co-ordinates (x, y) of P_2 are

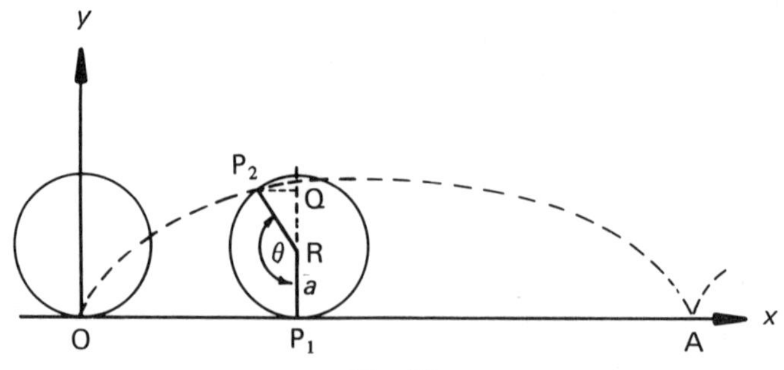

Fig. B1

given by

$$\begin{aligned} x &= OP_1 - P_2Q \\ &= a\theta - a \sin(\pi - \theta) \\ &= a(\theta - \sin\theta) \\ y &= P_1R + RQ \\ &= a + a\cos(\pi - \theta) \\ &= a(1 - \cos\theta) \end{aligned}$$

It is possible to obtain an equation in x and y, eliminating θ, but this would be quite involved and it is more convenient to leave the equation in the above form. Here θ is called a PARAMETER and $x = a(\theta - \sin\theta)$, $y = a(1 - \cos\theta)$ are referred to as the PARAMETRIC FORM of the equation of a cycloid.

8.4 Parametric Differentiation

If an equation is defined parametrically in terms of say t, then

$$\begin{aligned} \frac{dy}{dx} &= \frac{dy}{dt} \cdot \frac{dt}{dx} \\ &= \frac{dy}{dt}\bigg/\frac{dx}{dt} \end{aligned}$$

and

$$\begin{aligned} \frac{d^2y}{dx^2} &= \frac{d}{dx}\left(\frac{dy}{dx}\right) \\ &= \frac{d}{dx}\left(\frac{dy}{dt}\bigg/\frac{dx}{dt}\right) \end{aligned}$$

Using the method of Part 8.2 this can be obtained as

$$\left(\frac{dx}{dt} \cdot \frac{d^2y}{dt^2} - \frac{dy}{dt} \cdot \frac{d^2x}{dt^2}\right)\bigg/\left(\frac{dx}{dt}\right)^3$$

As long as the principle of implicit differentiation is understood there is no need to memorise this expression.

Example 8.4(a)
Show that the equation of the parabola $y^2 = 4x$ can be expressed parametrically as $x = t^2$, $y = 2t$. Obtain dy/dx by differentiating (i) parametrically, (ii) implicitly, (iii) the equation $y = \pm 2\sqrt{x}$.

If $y = 2t$, $x = t^2$ then $y^2 = 4x(= 4t^2)$.

(i) $dy/dt = 2$, $dx/dt = 2t$
$dy/dx = 2/2t = 1/t$
(ii) $y^2 = 4x$
$2y\ dy/dx = 4$
$dy/dx = 2/y$
(iii) $y = \pm 2\sqrt{x}$
$dy/dx = \pm 1/\sqrt{x}$

It is easily verified that all these results are equivalent.

Example 8.4(b)
Verify that the point P($\sqrt{5}$, 2) lies on the curve $4x^2 - y^2 = 16$. Using the forms (i) $y = 4 \sinh \theta$, $x = 2 \cosh \theta$ (ii) $y = 4 \tan t$, $x = 2 \sec t$, find the values of the parameter at P in each case. Show that the values of dy/dx and of d^2y/dx^2 at P are $2\sqrt{5}$ and -8 whichever parametric form is chosen.

When $x = \sqrt{5}$, $y = 2$, $4x^2 - y^2 = 4 \times 5 - 4 = 16$. Hence P lies on the curve.

(i) At P, $\sinh \theta = \frac{1}{2}$ and $\cosh \theta = \frac{1}{2}\sqrt{5}$
Using the method of Part 1.4,

$$e^{\theta} - e^{-\theta} = 1$$
$$e^{2\theta} - e^{\theta} - 1 = 0$$
$$e^{\theta} = (1 + \sqrt{5})/2 = 1.618$$

(This involves the formula method for solving the quadratic equation in e^{θ} but uses only a '+' instead of a '±' sign, since $e^{\theta} > 0$.)

$$\theta = \ln 1.618 = 0.4812$$
$$dy/d\theta = 4 \cosh \theta$$
$$dx/d\theta = 2 \sinh \theta$$
$$dy/dx = 2 \cosh \theta / \sinh \theta = 2\sqrt{5} \text{ at P}$$

$$d^2y/dx^2 = d/d\theta\,(dy/dx) \cdot d\theta/dx$$
$$= \frac{2 \sinh^2 \theta - 2 \cosh^2 \theta}{\sinh^2 \theta} \cdot \frac{1}{2 \sinh \theta}$$
$$= -8 \text{ at P}$$

(ii) At P, $\tan t = \frac{1}{2}$ (so that $\sin t = 1/\sqrt{5}$, $\cos t = 2/\sqrt{5}$), $t = 26°34'$

$$dy/dt = 4 \sec^2 t$$
$$dx/dt = 2 \sec t \tan t$$
$$dy/dx = 2 \sec t \cot t = 2/\sin t = 2\sqrt{5} \text{ at P}$$

$$
\begin{aligned}
d^2y/dx^2 &= d/dt\,(dy/dx) \cdot dt/dx \\
&= \frac{-2\cos t}{\sin^2 t} \cdot \frac{\cos t}{2\tan t} \\
&= -8 \text{ at P}
\end{aligned}
$$

8.5 Logarithmic Differentiation

Using implicit differentiation

$$
\begin{aligned}
d/dx\,(\ln y) &= d/dy\,(\ln y)\,dy/dx \\
&= \frac{1}{y}\frac{dy}{dx} \qquad (1)
\end{aligned}
$$

If $y = f(x)$, dy/dx is written $f'(x)$ and the above result can be expressed as

$$
\frac{d}{dx}(\ln f(x)) = \frac{f'(x)}{f(x)} \qquad (2)
$$

One use of these results occurs when differentiating a product. If, say, $y = x^2 \sin 3x$, the usual method of obtaining dy/dx is to use the product rule to obtain $3x^2 \cos 3x + 2x \sin 3x$.

An alternative method is to take the natural logarithm of both sides of the original equation

$$
\begin{aligned}
\ln y &= \ln (x^2 \sin 3x) \\
&= \ln x^2 + \ln \sin 3x
\end{aligned}
$$

since the logarithm of a product is the sum of the logarithms of its factors. In addition $\ln x^2$ can be expressed as $2 \ln x$ for the same reason

$$
\ln y = 2 \ln x + \ln \sin 3x
$$

Differentiating w.r.t. x, using results (1) and (2) given above, as appropriate

$$
\begin{aligned}
\frac{1}{y}\frac{dy}{dx} &= \frac{2}{x} + \frac{\cos 3x}{\sin 3x} \\
\frac{dy}{dx} &= \frac{2y}{x} + \frac{3y\cos 3x}{\sin 3x} \\
&= 2x \sin 3x + 3x^2 \cos 3x
\end{aligned}
$$

This process is generally referred to as LOGARITHMIC DIFFERENTIATION. Here it is obviously longer than using the product rule but when it is necessary to differentiate a product of three or more terms, logarithmic differentiation is likely to be shorter. If one of the terms involves e^{kx}, remember that $\ln e^{kx} = kx$.

Example 8.5(a)
Obtain $d/dx\ (3x^2e^{2x} \cos 4x)$ using (i) the product rule (ii) logarithmic differentiation.

(i) If $y = 3x^2e^{2x} \cos 4x$,

$$\begin{aligned} dy/dx &= 3x^2\ d/dx\ (e^{2x} \cos 4x) + 6xe^{2x} \cos 4x \\ &= 3x^2(-4e^{2x} \sin 4x + 2e^{2x} \cos 4x) + 6xe^{2x} \cos 4x \\ &= 2xe^{2x}(3x \cos 4x - 6x \sin 4x + 3 \cos 4x) \end{aligned}$$

(ii) If $y = 3x^2e^{2x} \cos 4x$,

$$\ln y = \ln 3 + 2 \ln x + 2x + \ln \cos 4x$$

$$\frac{1}{y}\frac{dy}{dx} = \frac{2}{x} + 2 - \frac{4 \sin 4x}{\cos 4x}$$

$$\frac{dy}{dx} = 2xe^{2x}(3 \cos 4x + 3x \cos 4x - 6x \sin 4x)$$

In (i), y was considered as the product of $2x^2$ and $e^{2x} \cos 4$. It then necessary to differentiate the product $e^{2x} \cos 4x$.

In (ii) $\ln 3x^2$ simplifies to $\ln 3 + \ln x^2 = \ln 3 + 2 \ln x$, of which $\ln 3$ is constant and so vanishes on being differentiated. The natural logarithm of e^{2x} is, of course, $2x$.

Example 8.5(b)
Find the value $d/dx\ (3 \ln (x + 1) \cos 4x\sqrt{(x + 4)})$ when $x = 0$.

Let $y = 3 \ln (x + 1) \cos 4x\sqrt{(x + 4)}$.

$$\ln y = \ln 3 + \ln \ln (x + 1) + \ln \cos 4x + \tfrac{1}{2} \ln (x + 4)$$

$$\frac{1}{y}\frac{dy}{dx} = \frac{1}{(x + 1) \ln (x + 1)} - \frac{4 \sin 4x}{\cos 4x} + \frac{1}{2(x + 4)}$$

$$\frac{dy}{dx} = \frac{3 \cos 4x\sqrt{(x + 4)}}{x + 1} - 12 \ln (x + 1) \sin 4x\sqrt{(x + 4)} + \frac{3 \ln (x + 1) \cos 4x}{2\sqrt{(x + 4)}}$$

When $x = 0$, $dy/dx = 6$.

Example 8.5(c)
Use logarithmic differentiation to devise a rule for differentiating the extended product $uvw\ldots$, where $u, v, w \ldots$ are functions of x.

Let $y = uvw \ldots$

$$\ln y = \ln u + \ln v + \ln w + \cdots$$

$$\frac{1}{y}\frac{dy}{dx} = \frac{1}{u}\frac{du}{dx} + \frac{1}{v}\frac{dv}{dx} + \frac{1}{w}\frac{dw}{dx} + \cdots$$

$$\frac{dy}{dx} = (vw \ldots)\frac{du}{dx} + (uw \ldots)\frac{dv}{dx} + (uv \ldots)\frac{dw}{dx} + \ldots$$

This rule, which is easily remembered, can be verified using Examples 8.5(a) and 8.5(b). It applies equally to quotients, using logarithmic differences.

If $y = u/v$,

$$\ln y = \ln u - \ln v$$

$$\frac{1}{y}\frac{dy}{dx} = \frac{1}{u}\frac{du}{dx} - \frac{1}{v}\frac{dv}{dx}$$

$$\frac{dy}{dx} = \frac{y}{u}\frac{du}{dx} - \frac{y}{v}\frac{dv}{dx}$$

$$= \frac{v\dfrac{du}{dx} - u\dfrac{dv}{dx}}{v^2}$$

8.6 Differentiation of Inverse Trigonometrical Functions

The result $\sin\theta = s$ can also be written in the form $\theta = \sin^{-1} s$, where the notation '$\sin^{-1}$', pronounced SINE INVERSE or ARC SINE, means 'the angle whose sine is'. Its value will lie in the range 0 to $\pi/2$ (or $-\pi/2$ to 0 if $s < 0$), this being the PRINCIPAL VALUE (compare with Part 2.5). A corresponding notation is used with the other trigonometrical functions.

Given a function $\sin^{-1} x/a$ where a is a constant, it is possible to differentiate w.r.t. x in the following way

$$\begin{aligned}
y &= \sin^{-1} x/a \\
x/a &= \sin y \\
1/a &= \cos y\, dy/dx \\
dy/dx &= 1/(a\cos y) \\
&= 1/a\surd(1 - \sin^2 y) \\
&= 1/a\surd(1 - x^2/a^2) \\
&= 1/\surd(a^2 - x^2)
\end{aligned}$$

The final three lines are added because it is usual to express dy/dx in terms of x. As is always the case when differentiating trigonometrical functions x is measured in radians.

Other inverse trigonometrical functions can be differentiated in a similar manner. To obtain $d/dx\ (\tan^{-1} x/a)$, the procedure is very similar

$$
\begin{aligned}
y &= \tan^{-1} x/a \\
x/a &= \tan y \\
1/a &= \sec^2 y\ dy/dx \\
dy/dx &= 1/a \sec^2 y \\
&= 1/a(1 + \tan^2 y) \\
&= a/(a^2 + a^2 \tan^2 y) \\
&= a/(a^2 + x^2)
\end{aligned}
$$

To obtain $d/dx\ (\text{cosec}^{-1} x/a)$ the same method is used but, as usual when attempting to differentiate trigonometrical functions secant, cosecant or cotangent, it is sensible to express results in terms of sine and cosine

$$
\begin{aligned}
y &= \text{cosec}^{-1} x/a \\
x/a &= \text{cosec}\ y = 1/\sin y \\
\frac{1}{a} &= -\frac{\cos y}{\sin^2 y}\frac{dy}{dx} \\
\frac{dy}{dx} &= -\frac{\sin^2 y}{a \cos y} \\
&= -\frac{\sin^2 y}{a\sqrt{(1 - \sin^2 y)}} \\
&= -\frac{a^2}{x^2}\Bigg/ a\sqrt{\left(1 - \frac{a^2}{x^2}\right)} \\
&= -a/x\sqrt{(x^2 - a^2)}
\end{aligned}
$$

Note that $\sin^{-1} x/a$ is *not* the reciprocal of $\text{cosec}^{-1} x/a$.

Examples 8.6(a) and 8.6(b) illustrate some slight variations on the procedure.

Example 8.6(a)
Show that $d/dx\ (\cot^{-1} x) = -1/(x^2 + 1)$

Let $y = \cot^{-1} x$.

$$
\begin{aligned}
x &= \cot y \\
1/x &= \tan y
\end{aligned}
$$

$$-1/x^2 = \sec^2 y \, dy/dx$$
$$= (1 + 1/x^2) \, dy/dx$$
$$dy/dx = -1/(x^2 + 1)$$

Example 8.6(b)
Find $d/dx\ (\cos^{-1} x/a)$ using the fact that dy/dx and dx/dy are reciprocals. Show that $\sin^{-1} x/a + \cos^{-1} x/a$ always totals $\pi/2$ and, using the result of $d/dx\ (\sin^{-1} x/a)$, verify the solution.

Let $y = \cos^{-1} x/a$.

$$x/a = \cos y$$
$$x = a \cos y$$
$$dx/dy = -a \sin y \text{ (differentiating w.r.t. } y\text{)}$$
$$dy/dx = -1/a \sin y$$
$$= -1/a\sqrt{(1 - \cos^2 y)}$$
$$= -1/a\sqrt{(1 - x^2/a^2)}$$
$$= -1/\sqrt{(a^2 - x^2)}$$

Let $\sin^{-1} x/a = p$.

$$\sin p = x/a = \cos (\pi/2 - p)$$
$$\therefore \cos^{-1} x/a = \pi/2 - p$$
$$\sin^{-1} x/a + \cos^{-1} x/a = \pi/2$$
$$\cos^{-1} x/a = \pi/2 - \sin^{-1} x/a$$
$$d/dx\ (\cos^{-1} x/a) = d/dx\ (\pi/2 - \sin^{-1} x/a)$$
$$= d/dx\ (\pi/2) - d/dx\ (\sin^{-1} x/a)$$
$$= -1/\sqrt{(a^2 - x^2)}$$

$d/dx\ (\pi/2) = 0$ since $\pi/2$ is a constant.

8.7 Differentiation of Inverse Hyperbolic Functions

The inverse notation used for trigonometrical functions can be extended to hyperbolic functions. If $\sinh y = x$, $y = \sinh^{-1} x$ but this time y is not an angle. The methods of Part 8.6 can be used to obtain $d/dx\ (\sinh^{-1} x)$, etc.

Let $y = \sinh^{-1} x$

$$x = \sinh y$$
$$1 = \cosh y \, dy/dx$$
$$dy/dx = 1/\cosh y$$
$$= 1/\sqrt{(\sinh^2 y + 1)}$$
$$= 1/\sqrt{(x^2 + 1)}$$

Example 8.7(a)
Show that $\sinh^{-1} x = \ln(x + \surd(x^2 + 1))$. Verify the result $d/dx(\sinh^{-1} x) = 1/\surd(x^2 + 1)$ by considering the differentiation of the logarithmic form.

Let $y = \sinh^{-1} x$.

$$x = \sinh y = \tfrac{1}{2}(e^y - e^{-y})$$
$$e^{2y} - 2xe^y - 1 = 0$$

Treating this as a quadratic equation in e^y,

$$e^y = x + \surd(x^2 + 1)$$

(Only the + sign need be considered since $e^y > 0$.)

$$y = \ln(x + \surd(x^2 + 1)) = \sinh^{-1} x$$

$$\frac{dy}{dx} = \frac{1 + \dfrac{x}{\surd(x^2 + 1)}}{x + \surd(x^2 + 1)}$$
$$= \frac{\surd(x^2 + 1) + x}{\surd(x^2 + 1)(x + \surd(x^2 + 1))}$$
$$= 1/\surd(x^2 + 1)$$

The value of dy/dx was obtained using Equation (2) of Part 8.5 with the function of a function rule. The reslt was simplified by multiplying numerator and denominator by $\surd(x^2 + 1)$ and cancelling the factor $x + \surd(x^2 + 1)$.

Unworked Examples 1

Parts 8.1 and 8.2

(1) Find dy/dx if $x^2 + 4y^2 = 4$.
(2) Find where the implicitly defined curve $3x^2y + e^x = \sin xy + y$ intersects the y-axis and show that dy/dx is zero at this point.
(3) Show that if $y = x^{1/3}$, $dy/dx = \frac{1}{3}x^{2/3}$ by (a) direct differentiation (b) implicit differentiation of $y^3 = x$.
(4) Find the points of intersection of the curves $C_1 \equiv x^2 + 3y^2 = 6$, $C_2 \equiv x^2 - y^2 = 2$ and show that, at these points, the tangents to the curve are at right angles.
(5) If $x^2 + y^2 + 2x + 4y + 1 = 0$ show that

$$\frac{dy}{dx} = -\frac{x + 1}{y + 2}$$

Parts 8.3 and 8.4

(6) Show that the general point ($2\cos t$, $\sin t$) represents a parametric form of the curve $x^2 + 4y^2 = 4$. Find $\mathrm{d}y/\mathrm{d}t$ and $\mathrm{d}x/\mathrm{d}t$ and hence verify the result of Question 1.

(7) Prove that if a curve is expressed parametrically in the form $x = f(t)$, $y = g(t)$, then $\mathrm{d}y/\mathrm{d}x = g'(t)/f'(t)$ and $\mathrm{d}^2y/\mathrm{d}x^2 = (f'(t)g''(t) - g'(t)f''(t))/(f't)^3$.

(8) If $x = 2\cos\theta$, $y = 3\sin\theta$, find the value of $\mathrm{d}y/\mathrm{d}x$ when $\theta = \pi/3$.

(9) If $x = t - 1/t$, $y = t^2 + 1/t^2$ show that (a) $x^2 = y - 2$ (b) $\mathrm{d}y/\mathrm{d}t = 2t - 2/t^3$ (c) $\mathrm{d}x/\mathrm{d}t = 1 + 1/t^2$. Show that obtaining $\mathrm{d}y/\mathrm{d}x$ from (a) leads to the same derivative as obtaining it from (b) and (c).

Part 8.5

(10) Verify the product rule for differentiation using natural logarithms.

(11) Find $\mathrm{d}/\mathrm{d}x\ (\sqrt{(x)}\mathrm{e}^{2x}/(x^2 + 2x + 4))$.

(12) If $y = a^x$ where a is a constant show that $\mathrm{d}y/\mathrm{d}x = a^x \ln a$ by taking logarithms.

(13) If $a = \mathrm{e}^b$ show that $b = \ln a$. Hence verify the result of Question 12 by differentiating $y = \mathrm{e}^{bx}$.

(14) Find the value of $\mathrm{d}/\mathrm{d}x\ (3x\mathrm{e}^{2x}\cos 3x)$ where $x = 0$.

(15) Show that if $y = \ln\cos x$, $\mathrm{d}y/\mathrm{d}x = -\tan x$.

(16) Obtain the value of $\mathrm{d}/\mathrm{d}x\ (x\sqrt{(x+1)}\sqrt{(x+2)}\sqrt{(x+3)})$ where $x = 1$.

Part 8.6

(17) Find $\mathrm{d}/\mathrm{d}x\ (\sin^{-1} 3x)$.

(18) Show that $\mathrm{d}/\mathrm{d}x\ (\mathrm{cosec}^{-1}\, x) = -1/x\sqrt{(x^2 - 1)}$.

(19) Find the value of $\mathrm{d}/\mathrm{d}x\ (\tan^{-1} \frac{1}{3}x)$ when $x = 4$.

(20) If $y = \cos^{-1} x$ find $\mathrm{d}y/\mathrm{d}x$ and $\mathrm{d}x/\mathrm{d}y$.

(21) Prove that if $y = x^2 \sin^{-1} \frac{1}{2}x$, $\mathrm{d}y/\mathrm{d}x = 2x\sin^{-1}\frac{1}{2}x + x^2/\sqrt{(4 - x^2)}$.

(22) Find $\mathrm{d}/\mathrm{d}x\ (\tan^{-1}\sqrt{x})$.

Part 8.7

(23) Find $\mathrm{d}/\mathrm{d}x\ (\tan^{-1} x)$ and $\mathrm{d}/\mathrm{d}x\ (\tanh^{-1} x)$.

(24) Prove that $\mathrm{d}/\mathrm{d}x\ (\coth^{-1} x) = -1/(x^2 - 1)$ by using the method illustrated in the text of Part 8.7. Show that

$$\coth^{-1} x = \frac{1}{2}\ln\frac{x+1}{x-1}$$

and hence verify this result.

(25) Obtain $\mathrm{d}/\mathrm{d}x\ (\sinh^{-1} x/\sinh x)$.
(26) Is $\mathrm{d}/\mathrm{d}x\ (\coth^{-1} x) = \mathrm{d}/\mathrm{d}x\ (\tanh^{-1} x)$?
(27) If $y = \sinh^{-1} (\tan x)$ use the result

$$\frac{\mathrm{d}}{\mathrm{d}x}(\sinh^{-1} x) = \frac{1}{\sqrt{(x^2 + 1)}}$$

to obtain $\mathrm{d}y/\mathrm{d}x$. Use the identity $\sinh^{-1} x = \ln (x + \sqrt{(x^2 + 1)})$ to prove that $\sinh^{-1} (\tan x) = \ln (\tan x + \sec x)$. Hence verify the derivative $\mathrm{d}y/\mathrm{d}x$.

PART 9: FURTHER INTEGRATION

9.1 Integration by Substitution

It was shown in Part 8.6 that $\mathrm{d}/\mathrm{d}x\ (\sin^{-1} x/a) = 1/\sqrt{(a^2 - x^2)}$. Since a can take any non-zero value we can simplify this result by choosing $a = 1$ so that $\mathrm{d}/\mathrm{d}x\ (\sin^{-1} x) = 1/\sqrt{(1 - x^2)}$.

As integration is the opposite process to differentiation, this result can be expressed in reverse as

$$\int \frac{\mathrm{d}x}{\sqrt{(1 - x^2)}} = \sin^{-1} x + C$$

C being the constant of integration.

Is it possible to carry out this integration without knowing the result? It certainly cannot be performed directly because of the nature of the denominator. However, the form of the denominator, $\sqrt{(1 - x^2)}$, suggests that it could be simplified by putting $x = \sin\theta$, making $\sqrt{(1 - x^2)} = \sqrt{(1 - \sin^2\theta)} = \cos\theta$. However if the x term in the denominator is to be replaced in this way, it also necessary to substitute for $\mathrm{d}x$. Since $x = \sin\theta$, $\mathrm{d}x/\mathrm{d}\theta = \cos\theta$ or, using derivatives $\mathrm{d}x = \cos\theta\ \mathrm{d}\theta$. Incorporating all these changes

$$\begin{aligned}\int \frac{\mathrm{d}x}{\sqrt{(1 - x^2)}} &= \int \frac{\cos\theta\ \mathrm{d}\theta}{\cos\theta} \\ &= \int \mathrm{d}\theta \\ &= \theta + C \\ &= \sin^{-1} x + C\end{aligned}$$

The final line is added since the result of the integration will normally be required in terms of the original variable. This process is known as INTEGRATION BY SUBSTITUTION.

If a definite integral is to be evaluated using this method, the limits are changed in the same way. To evaluate

$$\int_0^1 \frac{dx}{\sqrt{(1 - x^2)}}$$

using the same substitution, when $x = 1$, $\theta = \pi/2$ and when $x = 0$, $\theta = 0$. Hence

$$\begin{aligned}\int_0^1 \frac{dx}{\sqrt{(1 - x^2)}} &= \int_0^{\pi/2} d\theta \\ &= [\theta]_0^{\pi/2} \\ &= \pi/2\end{aligned}$$

When, as here, the replacement of the limits involves an inverse trigonometrical function, the principal value is selected. If the replacement involves a square root, the positive value is chosen.

Example 9.1(a)
By means of the substitution $u^2 = x^2 + 16$, evaluate

$$\int_0^3 \frac{2x\,dx}{\sqrt{(x^2 + 16)}}$$

If $u^2 = x^2 + 16$, $2u\,du = 2x\,dx$

$$\begin{aligned}\int_0^3 \frac{2x\,dx}{\sqrt{(x^2 + 16)}} &= \int_4^5 \frac{2u\,du}{u} \\ &= 2\int_4^5 du \\ &= 2[u]_4^5 \\ &= 2\end{aligned}$$

Example 9.1(b)
Integrate

$$\int \frac{4x\,dx}{(x^2 + 2)^5}$$

As we are not told what substitution to use, we must look at the integrand and use our judgement and, since the part of the integrand that looks the 'most difficult to deal with' is $(x^2 + 2)^5$, we could try simplify it by substituting for $x^2 + 2$.

Let $z = x^2 + 2$.

$$dz = 2x\,dx$$

$$\int \frac{4x\,dx}{(x^2+2)^5} = \int \frac{2\,dz}{z^5}$$

$$= 2\int z^{-5}\,dz$$

$$= -\frac{1}{2z^4} + C$$

$$= -1/2\,(x^2+4)^4 + C$$

Example 9.1(c)
Integrate $\int \sin^4 t \cos^3 t\,dt$

Let $s = \sin t$.

$$ds = \cos t\,dt$$

$$\int \sin^4 t \cos^3 t\,dt = \int \sin^4 t \cos^2 t \cos t\,dt$$

$$= \int s^4 (1 - s^2)\,ds$$

$$= \int (s^4 - s^6)\,ds$$

$$= \frac{1}{5}s^5 - \frac{1}{7}s^7 + C$$

$$= \frac{1}{5}\sin^5 t - \frac{1}{7}\sin^7 t + C$$

This illustrates a standard method of solving integrals of the form $\sin^m t \cos^n t\,dt$ where m and n are not both even. The substitution is $s = \sin t$ if n is odd, $c = \cos t$ if m is odd, either of these if both m and n are odd. If both m and n are even, the method of Part 3.2 or of Part 9.4 may be used.

Example 9.1(d)
Evaluate

$$\int_0^{\pi/2} \frac{2\,dx}{1 + 3\cos^2 x}$$

Integrals of this type involving only $\sin^2 x$, $\cos^2 x$ or $\tan^2 x$ are handled by the substitution $t = \tan x$. Then $dt = \sec^2 x\,dx = (t^2 + 1)\,dx$

$$dx = dt/(t^2+1)$$

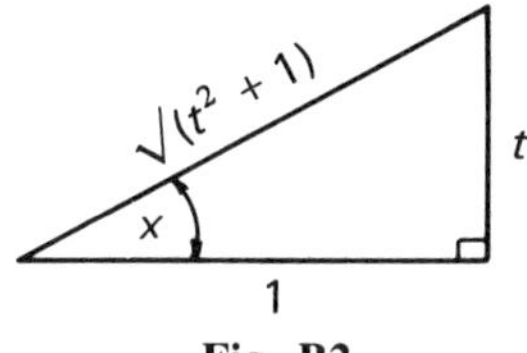

Fig. B2

The value of cos x (or sin x) is best obtained by reference to Figure B2. As tan $x = t$, the lengths of the sides of the right-angled triangle may be inserted

$$\cos x = 1/\sqrt{(t^2 + 1)}$$

$$\cos^2 x = 1/(t^2 + 1)$$

$$\int_0^{\pi/2} \frac{2\,dx}{1 + 3\cos^2 x} = \int_0^{\infty} \frac{2}{t^2 + 1} \cdot \frac{1}{1 + 3/(t^2 + 1)}\,dt = \int_0^{\infty} \frac{2\,dt}{t^2 + 4}$$

This integrand is one which was met in Part 8.6. If this has been forgotten it can be evaluated by another substitution $t = 2 \tan u$. Then $dt = 2 \sec^2 u \, du$ and $t^2 + 4 = 4 \sec^2 u$

$$\int_0^{\infty} \frac{2\,dt}{t^2 + 4} = \int_0^{\pi/2} du$$

$$= [u]_0^{\pi/2}$$

$$= \pi/2$$

Where a solution such as this involves a double substitution, it could have been arrived at directly by letting $\tan x = 2 \tan u$.

Example 9.1(e)
Show that

$$\int \frac{f'(x)}{f(x)}\,dx = \ln f(x) + C$$

Hence obtain

$$\int \frac{2x + 1}{x^2 + x + 1}\,dx$$

Let $u = f(x)$.

$$du = f'(x)\,dx$$

$$\int \frac{f'(x)}{f(x)}\,dx = \int \frac{du}{u}$$

$$= \ln u + C$$

$$= \ln f(x) + C$$

If $f(x) = x^2 + x + 1$,

$$f'(x) = 2x + 1$$

$$\int \frac{2x + 1}{x^2 + x + 1}\, dx = \ln (x^2 + x + 1) + C$$

Integrating an algebraic fraction, the numerator of which is the derivative of the denominator, gives the logarithm of the denominator. This is effectively Equation (2) of Part 8.5.

Example 9.1(f)
Evaluate $\int_0^2 p\sqrt{(4 - p^2)}\, dp$ using the substitutions (a) $u^2 = 4 - p^2$ (b) $p^2 = 4(1 - c^2)$.

(a) Let $u^2 = 4 - p^2$.

$$2u\, du = -2p\, dp$$

$$\int_0^2 p\sqrt{(4 - p^2)}\, dp = -\int_2^0 u^2\, du$$

$$= \int_0^2 u^2\, du$$

$$= \left[\frac{1}{3}u^3\right]_0^2$$

$$= \tfrac{8}{3}$$

(b) Let $p^2 = 4(1 - c^2)$.

$$p = 2\sqrt{(1 - c^2)}$$

$$dp = -2c/\sqrt{(1 - c^2)}\, dc$$

$$\sqrt{(4 - p^2)} = \sqrt{(4 - 4 + 4c^2)} = 2c$$

$$\int_0^2 p\sqrt{(4 - p^2)}\, dp = -\int_1^0 8c^2\, dc$$

$$= 8\int_0^1 c^2\, dc$$

$$= 8[\tfrac{1}{3}c^3]_0^1$$

$$= \tfrac{8}{3}$$

In both cases the change of variable makes the upper limit smaller than the lower limit. While there is no reason why the integral cannot be evaluated in this form, it is more usual to interchange them. This alters the sign of the integral and accounts for the disappearance of the minus sign in both parts.

Even in cases where the integration may be carried out by more than one substitution, the result is unaffected although the methods may vary in the amount of work involved.

9.2 Partial Fractions

Part 9.1 dealt with some methods of integrating algebraic fractions. This part will deal with some futher methods.

The highest power in a polynomial expression is called its DEGREE. If the degree of the numerator is equal to or greater than the degree of the denominator, it is possible to divide out until the situation is changed:

$$\int \frac{x^4 - 2x^3 + x^2 - 2x + 1}{x^2 + 1}\,dx = \int \left(x^2 - 2x + \frac{1}{x^2 + 1} \right) dx$$

$$= \tfrac{1}{3}x^3 - x^2 + \tan^{-1} x + C$$

We will therefore only be concerned with intergrating fractions in which the degree of the numerator is less than that of the denominator. These will be referred to as proper algebraic fractions.

Consider the addition of two such fractions.

$$\frac{2}{x+1} + \frac{3}{x-2} = \frac{2(x-2) + 3(x+1)}{(x+1)\,(x-2)}$$

$$= \frac{5x - 1}{x^2 - x - 2}$$

Two (or more) proper algebraic fractions when added (or subtracted) give a further proper algebraic fraction. In consequence, if we wish to reverse the process and express the result as the sum or difference of two simpler fractions then

(i) it is necessary to be able to factorise the denominator,

(ii) all numerators must be of one degree less than the corresponding denominator. In most cases, the denominator will be a linear factor and consequently the numerator will be a constant.

The process is called splitting into PARTIAL FRACTIONS.

To split $(5x - 1)/(x^2 - x - 2)$ into partial fractions it is written as follows:

$$\frac{5x - 1}{x^2 - x - 2} = \frac{5x - 1}{(x+1)(x-2)}$$

$$= \frac{A}{x+1} + \frac{B}{x-2}$$

where A and B are constants to be determined. The sign linking the fractions is optionally + or −. (If it is written + and the fractions are to be subtracted, B will be found to be negative.)

There are several methods of obtaining A and B.

(*a*) *Choice of Value of x*

If the numerator of the sum of

$$\frac{A}{x + 1} + \frac{B}{x - 2}$$

is expressed in terms of A and B, it can be identified to the original numerator giving

$$A(x - 2) + B(x + 1) \equiv 5x - 1$$

Choosing *any* two values of x will give two simultaneous equations in A and B, whose values can then be obtained. If, in particular, the values $x = -1$ and $x = 2$ are chosen, values of A and B can be obtained directly so that $A = 2$, $B = 3$. Choosing a third value of x will check the results.

(*b*) *Equating Coefficients*

Obtaining $A(x - 2) + B(x + 1) \equiv 5x - 1$ as in method (a), A and B can be found from the simultaneous equations obtained by equating the coefficients of x and the constant term on both sides of the equivalence relationship. In this case

$$A + B = 5$$
$$-2A + B = -1$$

giving the same values for A and B as method (a).

(*c*) *Cover-Up Rule*

The cover-up rule allows constants such a A or B to be obtained directly. From the original fraction with the denominator expressed as the product of linear factors, cover up one of the factors, say $(x + 1)$, to give

$$\frac{5x - 1}{(x - 2)}$$

If the value $x = -1$ (which would make the covered factor zero) is substituted in the uncovered fraction, the result

$$\frac{(5x - 1) - 1}{(-1 - 2)} = 2$$

is the value of A. In exactly the same way B is obtained as $(5x - 1)/(x + 1)$ when $x = 2$. Any combination of these methods may be used.

These methods allow any fraction, with non-repeated factors in the denominator, to be split into partial components. If the denominator contains a multiple factor, the splitting is into a series of fractions with

(i) all numerators as constants,
(ii) the denominators containing the factor in question with the degree of multiplicity reduced by one each time.

For example, $(x + 5)/(x - 2)^3$ would be split up as

$$\frac{x + 5}{(x - 2)^3} = \frac{A}{(x - 2)^3} + \frac{B}{(x - 2)^2} + \frac{C}{x - 2}$$

The first of these numerators only can be evaluated using the cover-up rule.

Example 9.2(a)
Split

$$\frac{3x^2 - 4x - 5}{x^3 + x^2 - 3x - 2}$$

into partial fractions.

$$\frac{3x^2 - 4x - 5}{x^3 + x^2 - 3x - 2} = \frac{3x^2 - 4x - 5}{(x + 2)(x^2 - x - 1)}$$

$$= \frac{3}{x + 2} + \frac{Ax + B}{x^2 - x - 1}$$

Here $3(x^2 - x - 1) + (Ax + B)(x + 2) \equiv 3x^2 - 4x - 5$. Equating coefficients,

$$x^2\text{:}\quad 3 + A = 3$$

$$x^\circ\text{:}\quad -3 + 2B = -5$$

$$A = 0,\ B = -1$$

$$\frac{3x^2 - 4x - 5}{x^3 + x^2 - 3x - 2} = \frac{3}{x + 2} - \frac{1}{x^2 - x - 1}$$

Here $x^3 + x^2 - 3x - 2 = (x + 2)(x^2 - x - 1)$. If it were possible to further factorise $x^2 - x - 1$ into two linear factors this must be carried out so that there would have been three partial fractions. As $x^2 - x - 1$ cannot be further factorised there are two partial fractions, with numerators 3 (obtained using the cover-up rule) and $Ax + B$ (to be of the right degree).

In equating the coefficients the notation x° indicates the equating of constant terms.

Example 9.2(b)
Prove that $3 + 6(x - 1) + (x - 1)^2 = x^2 + 4x - 2$. Hence deduce that

$$\frac{x^2 + 4x - 2}{(x - 1)^3}$$

will be split into partial fractions

$$\frac{3}{(x - 1)^3} + \frac{6}{(x - 1)^2} + \frac{1}{x - 1}$$

Verify the result using a conventional technique for splitting the fraction.

$$\begin{aligned} 3 + 6(x - 1) + (x - 1)^2 &= 3 + 6x - 6 + x^2 - 2x + 1 \\ &= x^2 + 4x - 2 \end{aligned}$$

$$\begin{aligned} \frac{x^2 + 4x - 2}{(x - 1)^3} &= \frac{3 + 6(x - 1) + (x - 1)^2}{(x - 1)^3} \\ &= \frac{3}{(x - 1)^3} + \frac{6}{(x - 1)^2} + \frac{1}{x - 1} \end{aligned}$$

By conventional technique

$$\frac{x^2 + 4x - 2}{(x - 1)^3} = \frac{3}{(x - 1)^3} + \frac{A}{(x - 1)^2} + \frac{B}{x - 1}$$

where $3 + A(x - 1) + B(x - 1)^2 \equiv x^2 + 4x - 2$.

Selecting values of x,

$x = 0$: $3 - A + B = -2$

$x = 2$: $3 + A + B = 10$

Hence $A = 6$, $B = 1$,

$$\frac{x^2 + 4x - 2}{(x - 1)^3} = \frac{3}{(x - 1)^3} + \frac{6}{(x - 1)^2} + \frac{1}{x - 1}$$

The numerator of the fraction with denominator $(x - 1)^3$ was obtained using the cover-up rule.

Example 9.2(c)
Integrate

$$\int \frac{x^2 - x - 1}{x^3 - x} \, dx$$

$$\int \frac{x^2 - x - 1}{x^3 - x}\,dx = \int \frac{x^2 - x - 1}{x(x + 1)(x - 1)}\,dx$$

$$= \int \left(\frac{1}{x} + \frac{\frac{1}{2}}{x + 1} - \frac{\frac{1}{2}}{x - 1}\right) dx$$

$$= \ln x + \tfrac{1}{2} \ln (x + 1) - \tfrac{1}{2} \ln (x - 1) + C$$

All the numerators of the partial fractions were obtained using the cover-up rule.

Example 9.2(d)
Evaluate

$$\int_2^3 \frac{2x^2 - 5x + 9}{(x - 1)^2(x + 2)}\,dx$$

$$\int_2^3 \frac{2x^2 - 5x + 9}{(x - 1)^2(x + 2)}\,dx = \int_2^3 \frac{2}{(x - 1)^2} + \frac{A}{x - 1} + \frac{3}{x + 2}\,dx$$

where $2(x + 2) + A(x - 1)(x + 2) + 3(x - 1)^2 \equiv 2x^2 - 5x + 9$. Putting $x = 0$,

$$4 - 2A + 3 = 9$$

$$A = -1$$

$$\int_2^3 \frac{2x^2 - 5x + 9}{(x - 1)^2(x + 2)}\,dx = \int_2^3 \left(\frac{2}{(x - 1)^2} - \frac{1}{x - 1} + \frac{3}{x + 2}\right) dx$$

$$= \left[-\frac{2}{x - 1} - \ln (x - 1) + 3 \ln (x + 2)\right]_2^3$$

$$= 1 + \ln \frac{125}{128}$$

$$\approx 0.976$$

Wherever a fractional integrand is presented with its denominator factorised, this is a strong suggestion that partial fractions may be needed in its evaluation.

Unworked Examples 2

Part 9.1

(1) Use the substitution $t = \tan \theta$ to show that

$$\int \frac{dt}{1 + t^2} = \tan^{-1} t + C$$

(2) Show that the substitution $x = 2 \sin \theta$ changes the integral, I, $\int \sqrt{(4 - x^2)}\, dx$ into the intergral $\int 4 \cos^2 \theta \, d\theta$. Use this substitution to intergrate I.

(3) Evaluate $\int_0^4 3x \sqrt{(x^2 + 9)}\, dx$ using the substitution $p^2 = x^2 + 9$.

(4) Use the substitution $t = \tan z$ to evaluate the definite integral

$$\int_{-\pi/4}^{\pi/4} 3 \tan^2 z \sec^2 z \, dz$$

(5) Use a suitable substitution to integrate

$$\int \frac{x\, dx}{\sqrt{(1 - x^2)}}$$

(6) Show that when using the substitution $t = \tan \frac{1}{2}\theta$, $\tan \theta = 2t/(1 - t^2)$, $\sin \theta = 2t/(1 + t^2)$, $\cos \theta = (1 - t^2)/(1 + t^2)$ and $d\,\theta = 2\, dt/(1 + t^2)$.

(7) Show that the substitution $x = \tan \theta$ makes the integral, I

$$\int \frac{dx}{(1 + x^2)^{3/2}}$$

$= \sin(\tan^{-1} x)$, whereas the substitution $x = \sinh u$ makes $I = \tanh(\sinh^{-1} x)$. Verify that $\sin(\tan^{-1} x) = \tanh(\sinh^{-1} x)$ when $x = 1.1752$.

(8) Integrate $\int 5 \sin^4 t \cos t \, dt$.

(9) Show that

$$\int_0^1 \frac{2x + 1}{x^2 + x + 1}\, dx = \ln 3$$

Part 9.2

(10) Show that

$$\int \frac{dx}{x^2 - 3x + 2} = \ln \frac{x - 2}{x - 1} + C$$

(11) Split into partial fractions

(a) $\dfrac{x + 1}{x^2 - 4x + 3}$

(b) $\dfrac{3x^2 + 10x + 11}{(x + 1)(x + 2)(x + 3)}$

(12) Show that, if a is a constant,

$$\frac{1}{x^2 - a^2} = \frac{1}{2a}\left(\frac{1}{x - a} - \frac{1}{x + a}\right)$$

Hence deduce that

$$\int \frac{dx}{x^2 - a^2} = \frac{1}{2a} \ln \frac{x - a}{x + a}$$

(13) Show that

$$\frac{2x^2 - 3x - 2}{(x + 2)(x^2 + 2)} = \frac{3}{x + 2} - \frac{x + 1}{x^2 + 2}$$

(14) As $d/dx\,(x^2 + 5x - 6) = 2x + 5$,

$$\int \frac{2x + 5}{x^2 + 5x - 6}\, dx = \ln (x^2 + 5x - 6)$$

Show that if $(2x + 5) / (x^2 + 5x - 6)$ is split into partial fractions, the same result is eventually reached.

(15) Split $x^3/(x + 1)^4$ into partial fractions.

(16) If

$$\frac{x^2 - 7x + 4}{(x + 4)(x + 1)(x + 2)} = \frac{A}{x + 4} + \frac{B}{x + 1} + \frac{C}{x + 2}$$

use the cover-up rule to find the values of A, B and C. Hence evaluate

$$\int_0^1 \frac{x^2 - 7x + 4}{(x + 4)(x + 1)(x + 2)}\, dx$$

correct to three places of decimals.

(17) Evaluate

$$\int_1^2 \frac{(3x + 1)}{(x^2 + x)}\, dx$$

(18) Using the factorisation process for the difference of two squares as far as possible, show that

$$t^4 - 1 = (t^2 + 1)(t + 1)(t - 1)$$

Hence show that

$$\int \frac{dt}{t^4 - 1} = \tfrac{1}{4}\left(\ln \frac{t - 1}{t + 1} - 2 \tan^{-1} t\right) + C$$

9.3 Integration by Parts

The product formula for differentiation states

$$\frac{d}{dx}(uv) = u \frac{dv}{dx} + v \frac{du}{dx}$$

Integrating both sides w.r.t. x,

$$uv = \int u \frac{dv}{dx}\, dx + \int v \frac{du}{dx}\, dx$$

$$= \int u\, dv + \int v\, du$$

Rearranging this in the form

$$\int u\, dv = uv - \int v\, du$$

gives the basic formula for the method of INTEGRATION BY PARTS. *This formula must be memorised.*

The method is generally used to integrate products, of which one term is equated with u and the other with dv. Since the result is expressed in terms of a second integral, progress is only made if $\int v\, du$ is simpler in form than $\int u\, dv$. In particular, as the term u is differentiated, it will usually be equated with powers of x if there are any.

To integrate $\int xe^{2x}\, dx$, u is equated with x and dv with $e^{2x}\, dx$. With du and v, these can be conveniently laid out as follows.

$$\int xe^{2x}\, dx = \tfrac{1}{2}xe^{2x} - \tfrac{1}{2}\int e^{2x}\, dx$$

$$= \tfrac{1}{2}xe^{2x} - \tfrac{1}{4}e^{2x} + C$$

$$u = x$$
$$dv = e^{2x}\, dx$$
$$v = \tfrac{1}{2}e^{2x}$$
$$du = dx$$

The first line is a direct application of the formula. The integral on the right can then be handled directly giving the final line.

Definite integrals are handled in exactly the same way except that the limits are inserted using the appropriate notation. Adding limits 1 and 0 to the integral above, it becomes

$$\int_0^1 xe^{2x}\, dx = \left[\tfrac{1}{2}xe^{2x}\right]_0^1 - \tfrac{1}{2}\int_0^1 e^{2x}\, dx$$

$$= \left[\tfrac{1}{2}xe^{2x} - \tfrac{1}{4}e^{2x}\right]_0^1$$

$$= \tfrac{1}{4}(e^2 + 1)$$

$$\approx 2.097$$

In some cases, the integral on the right, although simpler than the original, still cannot be integrated directly. This is illustrated with $\int x^2 \sin x\, dx$ which it is necessary to integrate by parts twice. At each stage the values equated with u and dv must be corresponding terms.

$$\int x^2 \sin x \, dx = -x^2 \cos x + 2 \int x \cos x \, dx$$

$u = x^2$
$dv = \sin x \, dx$
$v = -\cos x$
$du = 2x \, dx$

$$\int x \cos x \, dx = x \sin x - \int \sin x \, dx$$

$$= x \sin x + \cos x$$

Thus

$$\int x^2 \sin x \, dx = -x^2 \cos x + 2(x \sin x + \cos x) + C$$

$$= -x^2 \cos x + 2\, x \sin x + 2 \cos x + C$$

$u = x$
$dv = \cos x \, dx$
$v = \sin x$
$du = dx$

Occasionally integration by parts can be used to evaluate an integral such as $\ln x \, dx$, containing an integrand whose *derivative* is known. In such a case the integrand is equated with u and dv is equated with the derivative dx

$$\int \ln x \, dx = x \ln x - \int dx$$

$$= x \ln x - x + C$$

$u = \ln x$
$dv = dx$
$v = x$
$du = \frac{1}{x} \, dx$

Example 9.3(a)
Integrate $\int x \cos 3x \, dx$. Verify the result by differentiation.

$$\int x \cos 3x \, dx = \tfrac{1}{3}x \sin 3x - \tfrac{1}{3} \int \sin 3x \, dx$$

$$= \tfrac{1}{3}x \sin 3x + \tfrac{1}{9} \cos 3x + C$$

$u = x$
$dv = \cos 3x \, dx$
$v = \frac{1}{3} \sin 3x$
$du = dx$

$d/dx\ (\frac{1}{3}x \sin 3x + \frac{1}{9} \cos 3x + C)$
$\frac{1}{3} \sin 3x + x \cos 3x - \frac{1}{3} \sin 3x$
$= x \cos 3x$

Example 9.3(b)
Evaluate $\int_1^2 3x^2 \ln x \, dx$.

$$\int_1^2 3x^2 \ln x = [x^3 \ln x]_1^2 - \int_1^2 x^2 \, dx$$
$$= [x^3 \ln x - \tfrac{1}{3}x^3]_1^2$$
$$= (8 \ln 2 - \tfrac{8}{3}) - (-\tfrac{1}{3})$$
$$\approx 3.21$$

$$u = \ln x \qquad dv = 3x^2 \, dx \qquad v = x^3 \qquad du = \frac{1}{x} \, dx$$

Here, exceptionally, u is equated with the term other than a power of x. This is so that the result $d/dx\,(\ln x) = 1/x$ can be used.

Example 9.3(c)
Integrate $\int e^{2x} \sin x \, dx$.

Let

$$I = \int e^{2x} \sin x \, dx$$
$$= -e^{2x} \cos x + 2\int e^{2x} \cos x \, dx$$

$$u = e^{2x} \qquad dv = \sin x \, dx \qquad v = -\cos x \qquad du = 2e^{2x} \, dx$$

$$2\int e^{2x} \cos x dx = 2e^{2x} \sin x - 4\int e^{2x} \sin x \, dx$$
$$= 2e^{2x} \sin x - 4I$$

$$u = e^{2x} \qquad dv = \cos x \, dx \qquad v = \sin x \qquad du = 2e^{2x} \, dx$$

Hence

$$I = -e^{2x} \cos x + 2(e^{2x} \sin x - 2I)$$
$$5I = e^{2x}(2 \sin x - \cos x)$$
$$I = \tfrac{1}{5}e^{2x}(2 \sin x - \cos x) + C$$

Here there are no terms in the integrand involving powers of x and either of the factors e^{2x} or $\sin x$ may be equated to u. Since u was equated with e^{2x} for the first integration, it had to be equated with it again for the second integration. The method of solution used applies to any example where integration by parts twice leads to the original integrand.

Example 9.3(d)

Obtain $\int \tan^{-1} x \, dx$.

$$\int \tan^{-1} dx = x \tan^{-1} x - \int \frac{x \, dx}{1 + x^2}$$

$$= x \tan^{-1} x - \tfrac{1}{2} \int \frac{2x \, dx}{1 + x^2}$$

$$= x \tan^{-1} x - \tfrac{1}{2} \ln (1 + x^2) + C$$

$$u = \tan^{-1} x \qquad dv = dx \qquad v = x \qquad du = dx/(1 + x^2)$$

This is another example in which the integrand has a known derivative (see Part 8.6) and for which integration by parts is possible. The integrand on the right should be recognised as logarithmic — the numerator when multiplied by 2 is the derivative of the denominator — allowing the integration to be completed. This logarithmic integral could be solved using the substitution $u = 1 + x^2$ if its form were not recognised.

Example 9.3(e)

Show that $\int_0^\infty x^3 e^{-x} \, dx = 6$

$$\int_0^\infty x^3 e^{-x} \, dx = [x^3 e^{-x}]_0^\infty + 3 \int_0^\infty x^2 e^{-x} \, dx$$

$$u = x^3 \qquad dv = e^{-x} \, dx \qquad v = -e^{-x} \qquad du = 3x^2 \, dx$$

$$3 \int_0^\infty x^2 e^{-x} \, dx = [3x^2 e^{-x}]_0^\infty + 6 \int_0^\infty x e^{-x} \, dx$$

$$u = x^2 \qquad dv = e^{-x} \, dx \qquad v = -e^{-x} \qquad du = 2x \, dx$$

$$6 \int_0^\infty x e^{-x} \, dx = [6x e^{-x}]_0^\infty + 6 \int_0^\infty e^{-x} \, dx$$

$$= [6x e^{-x} - 6e^{-x}]_0^\infty$$

$$u = x \qquad dv = e^{-x} \, dx \qquad v = -e^{-x} \qquad du = dx$$

Hence

$$\int_0^\infty x^3 \, e^{-x} \, e^{-x} \, dx = [x^3 e^{-x} + 3x^2 e^{-x} + 6x e^{-x} - 6e^{-x}]_0^\infty$$

$$= 6$$

When substituting on the penultimate line, all terms involving e^{-x} become zero when the upper limit is substituted and all terms involving x become zero when the lower limit is substituted.

9.4 Reduction Formulae

When solving Example 9.3(e), it was necessary to carry out an integration by parts three times, the effect of the integration at each stage being to reduce the power of x in the integrand by 1. If we were asked to evaluate $\int_0^\infty x^{10} e^{-x} \, dx$, it could be done in the same way (given patience and a lot of paper!), but, in such a case, it would be more sensible to integrate

$\int_0^\infty x^n e^{-x}\, dx$ and obtain the answer as a formula in terms of n and into which the value $n = 10$ could be substituted. Used in this way, n is known as a PARAMETER of the integral.

If we write $I_n = \int_0^\infty x^n e^{-x}\, dx$ and carry out the integration by parts in terms of the parameter, we obtain an integral of the same form as I_n.

$$I_n = \int_0^\infty x^n e^{-x}\, dx \qquad u = x^n$$

$$dv = e^{-x}\, dx$$

$$= [-x^n e^{-x}]_0^\infty + n\int_0^\infty x^{n-1} e^{-x}\, dx \qquad v = -e^{-x}$$

$$du = nx^{n-1}\, dx$$

$$= 0 + nI_{n-1}$$

$$= nI_{n-1}$$

The expression $[-x^n e^{-x}]_0^\infty$ is equal to zero for the reasons explained in Example 9.3(e). The result $I_n = nI_{n-1}$ is called a REDUCTION FORMULA. To evaluate $\int_0^\infty x^{10} e^{-x}\, dx$ requires a value to be obtained for I_{10}. Using the reduction formula as many times as necessary.

$$\begin{aligned} I_{10} &= 10 \times I_9 \\ &= 10 \times (9 \times I_8) \\ &= 10 \times 9 \times (8 \times I_7), \text{ etc.} \end{aligned}$$

The simplest integral of this form will be I_0 which can be obtained directly:

$$I_0 = \int_0^\infty e^{-x}\, dx = [-e^{-x}]_0^\infty = 1$$

as $x^n = 1$ when $n = 0$. Hence

$$\begin{aligned} I_{10} &= 10!I_0 \\ &= 10! \end{aligned}$$

using factorial notation. (In general terms $I_n = n!$ Taking $n = 3$ verifies the result of Example 9.3(c).)

Sometimes the reduction formula expresses I_n in terms of I_{n-2} rather than in terms of I_{n-1}. If $I_n = \int \cos^n x\, dx$, working with indefinite integrals this time

$$I_n = \int \cos^n x\, dx \qquad u = \cos^{n-1} x$$

$$dv = \cos x\, dx$$

$$= \int \cos^{n-1} x \cos x\, dx \qquad v = \sin x$$

$$du = -(n-1)\cos^{n-2} x \sin x$$

$$= \cos^{n-1} x \sin x + (n-1)\int \cos^{n-2} x \sin^2 x\, dx$$

$$= \cos^{n-1} x \sin x + (n-1)\int \cos^{n-2} x\,(1 - \cos^2 x)\, dx$$

$$= \cos^{n-1} x \sin x + (n-1)(I_{n-2} - I_n)$$

which can be rearranged as

$$I_n = \frac{1}{n}(\cos^{n-1} x \sin x + [n - 1]I_{n-2})$$

A result I_n is obtained in terms of either I_0 (if n is even) or I_1 (if n is odd).

$$I_0 = \int dx = x$$

$$I_1 = \int \cos x \, dx = \sin x$$

Notice that the original integrand $\cos^n x$ has to be expressed as the product $\cos^{n-1} x \cos x$ in order to obtain a term which can be integrated.

The integral $\int \sin^n x \, dx$ is handled in the same way.

Some reduction formulae are expressed in terms of two parameters. To evaluate $\int_0^{\pi/2} \sin^m x \cos^n x \, dx$, it is necessary to start once again by rearranging the integrand.

$$u = \cos^{n-1} x$$

$$dv = \sin^m x \cos x \, dx$$

$$v = \frac{\sin^{m+1} x}{m + 1}$$

$$du = -(n - 1)\cos^{n-2} x \sin x$$

$$\begin{aligned}
I_{m,n} &= \int_0^{\pi/2} \sin^m x \cos^n x \, dx \\
&= \int_0^{\pi/2} \sin^m x \cos^{n-1} x \cos x \, dx \\
&= \left[\frac{\cos^{n-1} x \sin^{m+1} x}{m + 1}\right]_0^{\pi/2} + \frac{n - 1}{m + 1}\int_0^{\pi/2} \cos^{n-2} x \sin^{m+2} x \, dx \\
&= \frac{n - 1}{m + 1}\int_0^{\pi/2} \cos^{n-2} x \sin^{m+2} x \, dx
\end{aligned}$$

A reduction form can be obtained from this result but as one parameter has been increased, this is not particularly helpful. However writing $\sin^{m+2} x$ as $\sin^2 x$.

$\sin^2 x = \sin^m x\,(1 - \cos^2 x)$ we get

$$\begin{aligned}
I_{m,n} &= \frac{n - 1}{m + 1}\int_0^{\pi/2} \cos^{n-2} x \sin^{m+2} x \, dx \\
&= \frac{n - 1}{m + 1}\int_0^{\pi/2} \cos^{n-2} x \,(1 - \cos^2 x) \sin^m x \, dx \\
&= \frac{n - 1}{m + 1}\int_0^{\pi/2} (\cos^{n-2} x \sin^m x - \cos^n x \sin^m x) \, dx \\
&= \frac{n - 1}{m + 1}(I_{m,n-2} - I_{m,n})
\end{aligned}$$

This can be rearranged as

$$I_{m,n} = \frac{n-1}{m+n} I_{m,n-2}$$

$I_{m,n}$ will therefore be expressed in terms of $I_{m,0}$ or $I_{m,1}$ depending on whether n is even or odd. Since $I_{m,0} = \sin^m x \, dx$, this can be obtained as explained earlier. $I_{m,1} = \int_0^{\pi/2} \sin^m x \cos x \, dx$, which can be evaluated using the substitution $s = \sin x$. Example 9.1(c) illustrates the best method of solving integrals of this type when m and/or n are odd. If both parameters are even a reduction formula is needed.

Example 9.4(a)
Show that if $I_n = \int_0^\pi x^n \sin x \, dx$, then $I_n = \pi^n - n(n-1) I_{n-2}$. Hence evaluate I_4

$$I_n = \int_0^\pi x^n \sin x \, dx$$

$$= [-x^n \cos x]_0^\pi + n \int_0^\pi x^{n-1} \cos x \, dx$$

$$= \pi^n + n \int_0^\pi x^{n-1} \cos x \, dx$$

$u = x^n$
$dv = \sin x \, dx$
$v = -\cos x$
$du = nx^{n-1} \, dx$

$$n \int_0^\pi x^{n-1} \cos x \, dx = n[x^{n-1} \sin x]_0^\pi - n(n-1) \int_0^\pi x^{n-2} \sin x \, dx$$

$$= -n(n-1)I_{n-2}$$

Hence $I_n = \pi^n - n(n-1)I_{n-2}$.

$$I_4 = \pi^4 - 12I_2$$

$$= \pi^4 - 12(\pi^2 - 2I_0)$$

$$= \pi^4 - 12\pi^2 + 24I_0$$

$u = x^{n-1}$
$dv = \cos x \, dx$
$v = \sin x$
$du = (n-1)x^{n-2} \, dx$

But

$$I_0 = \int_0^\pi \sin x \, dx$$

$$= [-\cos x]_0^\pi$$

$$= 2$$

$$I_4 = \pi^4 - 12\pi^2 + 48 \approx 26.97$$

Example 9.4(b)
Find reduction formula for $I_n = \int \tan^n \theta \, d\theta$ and use it to integrate $\int \tan^6 \theta \, d\theta$.

$$I_n = \int \tan^n \theta \, d\theta$$

$$= \int \tan^{n-2} \theta \tan^2 \theta \, d\theta$$

$$= \int \tan^{n-2} \theta (\sec^2 \theta - 1) \, d\theta$$

$$= \int \tan^{n-2} \theta \sec^2 \theta \, d\theta - I_{n-2}$$

The integral $\int \tan^{n-2} \theta \sec^2 \theta \, d\theta$ is solved using the substitution $t = \tan \theta$ so that $dt = \sec^2 \theta \, d\theta$.

$$\int \tan^{n-2} \theta \sec^2 \theta \, d\theta = \int t^{n-2} \, dt$$

$$= \frac{1}{n-1} t^{n-1}$$

$$= \frac{1}{n-1} \tan^{n-1} \theta$$

The reduction formula becomes

$$I_n = \frac{1}{n-1} \tan^{n-1} \theta - I_{n-2}$$

Hence

$$I_6 = \tfrac{1}{5} \tan^5 \theta - I_4$$

$$= \tfrac{1}{5} \tan^5 \theta - \tfrac{1}{3} \tan^3 \theta + I_2$$

$$= \tfrac{1}{5} \tan^5 \theta - \tfrac{1}{3} \tan^3 \theta + \tan \theta - I_0$$

$$= \tfrac{1}{5} \tan^5 \theta - \tfrac{1}{3} \tan^3 \theta + \tan \theta - \theta + C$$

since $I_0 = \int d\theta = \theta + C$

Unworked Examples 3

Part 9.3

(1) Show that $\int xe^x \, dx = xe^x - e^x + C$ and that $\int x^2 e^x \, dx = x^2 e^x - 2xe^x + 2e^x + C$.

(2) Evaluate $\int_0^{\pi/4} e^x \sin x \, dx$.

(3) The result $\int \ln x \, dx = x \ln x - x$ was obtained in the text. Use this to verify the result of Example 9.3(b) taking $u = 3x^2$ and $dv = \ln x \, dx$.

(4) Integrate $\int 4x^2 \sin 2x \, dx$ by parts.

(5) Show that the result of Example 9.3(c) is unchanged if u is equated with $\sin x$ and $\mathrm{d}v$ with $\mathrm{e}^{2x}\,\mathrm{d}x$.

(6) Integrate $\int \sin^{-1} x\,\mathrm{d}x$.

(7) Prove that $\int_0^{\pi/2} x^2 \sin x\,\mathrm{d}x = \pi - 2$.

(8) Show that

$$\text{(a) } I_1 = \int \mathrm{e}^{ax} \sin bx\,\mathrm{d}x = \frac{1}{a^2 + b^2}\,\mathrm{e}^{ax}\,(a \sin bx - b \cos bx)$$

$$\text{(b) } I_2 = \int \mathrm{e}^{ax} \cos bx\,\mathrm{d}x = \frac{1}{a^2 + b^2}\,\mathrm{e}^{ax}\,(a \cos bx + b \sin bx)$$

by integrating by parts. Show further that
(i) $aI_1 + bI_2 = \mathrm{e}^{ax} \sin bx$,
(ii) $aI_2 - bI_1 = \mathrm{e}^{ax} \cos bx$.
Verify the results (a) and (b) by considering the real and imaginary parts of $\int \mathrm{e}^{(a+jb)x}\,\mathrm{d}x$.

Part 9.4

(9) Find a reduction formula for $\int x^n \cos x\,\mathrm{d}x$. Hence find $\int x^4 \cos x\,\mathrm{d}x$.

(10) Show that if $I_n = \int x^n \mathrm{e}^x\,\mathrm{d}x$ then $I_n = x^n \mathrm{e}^x - nI_{n-1}$. Using this formula, deduce
(a) $I_0 = \mathrm{e}^x$
(b) $I_1 = \mathrm{e}^x\,(x - 1)$
(c) $I_2 = \mathrm{e}^x\,(x^2 - 2x + 2)$
(d) $I_3 = \mathrm{e}^x\,(x^3 - 3x^2 + 6x - 6)$.
Neglecting the *alternating* signs of the terms in the bracket what do you notice about the terms themselves? Deduce an expression for I_4.

(11) Obtain a reduction formula for $\int \sin^m x\,\mathrm{d}x$.

(12) Using the method shown in Example 9.1(c) evaluate

$$\int_0^{\pi/2} \sin^5 x \cos^5 x\,\mathrm{d}x$$

Using the reduction formula for $\int_0^{\pi/2} \sin^m x \cos^n x\,\mathrm{d}x$,

$$I_{m,n} = \frac{n-1}{m+n} I_{m,n-2}$$

show that $I_{m,1} = 1/(m + 1)$ and hence verify the result when $m = n = 5$.

(13) If

$$I_n = \int_0^\infty \frac{\mathrm{d}x}{(1 + x^2)^n}$$

Show that $I_n = 2n(I_n - I_{n+1})$. Replacing n by $n - 1$ and rearranging, obtain the formula

$$I_n = \tfrac{1}{2}(2n - 3)/(n - 1)I_{n-1}.$$

Show that $I_1 = \pi/2$ and hence that

$$\int_0^\infty \frac{dx}{(1 + x^2)^3} = 3\pi/16.$$

PART 10: POWER SERIES

10.1 Maclaurin's Theorem

It is convenient to express certain functions such as $\sin x$ or e^x as a series of ascending powers of x, generally referred to as a POWER SERIES. We will assume that all the power series in which we have an interest can be expressed as an infinite series of the form

$$f(x) = a_0 + a_1 + a_2x^2 + a_3x^3 + \cdots$$
$$= \sum_{n=0}^{\infty} a_n x^n$$

This is a rather sweeping assumption and, although it may give valid results for the functions considered, it should be borne in mind that there are other functions which cannot be treated by these methods.

Assuming that

$$f(x) = \sum_{n=0}^{\infty} a_n x^n$$

and that the function can be differentiated indefinitely, we get

$$\begin{aligned}
f(x) &= a_0 + a_1x + a_2x^2 + a_3x^3 + a^4x^4 + \cdots \\
f'(x) &= a_1 + 2a_2x + 3a_3x^2 + 4a_4x^3 + \cdots \\
f''(x) &= 2.1a_2 + 3.2a_3x + 4.3a_4x^2 + \cdots \\
f'''(x) &= 3.2.1a_3 + 4.3.2a_4x + \cdots, \text{ etc.}
\end{aligned}$$

Putting $x = 0$ in each case, $f(0) = a_0$, $f'(0) = a_1$, $f''(0) = 2!\ a_2$ and $f'''(0) = 3!\ a_3$. Continuing in this way $f^n(0) = n!\ a_n$ for all values of n. (The notation $f^n(x)$ is used for the nth derivative of $f(x)$.) When $n = 0$, $f^n(x)$ means $f(x)$. Hence

$$f(x) = f(0) + xf'(0) + x^2\frac{f''(0)}{2!} + x^3\frac{f''(0)}{3!} + \cdots$$
$$= \sum_{n=0}^{\infty} x^n \frac{f^n(0)}{n}$$

This result is called MACLAURIN'S THEOREM.

Taking a simple application, when $f(x) = e^x$, $f'(x) = f''(x) = f'''(x) = \cdots = e^x$, so that $f^n(0) = 1$ for all values of n. Hence

$$e^x = 1 + x + x^2/2! + x^3/3! + x^4/4! + \cdots$$

Putting $x = 1$, gives the value of e as 2.7183. If $f(x) = \ln(1 + x)$, the following set of values is obtained

$$\begin{aligned} f(x) &= \ln(1 + x) & f(0) &= 0 \\ f'(x) &= 1/(1 + x) & f'(0) &= 1 \\ f''(x) &= -1/(1 + x)^2 & f''(0) &= -1 \\ f'''(x) &= 2.1/(1 + x)^3 & f'''(0) &= 2! \\ f^{iv}(x) &= -3.2/(1 + x)^4 & f^{iv}(0) &= -3! \end{aligned}$$

and so on, with

$$f^n(0) = (n - 1)!(-1)^{n-1}(n \neq 0)$$

(The term $(-1)^{n-1}$ makes the value negative when n is even and positive when n is odd.) Hence

$$\begin{aligned} \ln(1 + x) &= x - \frac{x^2}{2!} + \frac{2!x^3}{3!} - \frac{3!x^4}{4!} + \cdots \\ &= x - \tfrac{1}{2}x^2 + \tfrac{1}{3}x^3 - \tfrac{1}{4}x^4 + \cdots \end{aligned}$$

Example 10.1(a)

Use Maclaurin's theorem to obtain power series for $\sin x$ and $\cos x$. Show that the results $d/dx(\sin x) = \cos x$ and $d/dx(\cos x) = -\sin x$ hold when dealing with the series forms of the functions

For $f(x) = \sin x$

$$\begin{aligned} f(x) &= \sin x & f(0) &= 0 \\ f'(x) &= \cos x & f'(0) &= 1 \\ f''(x) &= -\sin x & f''(0) &= 0 \\ f'''(x) &= -\cos x & f'''(0) &= -1 \\ f^{iv}(x) &= \sin x & f^{iv}(0) &= 0, \text{ etc.} \end{aligned}$$

$$\sin x = x - x^3/3! + x^5/5! \cdots$$

For $f(x) = \cos x$

$$\begin{aligned} f(x) &= \cos x & f(0) &= 1 \\ f'(x) &= -\sin x & f'(0) &= 0 \\ f''(x) &= -\cos x & f''(0) &= -1 \\ f'''(x) &= \sin x & f'''(0) &= 0 \\ f^{iv}(x) &= \cos x & f^{iv}(0) &= 1, \text{ etc.} \end{aligned}$$

$\cos x = 1 - x^2/2! + x^4/4! - \cdots$

$$\begin{aligned} d/dx\,(\sin x) &= d/dx\,(x - x^3/3! + x^5/5! - \cdots \\ &= 1 - x^2/2! + x^4/4! - \cdots \\ &= \cos x \end{aligned}$$

$$\begin{aligned} d/dx\,(\cos x) &= d/dx\,(1 - x^2/2! + x^4/4! - x^6/6! + \cdots) \\ &= -x + x^3/3! - x^5/5! + \cdots \\ &= -\sin x \end{aligned}$$

Functions were described as odd or even in Part 1.3. Notice that odd functions have a series containing only odd powers of x while even functions have a series containing only even powers of x.

Example 10.1(b)
Show that the binomial expansion

$$(1 + x)^n = 1 + nx + \frac{n(n-1)}{2!}x^2 + \frac{n(n-1)(n-2)}{3!}x^3 + \cdots$$

can be obtained using Maclaurin's theorem.

$$\begin{aligned} \text{If } f(x) &= (1 + x)^n & f(0) &= 1 \\ f'(x) &= n(1 + x)^{n-1} & f'(0) &= n \\ f''(x) &= n(n-1)(1 + x)^{n-2} & f''(0) &= n(n-1) \\ f'''(x) &= n(n-1)(n-2)(1 + x)^{n-3} & f'''(0) &= n(n-1)(n-2), \\ & & & \text{etc.} \end{aligned}$$

$$(1 + x)^n = 1 + nx + \frac{n(n-1)}{2!}x^2 + \frac{n(n-1)(n-2)}{3!}x^3 + \cdots$$

10.2 Other Power Series

The power series for e^x, $\sin x$, $\cos x$, $\ln(1 + x)$ and $(1 + x)^n$ given in Part 10.1 all have a coefficient pattern, which makes them fairly easy to remember. Furthermore they are used frequently to obtain other power series and, for convenience they are summarised in Table 10.1.

Other series may be obtained from these by replacing, say, x by $-x$ so that the power series for e^{-x} is $1 - x + x^2/2! - x^3/3! + \cdots$ or by replacing x by $2x$ so that the power series for $\sin 2x = 2x - 8x^3/3! + 32x^5/5! - \cdots$. Another method is to combine two known series. Since

$\ln(1 + x) = x - \frac{1}{2}x^2 + \frac{1}{3}x^3 - \frac{1}{4}x^4 + \cdots$, it follows that

$\ln(1 - x) = -x - \frac{1}{2}x^2 - \frac{1}{3}x^3 - \frac{1}{4}x^4 - \cdots$

Table 10.1

Function	Power series
e^x	$1 + x + x^2/2! + x^3/3! + \cdots$
$\sin x$	$x - x^3/3! + x^5/5! - \cdots$
$\cos x$	$1 - x^2/2! + x^4/4! - \cdots$
$\ln(1 + x)$	$x - x^2/2 + x^3/3 - x^4/4 + \cdots$
$(1 + x)^n$	$1 + nx + n(n-1)x^2/2! + n(n-1)(n-2)x^3/3! + \cdots$

Thus

$$\ln \frac{1+x}{1-x} = \ln(1+x) - \ln(1-x)$$
$$= 2(x + \tfrac{1}{3}x^3 + \tfrac{1}{5}x^5 + \cdots)$$

Of course, Maclaurin's theorem can be used although in the cases of functions other than those in Table 10.1, this may not be practical for more than the first few terms of the series, as the higher derivatives of $f(x)$ become more involved. Taking, as an example $f(x) = \tan x$, we obtain

$$\begin{aligned}
f(x) &= \tan x & f(0) &= 0\\
f'(x) &= \tan^2 x + 1 & f'(0) &= 1\\
f''(x) &= 2\tan^3 x + 2\tan x & f''(0) &= 0\\
f'''(x) &= 6\tan^4 x + 8\tan^2 x + 2 & f'''(0) &= 2\\
f^{\text{iv}}(x) &= 24\tan^5 x + 40\tan^3 x + 16\tan x & f^{\text{iv}}(0) &= 0\\
f^{\text{v}}(x) &= 120\tan^6 x + 240\tan^4 x + 136\tan^2 x + 16 & f^{\text{v}}(0) &= 16
\end{aligned}$$

$$\tan x = x + \tfrac{1}{3}x^3 + \tfrac{2}{15}x^5 + \cdots$$

In each case the derivatives of $f(x)$ were obtained in terms of $\tan x$ and the result was simplified as far as possible.

Another method is to use a function, which when differentiated or integrated gives another function with a recognisable or easily obtainable series form. For example to obtain the power series for $\sin^{-1} x$ we have

$$\sin^{-1} x = \int \frac{dx}{\sqrt{(1-x^2)}} \qquad \text{(see Part 9.1)}$$

$$= \int (1 - x^2)^{-1/2}\, dx$$

$$= \int (1 + \tfrac{1}{2}x^2 + \tfrac{3}{8}x^4 + \tfrac{5}{16}x^6 + \cdots)\, dx$$

$$= x + \tfrac{1}{6}x^3 + \tfrac{3}{40}x^5 + \tfrac{5}{112}x^7 + \cdots + C$$

However, as $\sin^{-1} x$ is an odd function, $C = 0$

$$\sin^{-1} x = x + \tfrac{1}{6}x^3 + \tfrac{3}{40}x^5 + \tfrac{5}{112}x^7 + \ldots$$

$(1 - x^2)^{-1/2}$ has been expanded using the binomial theorem.

Example 10.2(a)
Find the first four terms taken in ascending powers of x in a series for $\ln(1 + \sin x)$.
Using the power series for $\ln(1 + x)$, replacing x by $\sin x$

$$\begin{aligned}\ln(1 + \sin x) &= \sin x - \tfrac{1}{2}\sin^2 x + \tfrac{1}{3}\sin^3 x - \tfrac{1}{4}\sin^4 x + \cdots \\ &= (x - \tfrac{1}{6}x^3) - \tfrac{1}{2}x^2 + \tfrac{1}{3}x^3 - \tfrac{1}{4}x^4 + \cdots \\ &= x - \tfrac{1}{2}x^2 + \tfrac{1}{6}x^3 - \tfrac{1}{4}x^4 + \cdots\end{aligned}$$

In each term of the expansion of $\ln(1 + \sin x)$, $\sin x$ and its powers are replaced by that part of its power series which will give powers of x up to x^4.

Example 10.2(b)
Use the result $\mathrm{d}/\mathrm{d}x\,(\tan^{-1} x) = 1/(1 + x^2)$ to obtain a power series for $\tan^{-1} x$. Deduce from the series that $\pi = 4\,(1 - \tfrac{1}{3} + \tfrac{1}{5} - \tfrac{1}{7} + \cdots)$

Since $\mathrm{d}/\mathrm{d}x\,(\tan^{-1} x) = 1/(1 + x^2)$,

$$\begin{aligned}\tan^{-1} x &= \int \frac{\mathrm{d}x}{1 + x^2} \\ &= \int (1 + x^2)^{-1}\,\mathrm{d}x \\ &= \int (1 - x^2 + x^4 - x^6 + \cdots)\,\mathrm{d}x \\ &= x - \tfrac{1}{3}x^3 + \tfrac{1}{5}x^5 - \tfrac{1}{7}x^7 + \ldots + C\end{aligned}$$

But $\tan^{-1} x$ is an odd function and hence $C = 0$

$$\tan^{-1} x = x - \tfrac{1}{3}x^3 + \tfrac{1}{5}x^5 - \tfrac{1}{7}x^7 + \ldots$$

When $x = 1$, $\tan^{-1} x = \pi/4$

$$\begin{aligned}\pi/4 &= 1 - \tfrac{1}{3} + \tfrac{1}{5} - \tfrac{1}{7} + \cdots \\ \pi &= 4(1 - \tfrac{1}{3} + \tfrac{1}{5} - \tfrac{1}{7} + \cdots)\end{aligned}$$

Example 10.2(c)
State the power series for $\sin x$ and $\cos x$ and hence obtain one for $\sin 2x$. Show that the identity $\sin 2x = 2 \sin x \cos x$ is at least reasonable by

obtaining the first three terms of the power series of both

$$\sin x = x - \frac{x^3}{3!} + \frac{x^5}{5!} - \cdots$$

$$= x - \tfrac{1}{6}x^3 + \tfrac{1}{120}x^5 - \cdots$$

$$\cos x = 1 - \frac{x^2}{2!} + \frac{x^4}{4!} - \cdots$$

$$= 1 - \tfrac{1}{2}x^2 + \tfrac{1}{24}x^4 - \cdots$$

$$\sin 2x = 2x - \frac{(2x)^3}{3!} + \frac{(2x)^5}{5!} - \cdots$$

$$= 2x - \tfrac{4}{3}x^3 + \tfrac{2}{15}x^5 \cdots$$

$$2 \sin x \cos x = 2(x - \tfrac{1}{6}x^3 + \tfrac{1}{120}x^5 - \cdots)(1 - \tfrac{1}{2}x^2 + \tfrac{1}{24}x^4 - \cdots)$$

$$= 2(x - [\tfrac{1}{6} + \tfrac{1}{2}]x^3 + [\tfrac{1}{120} + \tfrac{1}{12} + \tfrac{1}{24}]x^5 + \cdots)$$

$$= 2x - \tfrac{4}{3}x^3 + \tfrac{2}{15}x^5 - \cdots$$

Showing that two functions have power series with the first few terms equal does not make the functions equal.

10.3 Taylor's Theorem

This gives a formula for the expansion of $f(x + h)$ in terms of ascending powers of h. It can be obtained in a similar manner to that used for Maclaurin's theorem but will simply be stated here

$$f(x + h) = f(x) + h(f'(x) + \frac{h^2}{2!}f''(x) + \frac{h^3}{3!}f'''(x) + \cdots$$

$$= \sum_{n=0}^{\infty} \frac{h^n}{n!} f^n(x)$$

Putting $x = 0$ gives

$$f(h) = \sum_{n=0}^{\infty} h^n \frac{f^n(0)}{n!}$$

which is Maclaurin's theorem expressed in terms of h, so that it can be regarded as a special case of Taylor's theorem.

Example 10.3(a)
Use Taylor's theorem to obtain the expansion of $\ln(x + h)$. By putting $h = x^2$, verify the fourth entry of Table 10.1.

If $f(x) = \ln x$, $f(x + h) = \ln (x + h)$. Using Taylor's theorem,

$$\ln (x + h) = \ln x + h\frac{1}{x} - \frac{h^2}{2!}\frac{1}{x^2} + \frac{h^3}{3!}\cdot\frac{2!}{x^3} - \frac{h^4}{4!}\cdot\frac{3!}{x^4} + \cdots$$

$$= \ln x + \frac{h}{x} - \tfrac{1}{2}\left(\frac{h}{x}\right)^2 + \tfrac{1}{3}\left(\frac{h^3}{x}\right) - \cdots$$

Putting $h = x^2$,

$$\ln (x + x^2) - \ln x = x - \tfrac{1}{2}x^2 + \tfrac{1}{3}x^3 - \cdots$$

But $\ln (x + x^2) - \ln x = \ln (x + x^2)/x = \ln (1 + x)$.

$$\ln (1 + x) = x - \tfrac{1}{2}x^2 + \tfrac{1}{3}x^3 - \cdots$$

Example 10.3(b)
Obtain sin 31° correct to four places of decimals using Taylor's theorem.

Using Taylor's theorem,

$$\sin (x + h) = \sin x + h \cos x - \frac{h^2}{2!}\sin x - \frac{h^3}{3!}\cos x + \cdots$$

Here $x = \pi/6$, $h = 0.0175$ since angles must be measured in radians.

$$\sin 31° = 0.5 + 0.0175 \times 0.866 - \tfrac{1}{2} \times 0.0175^2 \times 0.5 - \cdots$$
$$\approx 0.5151$$

Unworked Examples 4

Parts 10.1 and 10.2

(1) Use Maclaurin's theorem to obtain a power series for $\sinh x$.
(2) Show that $d/dx\,(\sec x) = \sec x \tan x$ and hence that all higher derivatives of $\sec x$ can be expressed in terms of $\sec x$ and $\tan x$. Obtain the first three terms of the expansion of $\sec x$.
(3) Obtain a power series for e^{-x} by (i) direct application of Maclaurin's theorem, (ii) using the series for e^x.
(4) Obtain the expansion of $(1 + x)^n$. Use the result to calculate $\sqrt{(0.996)}$ correct to three places of decimals.
(5) Using the series for $\ln (1 + x)$, show that

$$\ln\frac{1 + x}{1 - x} = 2(x + x^3/3 + x^5/5 + \cdots)$$

By putting $z = (1 + x)/(1 - x)$, show that $x = (z - 1)/(z + 1)$ and that

$$\ln z = 2\left(\frac{z - 1}{z + 1} + \tfrac{1}{3}\left(\frac{z - 1}{z - 1}\right)^3 + \tfrac{1}{5}\left(\frac{z - 1}{z + 1}\right)^5 + \cdots\right)$$

Hence obtain ln 2 and ln 3 correct to three places of decimals.

(6) By using the power series for $\tan x$ or otherwise find a power series for $\ln \sec x$.

(7) Use the result $d/dx\,(\sinh^{-1} x) = 1/\sqrt{(1 + x^2)}$ to obtain a power series for $\sinh^{-1} x$. Hence obtain $\sinh^{-1} 0.5$ correct to three places of decimals.

(8) Show that $\tan x$ is an odd function. Assuming that $\tan x$ can be expanded in the form

$$ax + bx^3 + cx^5 + \cdots$$

use the result $\sin x = \cos x \tan x$ to determine coefficients a, b and c.

Part 10.3

(9) Obtain the expansion of $(x + 1)^n$ using Taylor's theorem taking $f(x) = x^n$, $h = 1$.

(10) Show that $\tan^{-1}(1 + x) = \pi/4 + \frac{1}{2}x - \frac{1}{4}x^2 + \frac{1}{12}x^3 - \cdots$ Hence obtain $\tan^{-1} 1.1$ in degrees and minutes.

(11) Use Taylor's theorem to obtain (a) tan 44°, (b) tan 46° correct to three places of decimals.

(12) Taking $f(x) = \sin x$, use Taylor's theorem to obtain $\sin(x + y)$. Hence show that

$$\sin(x + y) = \sin x \cos y + \cos x \sin y.$$

(13) Show that by using Taylor's theorem, $\cos(x + h)$ can be expanded as

$$\cos(x + h) = \cos x - h \sin x - \frac{h^2}{2!}\cos x + \frac{h^3}{3!}\sin x + \cdots$$

Hence show that $\cos(\pi/4 + h) = 1/\sqrt{2}(1 - h - h^2/2! + h^3/3! - \cdots) = 1\sqrt{2}(\cos h - \sin h)$.

PART 11: INDETERMINATE FORMS

11.1 The Limit Notation

Consider the series of fractions

$\frac{1}{2}, \frac{2}{3}, \frac{3}{4}, \frac{4}{5} \cdots$

The nth term of the series is equal to $n/(n+1)$ and there is no limit to the value of n which can be used in this formula (except that it must be a positive integer). However, successive terms get nearer and nearer to 1, without ever equalling it (since the numerator will always be smaller than the denominator) but getting as near to it as we choose. The value 1 is called the LIMIT (or LIMITING VALUE) of the series and the notation for this is

$$\lim_{n \to \infty} \frac{n}{n+1} = 1$$

Notice that attempting to substitute ∞ gives the meaningless result ∞/∞. This, like its counterpart 0/0 is called an INDETERMINATE FORM (and can be compared with the solution of the linearly dependent set of equations met in Part 5.1).

There are various methods of handling indeterminate forms. In the example above, it is possible to rearrange the fraction algebraically so that

$$\lim_{n \to \infty} \frac{n}{n+1} = \lim_{n \to \infty} \left(1 - \frac{1}{n+1}\right)$$
$$= 1$$

since the value of $1/(n+1)$ is meaningful for infinitely large values of n. Other simple methods that may be used to obtain the limiting value of an indeterminate form are illustrated in Examples 11.1(a) to 11.1(c). In more involved cases the methods of Parts 11.2 and 11.3 can be used.

Example 11.1(a)
Obtain

$$\lim_{x \to 0} (x^2 + x^3)/x^2$$

$$\lim_{x \to 0} (x^2 + x^3)/x^2 = \lim_{x \to 0} (1 + x) = 1$$

Example 11.1(b)
Show that as $x \to 1$, $(x^2 - 1)/(x - 1) \to 2$

We require

$$\lim_{x \to 1} \frac{x^2 - 1}{x - 1}$$

$$\lim_{x \to 1} \frac{x^2 - 1}{x - 1} = \lim_{x \to 1} (x + 1)$$
$$= 2$$

Example 11.1(c)
Show that

$$\lim_{z\to\infty} \frac{a_1 z^2 + b_1 z + c_1}{a_2 z^2 + b_2 z + c_2} = \frac{a_1}{a_2}$$

$$\lim_{z\to\infty} \frac{a_1 z^2 + b_1 z + c_1}{a_2 z^2 + b_2 z + c_2}$$

$$= \lim_{z\to\infty} \frac{a_1 + \dfrac{b_1}{b_z} + \dfrac{c_1}{c_{z^2}}}{a_1 + \dfrac{z}{z} + \dfrac{z^2}{z^2}}$$

$$= \frac{a_1}{a_2}$$

11.2 Use of Series

Another method of evaluating indeterminate forms is to use a power series. For example

$$\lim_{x\to 0} \frac{e^x - 1}{x}$$

cannot be evaluated directly nor can $(e^x - 1)/x$ be simplified algebraically. However, using the power series for e^x,

$$\lim_{x\to 0} \frac{e^x - 1}{x}$$

$$= \lim_{x\to 0} \frac{x + x^2/2! + x^3/3! + \cdots}{x}$$

$$= \lim_{x\to 0} 1 + x/2! + x^2/3! + \cdots$$

$$= 1$$

Example 11.2(a)
Show that

$$\lim_{x\to 0} \frac{\sin x}{x} = 1$$

where x is measured in radians. Hence evaluate sin 2° correct to four places of decimals

$$\lim_{x\to 0} \frac{\sin x}{x}$$

$$= \lim_{x\to 0} \frac{1}{x}(x - x^3/3! + x^5/5! - \cdots)$$

$$= \lim_{x\to 0} (1 - x^2/3! + x^4/5! - \cdots)$$

$$= 1$$

Hence $\sin x \to x$ when x is small and measured in radians. This allows some ratios to be evaluated.

$$\sin 2^\circ \approx \pi/90$$
$$= 0.0349 \text{ (corr. 4 pl. dec.)}$$

Example 11.2(b)
Show that

$$\lim_{x\to 0} \frac{1 - \cos x}{x^2} = \frac{1}{2}$$

$$\lim_{x\to 0} \frac{1 - (1 - x^2/2! + x^4/4! \cdots)}{x^2}$$

$$= \lim_{x\to 0} \left(\frac{1}{2} - \frac{x^2}{4!} + \frac{x^4}{6!} - \cdots\right)$$

$$= \frac{1}{2}$$

11.3 L'Hospital's Rule

A third method of evaluating limits of indeterminate form, known a L'HOSPITAL'S RULE, depends on Taylor's theorem.

Suppose we required

$$\lim_{x\to a} \frac{f(x)}{g(x)}$$

where $f(a) = g(a) = 0$. Let $x = a + h$.

$$\lim_{x \to a} \frac{f(x)}{g(x)} = \lim_{h \to 0} \frac{f(a+h)}{g(a+h)}$$

$$= \lim_{h \to 0} \frac{f'(a) + hf'(a) + \dfrac{h^2}{2!} f''(a) + \cdots}{g(a) + hg'(a) + \dfrac{h^2}{2!} g''(a) + \cdots}$$

$$= \lim_{h \to 0} \frac{hf'(a) + \dfrac{h^2}{2!} f''(a) + \cdots}{hg'(a) + \dfrac{h^2}{2!} g''(a) + \cdots}$$

$$= \lim_{h \to 0} \frac{f'(a) + \dfrac{h}{2!} f''(a) + \cdots}{g'(a) + \dfrac{h}{2!} g''(a) + \cdots}$$

$$= \frac{f'(a)}{g'(a)}$$

If $f'(a) = g'(a) = 0$, the limiting value is equal to $[f''(a)]/[g''(a)]$ and so on. Hence

$$\lim_{x \to a} \frac{f(x)}{g(x)} = \lim_{x \to a} \frac{f'(x)}{g'(x)}$$

$$= \lim_{x \to a} \frac{f''(x)}{g''(x)}, \text{ etc.}$$

Example 11.3(a)
Show that

$$\lim_{x \to 0} \frac{\tan ax}{\tan bx} = \frac{a}{b}$$

$$\lim_{x \to 0} \frac{\tan ax}{\tan bx} = \lim_{x \to 0} \frac{a \sec^2 ax}{\mathrm{b} \sec^2 bx}$$

$$= \frac{a}{b}$$

This result could have been obtained using the power series for tangent, but this would be more involved.

Example 11.3(b)
Verify the result in the text of Part 11.2 using L'Hospital's rule.

$$\lim_{x\to 0} \frac{e^x - 1}{x} = \lim_{x\to 0} \frac{e^x}{1}$$
$$= \lim_{x\to 0} e^x$$
$$= 1$$

Example 11.3(c)
Verify the result of Example 11.2(b) by applying L'Hospital's rule.

$$\lim_{x\to 0} \frac{1 - \cos x}{x^2}$$
$$= \lim_{x\to 0} \frac{\sin x}{2x}$$
$$= \lim_{x\to 0} \frac{\cos x}{2}$$
$$= \frac{1}{2}$$

Unworked Examples 5

Part 11.1

(1) Obtain

$$\lim_{x\to 1} \frac{x^2 - 2x + 1}{x - 1}$$

(2) Show that

$$\lim_{x\to 0} \frac{2x + 1}{3x + 1} = 1$$

but that

$$\lim_{x\to \infty} \frac{2x + 1}{3x + 1} = \frac{2}{3}$$

(3) Find

$$\lim_{p \to \infty} \frac{ap + b}{cp + d}$$

where a, b, c and d are constants.

(4) Find

$$\lim_{x \to 0} \frac{x + 1/x}{x - 1/x}$$

(5) Find

$$\lim_{x \to 1} \frac{x^2 + x - 2}{x^2 - 1}$$

(6) Show that

$$\lim_{x \to 0} \frac{\sin 2x}{\sin x} = 2$$

Use the formula $\sin 3x = 3 \sin x - 4 \sin^3 x$ to obtain the value of

$$\lim_{x \to 0} \frac{\sin 3x}{\sin x}$$

Parts 11.2 and 11.3

Use a suitable method to obtain the limiting value in Question 7 to Question 11.

(7) $\lim_{x \to 0} \tan 3x \cot x$.

(8) $\lim_{x \to 0} x \cot x$.

(9) $\lim_{x \to 0} \dfrac{\cos px}{\cos x}$ and $\lim_{x \to 0} \dfrac{\sin px}{\sin x}$

where p is a constant. Compare your answer with that given for Question 6.

(10) $\lim_{x \to 0} \dfrac{1}{x^2} (x - \sin x)$

$$\lim_{x \to 0} \frac{1}{x^3} (x - \sin x)$$

$$\lim_{x \to 0} \frac{1}{x^3} (x - \sinh x)$$

(11) $\lim_{x\to 0} x^n e^{-x}$.

(12) Show that d/dx (cosec x) = $-\cos x \operatorname{cosec}^2 x$ and that d/dx (cot x) = $-\operatorname{cosec}^2 x$. Hence show that

$$\lim_{x\to 0} \frac{\operatorname{cosec} x - \cot x}{x} = \lim_{x\to 0} \frac{1 - \cos x}{\sin^2 x} = \frac{1}{2}$$

PART 12: CONVERGENCE AND DIVERGENCE OF SERIES

12.1 Convergence of a Power Series

In Part 10 various power series were obtained in the form $a_0 + a_1x + a_2x^2 + a_3x^3 + \cdots$. In Example 10.2(b), having obtained the coefficients, a value was substituted for x and the impression may have been gained that once the series had been formed, the variable x can take any value.

This is not true in all cases. For example, if $a_0 = a_1 = a_2 = a_3 = \cdots = 1$, the resulting power series $1 + x + x^2 + x^3 + \cdots$ may be recognised as a geometric progression. In this case, if $x > 1$, individual terms and consequently their sum, increase indefinitely. Such a series is said to be DIVERGENT. On the other hand if $-1 < x < 1$, it will be seen that $\lim_{x\to\infty} x^n = 0$ and, moreover, the sum itself approaches a limiting value, which is given by the formula $1/(1 - x)$. A series of this type is said to be CONVERGENT and the limiting value is called the SUM TO INFINITY. (A third, exceptional, possibility occurs when $x = -1$. The resulting series $1 - 1 + 1 - 1 + 1 - \cdots$ will have sum 1 or 0 depending on whether an odd or even number of terms is considered. A series of this type is said to be OSCILLATORY.)

For an infinite series $u_1 + u_2 + u_3 + \cdots$ the following rules of convergency apply

(i) A series can only be convergent if its terms ultimately approach zero, i.e.
$$\lim_{n\to\infty} u_n = 0$$
However, this condition in itself does not guarantee convergency (see Example 12.1(a) (ii)).

(ii) The nature of a series is unchanged if any finite number of terms are removed.

(iii) If all terms are positive, the series is convergent if
$$\lim_{n\to\infty} (u_{n+1}/u_n) < 1$$
If the value of this ratio exceeds 1, the series is divergent while if the value is exactly 1 this RATIO TEST is inconclusive.

(iv) If the terms of the series are alternatively positive and negative and, beyond some arbitrary limit, and each term is numerically smaller than its predecessor, the series is convergent.
(v) If each term of a given series is less than the corresponding term in another series known to be convergent, the given series is also convergent. Conversely if each term is greater than the corresponding term of a divergent series, the given series is divergent.

For an infinite power series, $a_0 + a_1x + a_2x^2 + \cdots$, the ratio test for convergency becomes

$$\lim_{n\to\infty} a_{n+1}\, x/a_n < 1$$

If

$$\lim_{n\to\infty} a_n/a_{n+1} = r$$

the power series converges if $-r < x < r$ and may converge if $x = \pm r$. The value of r is called the RADIUS OF CONVERGENCE of the series.

Example 12.1(a)
Discuss the convergence of the series
(i) $S_1 = 1 - \frac{1}{2} + \frac{1}{3} - \frac{1}{4} + \frac{1}{5} - \cdots$
(ii) $S_2 = 1 + \frac{1}{2} + \frac{1}{3} + \frac{1}{4} + \frac{1}{5} + \cdots$

(i) Referring to rule (d) in the text, S_1 must be convergent
(ii) The nature of S_2 is less clear. Attempting the ratio test, we have that $u_n = 1/n$ and so

$$\begin{aligned}\lim_{n\to\infty} u_{n+1}/u_n &= \lim_{n\to\infty} n/(n+1) \\ &= \lim_{n\to\infty} (1 - 1/(n+1)) \\ &= 1\end{aligned}$$

The test is inconclusive. However, writing S_2 as $1 + \frac{1}{2} + (\frac{1}{3} + \frac{1}{4}) + (\frac{1}{5} + \frac{1}{6} + \frac{1}{7} + \frac{1}{8}) + \cdots$ each term or group of bracketed terms (after the first) is greater than the corresponding term of

$$1 + \tfrac{1}{2} + (\tfrac{1}{4} + \tfrac{1}{4}) + (\tfrac{1}{8} + \tfrac{1}{8} + \tfrac{1}{8} + \tfrac{1}{8}) + \cdots$$

This latter series is equal to $1 + \frac{1}{2} + \frac{1}{2} + \frac{1}{2} + \cdots$ and is therefore divergent. So therefore is S_2. S_2 is called the HARMONIC SERIES and is one of the more important divergent series.

Example 12.1(b)
Show that the power series for e^x is convergent for all values of x

$$e^x = \sum_{n=0}^{\infty} x^n/n!$$

The radius of convergence

$$r = \lim_{n\to\infty} (n + 1)!/n!$$
$$= \lim_{n\to\infty} (n + 1) \to \infty$$

Hence the series for e^x is convergent for all values of x.

Example 12.1(c)
Show that the binomial expansion of $(1 + x)^m$, where m is not a positive integer, converges if $-1 < x < 1$. What happens when m is a positive integer?

$$(1 + x)^m = 1 + mx + m(m - 1)x^2/2! + m(m - 1)(m - 2)x^3/3! + \cdots$$

If m is not a positive integer terms eventually become all the same sign if x is negative or, alternatively, positive and negative if x is positive. In either case we require

$$\lim_{n\to\infty} |u_{n+1}/u_n| < 1$$

Radius of convergence

$$r = \lim_{n\to\infty} \frac{m!}{n!(m - n)!} \times \left.\frac{(n + 1)!(m - n - 1)!}{m!}\right|$$
$$= \lim_{n\to\infty} \left|\frac{n + 1}{m - n}\right| = 1$$

The series converges if $-1 < x < 1$.
If m is a positive integer, the series is of finite length.

Unworked Examples 6

Part 12.1

(1) For what values of x do the following series converge?
(a) $x + x^2/4 + x^3/9 + x^4/16 + \cdots$
(b) $1 + x/2 + x^2/4 + x^3/8 + \cdots$
(c) $1 + \frac{1}{3}x + 1/2!(\frac{1}{3}x)^2 + 1/3!(\frac{1}{3}x)^3 + \cdots$
(2) By comparing terms of the series

$$S_1 \equiv \tfrac{2}{1} + \tfrac{3}{4} + \tfrac{4}{9} + \tfrac{5}{16} + \tfrac{6}{25} + \cdots$$

with the corresponding terms of the harmonic series, show that S_1 is divergent. Is the series

$$S_2 = \tfrac{2}{1} - \tfrac{3}{4} + \tfrac{4}{9} - \tfrac{5}{16} + \tfrac{6}{25} - \cdots$$

convergent or divergent?

(3) Show that the power series for $\sin x$ and $\cos x$ are convergent for all values of x. Use the series to obtain $\sin 5° 44'$ and $\cos 5° 44'$ both correct to four places of decimals. (Note $5° 44'$ can be taken as 0.1 rad.)

(4) Why is the series $\frac{1}{2} + \frac{2}{3} + \frac{3}{4} + \frac{4}{5} + \frac{5}{6} + \cdots$ divergent?

(5) If $S \equiv 1 + 2x + 3x^2 + 4x^3 + \cdots$ show that $S - xS = 1 + x + x^2 + x^3 + \cdots$ Obtain the sum to infinity of the series $S - xS$ and hence of S. For what range of values of S convergent?

(6) Show that the series for $\ln(1 + x)$ is convergent if $x < 1$ by considering its radius of convergence. Is this series convergent if (a) $x = 1$ (b) $x = -1$. Justify these two final answers.

PART 13: PARTIAL DIFFERENTIATION

13.1 Partial Derivative Notation

The formula for the volume V of a cylinder of radius r and height h is $V = \pi r^2 h$. This means that changing r or h, or both, affects V.

Since it is too involved to consider the effect on the volume of changing both the height and the radius, let us increase the height by an amount δh but keep the radius constant (Figure B3). Suppose the corresponding change in volume is δV_h. Since $V = \pi r^2 h$

$$V + \delta V_h = \pi r^2 (h + \delta h)$$

$$\delta V_h = \pi r^2 \delta h$$

$$\frac{\delta V_h}{\delta h} = \pi r^2$$

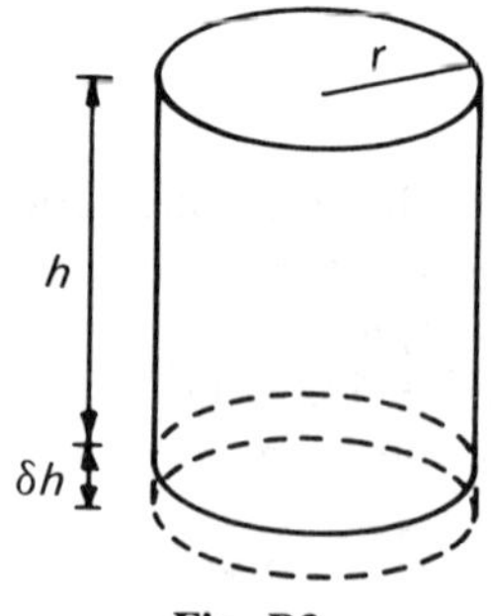

Fig. B3

To obtain a derivative, we consider

$$\lim_{h \to 0} \frac{\delta V_h}{\delta h}$$

It is wrong to write this as dV_h/dh since, normally, V will be affected by changes in r as well as h and so a notation must be devised to allow for this special situation in which a factor, normally variable, is held constant. In this case we write

$$\lim_{h \to 0} \frac{\delta V_h}{\delta h} = \frac{\partial V}{\partial h}$$

so that if $V = \pi r^2 h$, $\partial V/\partial h = \pi r^2$. A derivative of this type is called a PARTIAL DERIVATIVE.

Next consider the alternative possibility, keeping h constant and letting r increase by an amount δr (Figure B4). Suppose the volume increase is now δV_r

$$V = \pi r^2 h$$

$$V + \delta V_r = \pi(r + \delta r)^2 h$$

$$\delta V_r = 2\pi r h \delta r + \pi h (\delta r)^2$$

$$\frac{\delta V_r}{\delta r} = 2\pi r h + \pi h \delta r$$

This time

$$\lim_{r \to 0} = \frac{\delta V_r}{\delta r} = \frac{\partial V}{\partial r}$$

$$\partial V/\partial r = 2\pi r h$$

In both cases, the partial derivatives could have been obtained from the original formula $V = \pi r^2 h$, by simply following the normal rules of differential calculus except that one of the variables on the right is treated as if it were a constant.

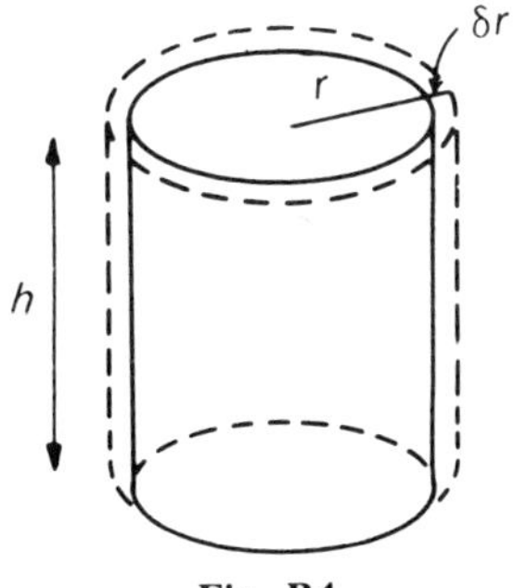

Fig. B4

More formally if z is a function of two variables x and y, written $z = f(x, y)$ then

$$\frac{\partial z}{\partial x} = \lim_{\delta x \to 0} \frac{f(x + \delta x, y) - f(x, y)}{\delta x}$$

and

$$\frac{\partial z}{\partial y} = \lim_{\delta y \to 0} \frac{f(x, y + \delta y) - f(x, y)}{\delta y}$$

The abbreviations z_x and z_y are often used for $\partial z/\partial x$ and $\partial z/\partial y$.

Example 13.1(a)
If $z = 3x^2y - 2x$, find z_x and z_y

$$z_x = 6xy - 2$$
$$z_y = 3x^2$$

In obtaining z_y the term $-2x$ is treated as a constant and, as such, vanishes on being differentiated.

Example 13.1(b)
Show that $\partial/\partial x\ (xye^{xy}) = ye^{xy}(1 + xy)$ and obtain a corresponding expression for $\partial/\partial y\ (xye^{xy})$

$$\begin{aligned} \partial/\partial x\ (xye^{xy}) &= ye^{xy} + xy^2e^{xy} \\ &= ye^{xy}(1 + xy) \\ \partial/\partial y\ (xye^{xy}) &= xe^{xy} + x^2ye^{xy} \\ &= xe^{xy}(1 + xy) \end{aligned}$$

In this example the partial derivatives involve both the product rule and the function of a function rule $\partial/\partial x\ (e^{xy}) = ye^{xy}$, etc.

Note that since interchanging x for y and y for x leaves xye^{xy} unaltered, the same change can be used to obtain $\partial/\partial y\ (xye^{xy})$ from $\partial/\partial x\ (xye^{xy})$

Example 13.1(c)
If $z = 2x^2 + y^2 + 3xy + 2x + y + 1$, find values of x, y and z at which $z_x = z_y = 0$.

$$z_x = 4x + 3y + 2$$
$$z_y = 2y + 3x + 1$$

The equations

$$4x + 3y = -2$$
$$3x + 2y = -1$$

can be solved by any of the usual methods, including those discussed in Parts 4, 5 and 6 and will lead to the solutions $x = 1$, $y = -2$. Using these values, $z = 1$.

13.2 Second Partial Derivatives

If $z = f(x, y)$, z_x and z_y themselves are functions of both x and y and it is possible to obtain four partial derivatives

(a) $\frac{\partial}{\partial x}\left(\frac{\partial z}{\partial x}\right)$ or $\frac{\partial^2 z}{\partial x^2}$ or z_{xx}

(b) $\frac{\partial}{\partial y}\left(\frac{\partial z}{\partial y}\right)$ or $\frac{\partial^2 z}{\partial y^2}$ or z_{yy}

(c) $\frac{\partial}{\partial x}\left(\frac{\partial z}{\partial y}\right)$ or $\frac{\partial^2 z}{\partial x \partial y}$ or z_{xy}

(d) $\frac{\partial}{\partial y}\left(\frac{\partial z}{\partial x}\right)$ or $\frac{\partial^2 z}{\partial y \partial x}$ or z_{yx}

It will be found that for all functions dealt with in this book $z_{xy} = z_{yx}$.

Example 13.2(a)
Find all the first and second partial derivatives of $z = x^2y - \cos 3xy^2$.

$$\begin{aligned}
z &= x^2y - \cos 3xy^2 \\
z_x &= 2xy + 3y^2 \sin 3xy^2 \\
z_y &= x^2 + 6xy \sin 3xy^2 \\
z_{xx} &= 2y + 9y^4 \cos 3xy^2 \\
z_{yy} &= 6x \sin 3xy^2 + 36x^2\, y^2 \cos 3xy^2 \\
z_{xy} &= 2x + 6y \sin 3xy^2 + 18xy^3 \cos 3xy^2 \\
z_{yx} &= 2x + 6y \sin 3xy^2 + 18xy^3 \cos 3xy^2
\end{aligned}$$

In obtaining z_{yy} from z_y, the term $\partial/\partial y\,(x^2)$ vanishes but $\partial/\partial y\,(6xy \sin 3xy^2)$ is treated as a product.

In this example z_{xy} and z_{yx} have been obtained independently although it is know that they are equal. This acts as a check.

Example 13.2(b)
Show that $u = e^x(\sin y + \cos y)$ satisfies the equation

$$\frac{\partial^2 u}{\partial x^2} + \frac{\partial^2 u}{\partial y^2} = 0$$

$$
\begin{aligned}
u &= e^x(\sin y + \cos y) \\
u_x &= u_{xx} = e^x(\sin y + \cos y) \\
u_y &= e^x(\cos y - \sin y) \\
u_{yy} &= e^x(-\sin y - \cos y) \\
&= -e^x(\sin y + \cos y)
\end{aligned}
$$

Hence

$$\frac{\partial^2 u}{\partial x^2} + \frac{\partial^2 u}{\partial y^2} = 0$$

13.3 Total Differential

Returning to the volume of the cylinder discussed in Part 13.1, we obtained the results

$$\frac{\delta V_h}{\delta h} = \pi r^2$$

and

$$
\begin{aligned}
\frac{\delta V_r}{\delta r} &= 2\pi rh + \pi h\delta r \\
&\approx 2\pi rh
\end{aligned}
$$

since, if r is small, the term $\pi h\delta r$ contributes little to $\delta V_r/\delta r$. The changes in volume, δV_h and δV_r, will therefore be given, approximately, by the formula

$$
\begin{aligned}
\delta V_h &= \pi r^2 \delta h \\
&= \frac{\partial V}{\partial h}\,\delta h
\end{aligned}
$$

and

$$
\begin{aligned}
\delta V_r &= 2\pi rh\delta r \\
&= \frac{\partial V}{\partial r}\,\delta r
\end{aligned}
$$

As these were the effects of (a) changing h but keeping r constant and (b) changing r but keeping h constant, it is reasonable to suppose that changing both h and r, would lead to a total volume change which we can denote by δV equal to $\delta V_h + \delta V_r$

$$\delta V = \frac{\partial V}{\partial h}\,\delta h + \frac{\partial V}{\partial r}\,\delta r$$

The right-hand side is called the TOTAL DIFFERENTIAL of V and its components, which are all of the same form are called its partial differentials. There are two of these because in the original formula, $V = \pi r^2 h$, there were two variable terms on the right, r and h. In a formula such as $w = f(x, y, z)$ where w is dependent on three variables, the total differential formula would be extended to

$$\delta w = \frac{\partial w}{\partial x}\,\delta x + \frac{\partial w}{\partial y}\,\delta y + \frac{\partial w}{\partial z}\,\delta z$$

and so on.

If x, y and z are all functions of a further variable t, w itself must also be a function of t, so that it is possible to modify this formula to

$$\frac{\delta w}{\delta t} = \frac{\partial w}{\partial x}\frac{\delta x}{\delta t} + \frac{\partial w}{\partial y}\frac{\delta y}{\delta t} + \frac{\partial w}{\partial z}\frac{\delta z}{\delta t}$$

which, as $t \to 0$, becomes

$$\frac{\mathrm{d}w}{\mathrm{d}t} = \frac{\partial w}{\partial x}\cdot\frac{\mathrm{d}x}{\mathrm{d}t} + \frac{\partial w}{\partial y}\cdot\frac{\mathrm{d}y}{\mathrm{d}t} + \frac{\partial w}{\partial z}\cdot\frac{\mathrm{d}z}{\mathrm{d}t}$$

(Since w, x, y and z are functions of a single variable t, only ordinary derivatives $\mathrm{d}w/\mathrm{d}t$, etc. are required.)

Example 13.3(a)

State the formula for the volume V of a right pyramid with a square base of side a and vertical height h. If a is equal to 4.9 cm and h to 6.3 cm, find the error in V and hence the true volume, if a is taken as 5 cm and h as 6 cm

$$V = \tfrac{1}{3}a^2h$$

If $a = 5$, $h = 6$, $V = 50$. The error in V, δV, is obtained using the total differential formula with $\delta a = -0.1$, $\delta h = 0.3$.

$$\begin{aligned}\delta V &= \frac{\partial V}{\partial a}\,\delta a + \frac{\partial V}{\partial h}\,\delta h \\ &= \tfrac{2}{3}ah\delta a + \tfrac{1}{3}a^2\delta h \\ &= \tfrac{1}{3}a(2h\delta a + a\delta h) \\ &= \tfrac{1}{3} \times 5(2 \times 6 \times (-0.1) + 5 \times 0.3) \\ &= 0.5\end{aligned}$$

Error $= 0.5\ \mathrm{cm}^2$. True volume $= 50.5\ \mathrm{cm}^2$.

In this example δa is taken negative because taking the true value of a, rather than the assumed value 5, will reduce V and δh is taken positive for the opposite reason.

Notice that the partial differentials were of opposite sign and therefore partially cancelled each other out. However, the net effect is an increase in volume, recognisable by the fact that δV was positive.

Example 13.3(b)
Calculate the result of $\sqrt{(9.95^2 - 6.01^2)}$ correct to two places of decimals without the help of a calculating aid.

If $c = \sqrt{(a^2 - b^2)}$, when $a = 10$, $b = 6$, $c = 8$, taking $\delta a = -0.05$, $\delta b = 0.01$,

$$
\begin{aligned}
\delta c &= \frac{\partial c}{\partial a}\delta a + \frac{\partial c}{\partial b}\delta b \\
&= \frac{a}{\sqrt{(a^2 - b^2)}}\delta a - \frac{b}{\sqrt{(a^2 - b^2)}}\delta b \\
&= \frac{1}{c}(a\delta a - b\delta b) \\
&= \tfrac{1}{8}(10 \times (-0.05) - 6 \times (0.01)) \\
&= -0.07
\end{aligned}
$$

$$c = 8 - 0.07 = 7.93 \text{ (corr. 2 pl. dec.)}$$

If the question does not actually quote a formula, one must be introduced in order to obtain the required partial derivatives.

Example 13.3(c)
The angular acceleration of a flywheel of mass m and radius of gyration K is given by the formula

$$\alpha = \frac{T}{mK^2}$$

where T is the applied torque. Find α if $T = 39.95$, $m = 499$ and $K = 0.41$ using the total differential formula.

Taking $T = 40$, $m = 500$, $K = 0.4$, $\alpha = 0.5$, when $\delta T = -0.05$, $\delta m = -1$, $\delta K = 0.01$,

$$
\begin{aligned}
\delta\alpha &= \frac{\partial\alpha}{\partial T}\delta T + \frac{\partial\alpha}{\partial m}\delta m + \frac{\partial\alpha}{\partial K}\delta K \\
&= \frac{1}{mK^2}\delta T - \frac{T}{m^2K^2}\delta m - \frac{2T}{mK^3}\delta K \\
&= \frac{1}{m^2K^3}(mK\delta T - TK\delta m - 2mT\delta K)
\end{aligned}
$$

$$= \frac{(500 \times 0.4 \times (-0.05) - 40 \times 0.4 \times (-1) - 2 \times 500 \times 40 \times 0.01)}{500^2 \times 0.4^3}$$

$$\approx -0.025$$

Required value of $\alpha = 0.5 - 0.025 = 0.475$

Example 13.3(d)
Two bodies A and B are projected at a steady rate from a point O in straight lines inclined at 60°. Show that when OA $= x$, OB $= y$, the rate of increase of the distance AB is given by

$$\frac{x - \frac{1}{2}y}{\sqrt{(x^2 + y^2 - xy)}} \frac{dx}{dt} - \frac{\frac{1}{2}x - y}{\sqrt{(x^2 + y^2 - xy)}} \frac{dy}{dt}$$

Evaluate the rate of increase if x = 160 cm and y = 60 cm after 2 s.

Referring to Figure B5, if AB $= Z$

$$Z^2 = x^2 + y^2 - 2xy \cos 60°$$

$$= x^2 + y^2 - xy$$

$$2Z \frac{\partial Z}{\partial x} = 2x - y$$

$$\frac{\partial Z}{\partial x} = \frac{2x - y}{2Z}$$

$$= \frac{x - \frac{1}{2}y}{Z}$$

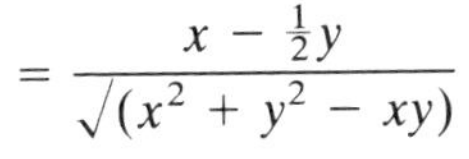

$$= \frac{x - \frac{1}{2}y}{\sqrt{(x^2 + y^2 - xy)}}$$

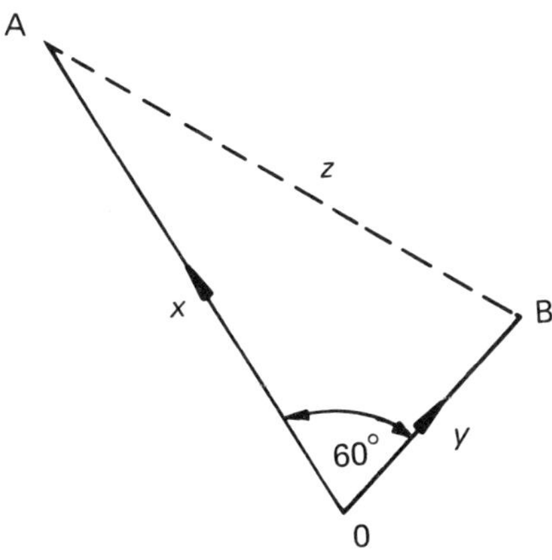

Fig. B5

Similarly

$$\frac{\partial Z}{\partial y} = \frac{y - \frac{1}{2}x}{\sqrt{(x^2 + y^2 - xy)}}$$

$$\frac{dZ}{dt} = \frac{\partial Z}{\partial x}\frac{dx}{dt} + \frac{\partial Z}{\partial y}\frac{dy}{dt}$$

$$= \frac{x - \frac{1}{2}y}{\sqrt{(x^2 + y^2 - xy)}}\frac{dx}{dt} + \frac{y - \frac{1}{2}x}{(x^2 + y^2 - xy)}\frac{dy}{dt}$$

$$= \frac{x - \frac{1}{2}y}{(x^2 + y^2 - xy)}\frac{dx}{dt} - \frac{\frac{1}{2}x - y}{\sqrt{(x^2 + y^2 - xy)}}\frac{dy}{dt}$$

When $x = 160$, $y = 60$, $t = 2$,

$$x - \tfrac{1}{2}y = 130$$

$$\tfrac{1}{2}x - y = 20$$

$$\sqrt{(x^2 + y^2 - xy)} = 140$$

$$\frac{dx}{dt} = 80$$

and

$$\frac{dy}{dt} = 30$$

$$\frac{dZ}{dt} = \frac{130}{140} \times 80 - \frac{20}{140} \times 30$$

$$= 70 \text{ cm s}^{-1}$$

13.4 Percentage Changes

In dealing with formulae in which there are no + or − signs, the resulting changes can be easily calculated in relative or percentage terms.

Suppose $A = B^k C^l \cdots$ where k and l are powers which may be positive or negative and need not be integral.

$$\delta A = \frac{\partial A}{\partial B}\delta B + \frac{\partial A}{\partial C}\delta C + \cdots$$

$$= (kB^{k-1}C^l \cdots)\delta B + (lB^k C^{l-1} \cdots)\delta C + \cdots$$

$$= (kB^k C^l \cdots)\frac{\delta B}{B} + (lB^k C^l \cdots)\frac{\delta C}{C} + \cdots$$

$$= kA\frac{\delta B}{B} + lA\frac{\delta C}{C} + \cdots$$

$$\frac{\delta A}{A} = k\frac{\delta B}{B} + l\frac{\delta C}{C} + \cdots$$

Hence the relative or percentage change in A is equal to k times the corresponding change in B plus l times the corresponding change in C and so on.

Example 13.4(a)
Find the volume of a cylinder, radius 5 cm, height 8 cm. Find the percentage increase in volume if the radius increases by 2% and the height decreases by 1% using (i) the actual figures provided, (ii) the percentage change formula. Leave all answers in terms of π. (Volume $= \pi r^2 h = 200\pi$ cm^3.)

(i) With the percentage changes quoted, the radius is increased to 5.1 cm and the height is decreased to 7.92 cm. New volume (obtainable using the total differential formula or otherwise) is 206π cm^3.

Percentage increase in volume $= 6/200 \times 100 = 3\%$

(ii) Percentage increase in volume

$$= 2 \times \frac{\delta r}{r} + 1 \times \frac{\delta h}{h}$$

$$= 2 \times 2\% + 1 \times (-1\%)$$

$$= 3\%$$

Example 13.4(b)
The result of an experiment is calculated from the quotient A/B where A and B are two measured quantities. If these may each be up to 1% in error, show that the maximum error in the result is 2%. Verify this figure if $A = 800$ and $B = 16$.

Calling the result R, so that $R = A/B$, $\delta R/R = \delta A/A - \delta B/B$. The maximum error occurs when A is overestimated by 1% and B is underestimated by the same amount (or vice versa). In either case the percentage error in R is 2%.

If $A = 800$ and $B = 16$ then $R = 50$.

If $A = 800 + 1\%$, i.e. 808 and $B = 16 - 1\%$, i.c. 15.84, $R = 808/15.84 \approx 51$.

If $A = 800 - 1\%$, i.e. 792 and $B = 16 - 1\%$ i.e. 16.16, $R = 792/16.16 \approx 49$.

In each case the error in R is 2%

13.5 Stationary Points

The normal process for obtaining the maximum or minimum value of a function of one variable involves equating the first derivative to zero. The roots of this equation lead to the stationary points and further tests distinguish between a maximum, a minimum and other forms of stationary point.

A similar process is used to obtain the maximum or minimum value, say z, of a function of two variables, x and y.

If $z = f(x, y)$ then at a stationary point $\partial z/\partial x = \partial z/\partial y = 0$. Solving these simultaneous equations leads to the stationary point or points and further tests indicate their nature. The complexity of these subsequent tests increase with the number of variables involved. In certain examples, physical or graphical considerations enable the nature of the stationary point to be determined. As an example, consider an open rectangular tank (see Figure B6). Suppose that its capacity is 32 litres and we are asked to determine the dimensions needed to minimise the amount of material needed in its construction. Taking the base dimensions as a cm by b cm, its height, in centimetres, will be $32\,000/ab$.

The amount of material needed, M, taken in cm^2, is given by

$$M = ab + 64\,000/a + 64\,000/b$$

$$\partial M/\partial a = b - 64\,000/a^2 \qquad \partial M/\partial b = a - 64\,000/b^2$$

At a stationary point, $b - 64\,000/a^2 = a - 64\,000/b^2 = 0$. These equations are easily solved to give $a = b = 40$ so that the base of the tank is a square of side 40 cm and the height is 20 cm. Since the amount of material needed could increase without limit by extending one of the dimensions and reducing the other two (keeping the capacity unchanged of course), these figures can only be the lengths needed for a minimum quantity of material.

Example 13.5(a)
Find the stationary point of the function $z = x^2 - 2y^2 + 3xy - 9x - 5y + 12$.

$$\partial z/\partial x = 2x + 3y - 9$$

$$\partial z/\partial y = -4y + 3x - 5$$

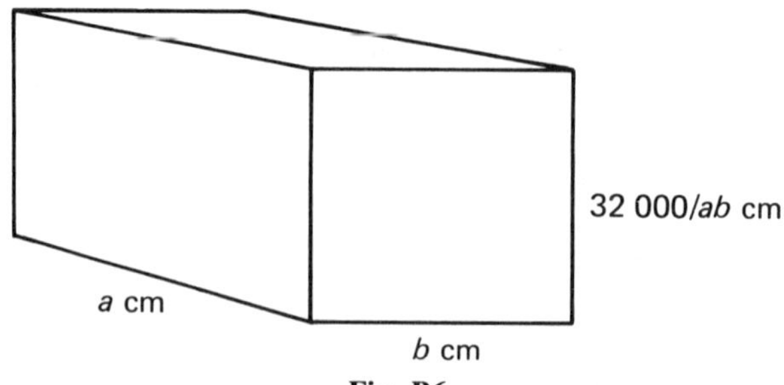

Fig. B6

At a stationary point,

$$2x + 3y = 9$$
$$-4y + 3x = 5$$

so that $x = 3$ and $y = 1$. This means that $z = -4$ and the required stationary point has co-ordinates $(3, 1, -4)$.

Unworked Examples 7

Part 13.1

(1) If $z = 3xy - x^2y + 4x/y^3$, find $\partial z/\partial x$ and $\partial z/\partial y$.
(2) Find $\partial u/\partial x$ and $\partial u/\partial y$ when
   (a) $u = \ln x^2y - ye^x$
   (b) $u = x^3y + x^2y^2 + 3xy^5$.
(3) Show that $z = (x + y) \ln (x/y)$ satisfies the equation

$$x\frac{\partial z}{\partial x} + y\frac{\partial z}{\partial y} = z$$

(4) If $z = 1/\sqrt{(x^2 + y^2)}$ show that (a) $z_x = -x/z^3$ (b) $z_y = -y/z^3$. Hence prove that $xz_x + yz_y + z = 0$.
(5) If $u = 4x^2 + 2y^2 + 3xy - 2x + 5y + 7$, find values of x and y and hence of u at which $u_x = u_y = 0$.
(6) Show that if $z = 2x^2 + 2y^2 + 4xy - x - y$ then z_x and z_y are equal.

Part 13.2

(7) Find all the first and second partial derivatives of
   (a) $u = 3xy^3 + \cos 2xy - 2$,
   (b) $u = e^{3xy} + \sin x^2y^2$.
(8) Show that, if $z = ye^{2x} + xe^{2y}$, $\partial^2 z/\partial x^2 + \partial^2 z/\partial y^2 = 4z$.
(9) Differentiate $u = 3x^3y - 2y/x + 2y^3e^x$ partially and show that u_{xy} and u_{yx} are equal.
(10) If $z = x^2y - 4e^{2xy} - y \sin x$, find the values of z_{xx} and z_{xy} when $x = y = 0$.
(11) Show that if $z = ye^{2x} + x^2y^3$
   (a) $\partial^2 z/\partial x^2 - 2\partial z/\partial x = 2y^3(1 - 2x)$
   (b) $\partial^2 z/\partial y^2$ has a value zero and $\partial^2 z/\partial x\partial y$ has a value 2 when $x = 0$ irrespective of the value of y.
(12) Show that $u = x^3e^{xy}$ satisfies the partial differential equation $u_{xy} = u + xu_x$.

(13) If $T = a^3b^2/\sqrt{c}$ find T without calculating aids if $a = 1.98$, $b = 14.85$ and $c = 25.25$ using (a) the total differential formula, (b) percentage changes.

(14) For a perfect gas, the law $PV = RT$ is applicable where P, V and T are pressure, volume and temperature and R is a constant. If T increases from 400 K to 402 K and V increases from 5 m^3 to 5.05 m^3, find the change in P if its value is initially 0.1 N cm^{-2}.

(15) Find the hypotenuse of a right-angled triangle using the total differential formula if its smaller sides are of length 7.94 cm and 14.98 cm.

(16) Find the percentage error in calculating the volume of a sphere based on a 1% underestimate of the radius.

(17) If $T = 2\pi\sqrt{(l/g)}$ show that $\delta T/T = \frac{1}{2}(\delta l/l - \delta g/g)$ using the usual notation.

(18) The angles of a triangle are to be calculated from sides a, b and c using the cosine rule. If the lengths are subject to errors δa, δb and δc show that

$$\delta A = \frac{a}{2\Delta}(\delta a - \delta b \cos \mathrm{C} - \delta c \cos \mathrm{B})$$

where A, B and C are the angles and Δ is the area of the triangle. Calculate A if $a = 99.5$, $b = 100.2$ and $c = 100.7$. (Remember that A will be measured in radians.)

(19) Show that the function $3x^2 + 4y^2 + 7xy - 5x - 6y + 10$ has just one stationary point which occurs where $x = 2$ and $y = -1$.

(20) Show that $xy + 1/x + 1/y$ has a stationary value of 3 occurring when $x = y = 1$.

(21) A sheet of metal 24 cm wide is to be made into a gutter by bending equal lengths x cm through equal angles θ (Figure B7). Find the values of x and θ which make its capacity a maximum.

Fig. B7

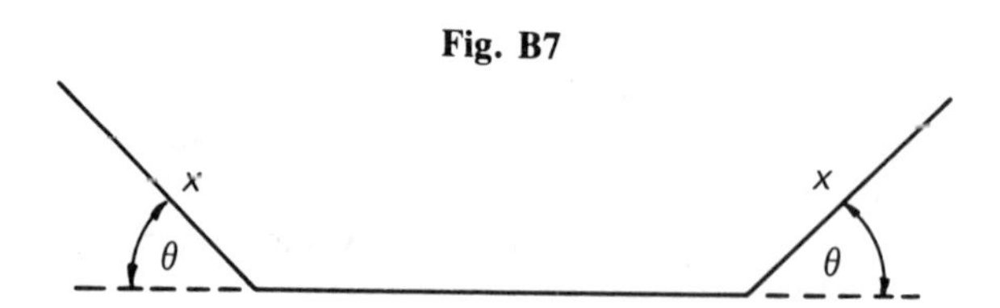

SECTION C
Differential Equations

PART 14: FORMATION OF DIFFERENTIAL EQUATIONS

14.1 Terminology

A DIFFERENTIAL EQUATION is an equation involving at least one derivative. A simple example is

$$dy/dx = 2x \tag{1}$$

In this equation x is the INDEPENDENT VARIABLE and y is the DEPENDENT VARIABLE. An ANALYTICAL SOLUTION of the equation is another equation linking the variables in such a way that derivatives are eliminated. This involves either direct or indirect integration, so that the solution of Equation (1) is $y = x^2 + C$, with C, actually the constant of integration, being referred to as an ARBITRARY CONSTANT.

If the differential equation describes mathematically either a physical or an engineering system, a CONDITION under which it is set up, will be given or implied and this will allow the arbitrary constant to be evaluated. The solution of the equation could therefore be the GENERAL SOLUTION, expressed in terms of the arbitrary constant, or a PARTICULAR SOLUTION, if this constant has been evaluated.

An equation such as $d^2y/dx^2 + 2dy/dx - 5y = 0$ is said to be of ORDER 2, a reference to the highest derivative. To solve an nth order differential equation, n direct or indirect integrations are required and the general solution will contain n arbitrary constants. It is usual to denote these by upper case letters A, B, C . . . . A particular solution of such an equation will require n conditions to be supplied.

The highest power of any derivative is called the DEGREE of the differential equation so that $d^2y/dx^2 + 3(dy/dx)^4 - 2y = e^{2x} \cos 3x$ is an example of a second order equation of degree 4.

14.2 Differential Equations Dealing with Geometrical Problems

Differential equations may arise from a variety of situations. Those arising from geometrical problems are usually quite straightforward and it is possible to use simple examples to illustrate some of the terms introduced in Part 14.1.

Figure C1 shows a set of four identical curves separated vertically by a distance 1. These are part of a FAMILY OF CURVES, the set of which contains an infinite number of members. A general equation for the gradient of the curves, $dy/dx = 2x$, applies to all members, each of which can be identified by its equation. In order words, the differential Equation (1) of Part 14.1 refers to the family of curves. A general solution could refer to any member but a condition, say $y = 0$ when $x = 0$, identifies a particular member.

Example 14.2(a)
Find which member of the family of curves $dy/dx = 3x^2 - 2x + 1$ passes through the point (1, 2).

$$dy/dx = 3x^2 - 2x + 1$$
$$y = x^3 - x^2 + x + A$$

Since the curve passes through (1, 2), this implies the condition $y = 2$ when $x = 1$. Using this condition $A = 1$. It follows that the required member is $y = x^3 - x^2 + x + 1$.

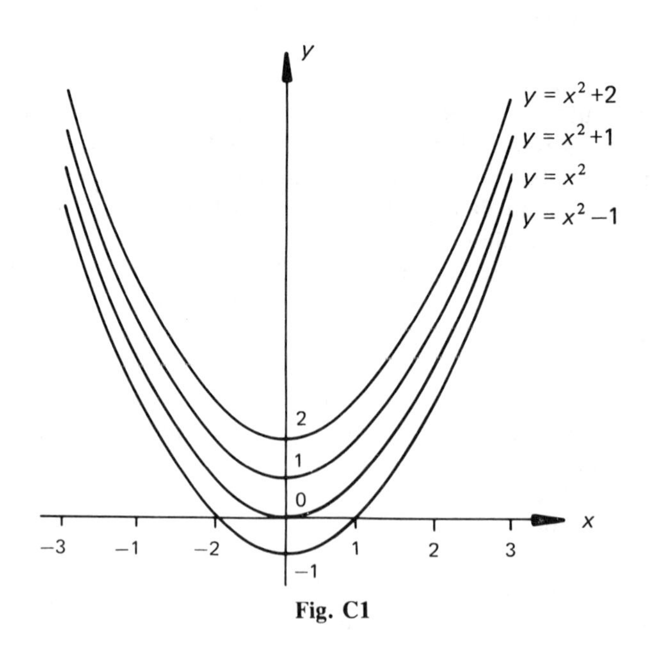

Fig. C1

Example 14.2(b)
Find the equation of the curve whose gradient is given by the formula $dy/dx = -x/\sqrt{(a^2 - x^2)}$, where a is a constant, if $x = a$ when $y = 0$

$$\frac{dy}{dx} = \frac{-x}{\sqrt{(a^2 - x^2)}}$$

$$y = -\int \frac{x\,dx}{\sqrt{(a^2 - x^2)}}$$

$$= \sqrt{(a^2 - x^2)} + A$$

From the condition $x = a$ when $y = 0$, $A = 0$ so that $y = \sqrt{(a^2 - x^2)}$. The integration could be carried out using the substitution $z^2 = a^2 - x^2$.

14.3 Method of Isoclines

A method of solving differential equations graphically uses ISOCLINES. These are lines along which all members of a family of curves have equal gradients. Figure C2 shows dotted isoclines for the equation $dy/dx = 2x$ with small line segments marking the gradients. From these, the family of curves

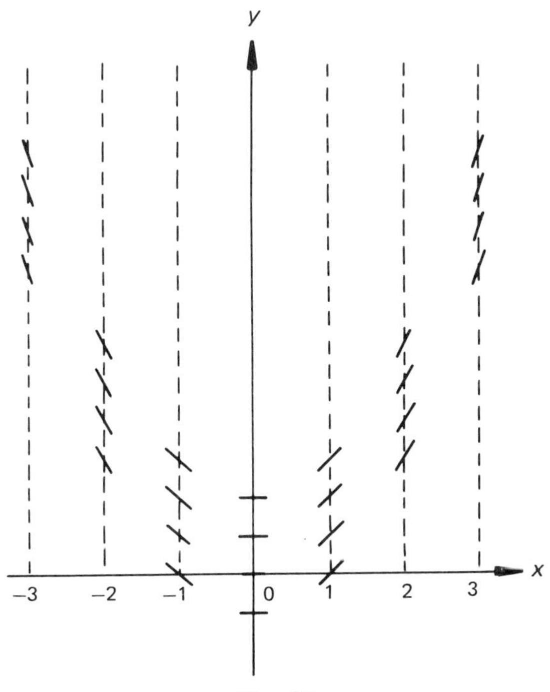

Fig. C2

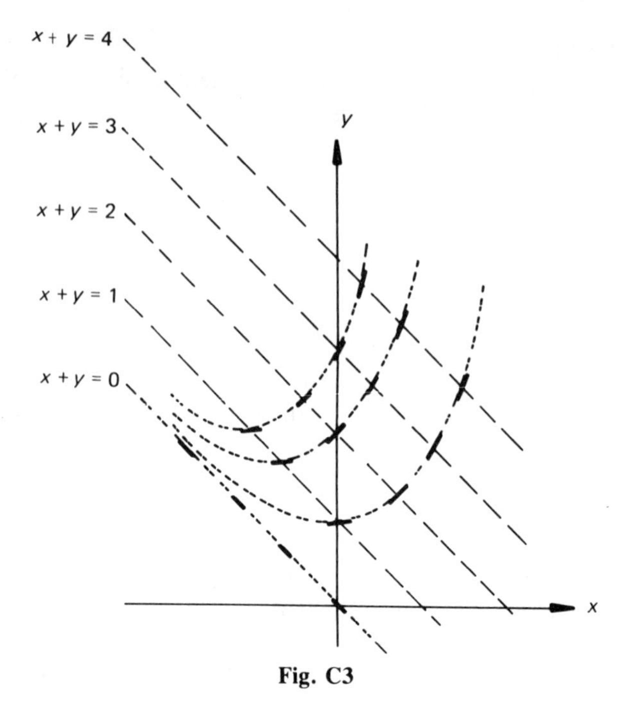

Fig. C3

themselves can be reconstituted but not their equations. The graphs are therefore another form of solution of the differential equation.

In this example, since dy/dx is expressed in terms of x, all isoclines are vertical line but if dy/dx is expressed in terms of x and y, this is not the case and isoclines must be constructed as in Example 14.3(a).

Example 14.3(a)
Obtain a graphical solution of the differential equation $dy/dx = x + y - 1$ using isoclines.

The isoclines are lines parallel to $x + y = 0$ as shown in Figure C3. Some of the family of curves have been drawn.

14.4 Types of Condition

In Examples 14.2(a) and 14.2(b), the differential equation linked two displacement variables and the conditions were expressed in terms of these. This type of condition is a BOUNDARY CONDITION. In contrast, when differential equations are formulated from engineering systems, in almost every case, the independent variable is time and the associated condition is referred to as an INITIAL CONDITION. This terminology is by no means standard but will be adhered to in this book.

14.5 Formulating Differential Equations from Engineering Problems

Differential equations arise in engineering from the application of physical laws, expressing the factors involved in terms of derivatives. In a mechanical problem, it is quite likely to arise from an application of Newton's second law, with acceleration expressed as $\mathrm{d}v/\mathrm{d}t$ or as $\mathrm{d}^2s/\mathrm{d}t^2$. In a structural case, the derivative may arise from a consideration of how vertical displacement of a loaded member varies with horizontal displacement, while in electrical engineering it may be due to the potential drop across resistors, inductors and capacitors.

In each case, in order to obtain the differential equation, a MATHEMATICAL MODEL is chosen which approximates the situation but simplifies it, so that, for example, in the case of a body falling freely, air resistance might be ignored and gravitational acceleration treated as a constant. On this basis, a simple differential equation can be obtained and, if the solution calculated from it is in good agreement with results obtained by observation or experiment, the model is considered to be a good one. If the agreement is poor, the model is unsatisfactory and the assumptions made about it must be reconsidered.

As an example, let us consider the model of freely falling body in a little more detail. If it has a mass m, with the assumptions made above, the sole force acting upon it is its weight mg. If distance is measured from the ground upwards, as illustrated in Figure C4, at a general point P, distance s above ground level, Newton's second law states that $mg = -m\,\mathrm{d}^2s/\mathrm{d}t^2$, the minus sign indicating that distance—and consequently acceleration—and force are measured in opposing directions. The equation can be simplified to $\mathrm{d}^2s/\mathrm{d}t^2 = -g$. As this is a second order equation, two initial conditions are required. If the body had been *dropped* from a height s_0 and timed from the moment of release, the conditions are $s = s_0$ and $\mathrm{d}s/\mathrm{d}t = 0$ when $t = 0$. As $\mathrm{d}^2s/\mathrm{d}t^2 = -g$ with g constant, the solution can be obtained by integrating twice.

$$\mathrm{d}s/\mathrm{d}t = -gt + A$$

$$s = -\tfrac{1}{2}gt^2 + At + B$$

From the conditions $A = 0$, $B = s_0$ making $s = s_0 - \tfrac{1}{2}gt^2$.

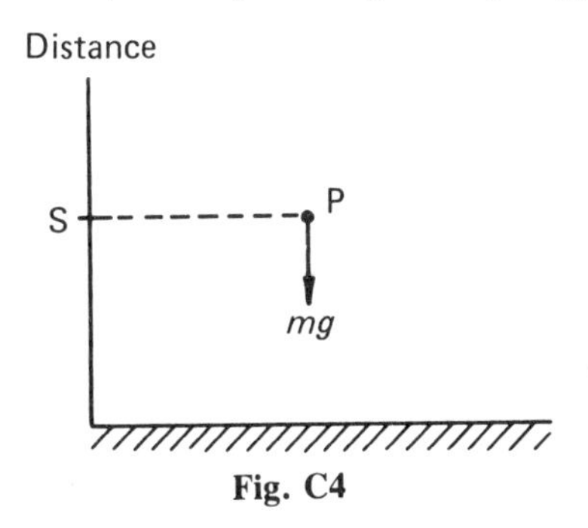

Fig. C4

We can use this solution to determine the time taken for the body to reach ground level, which will be the value of t when $s = 0$. This is obtained as $\sqrt{(2s_0/g)}$ and, using the intermediate result $\mathrm{d}s/\mathrm{d}t = -gt$, the velocity of impact is $-\sqrt{(2gs_0)}$, the minus sign being purely directional.

Unworked Examples 1

Parts 14.1 and 14.2

(1) Verify that $y = 3\mathrm{e}^x + x + 1$ is the particular solution of the differential equation $\mathrm{d}y/\mathrm{d}x = y - x$ with conditions $y = 4$ when $x = 0$.
(2) Taking the differential equation of Question 1 state
 (a) which is the independent variable
 (b) which is the dependent variable
 (c) the order of the equation
 (d) the degree of the equation.
(3) Which member of the family of curves $\mathrm{d}y/\mathrm{d}x = 6x^2 + 4x - 3$ passes through the origin?
(4) Obtain a general solution of the differential equation $\mathrm{d}y/\mathrm{d}x = x$.
(5) Obtain the general solution of the differential equation $\mathrm{d}y/\mathrm{d}x = x^4 - 1$. Evaluate the arbitrary constant given the conditions $y = 2$ when $x = 0$.
(6) Show that for various values of m, $y = mx$ represents a family of straight lines passing through the origin. Find a differential equation which represents this family in which m has been eliminated.
(7) State the order and degree of the differential equation

$$\left(\frac{\mathrm{d}y}{\mathrm{d}x}\right)^2 = \frac{1}{1 - x^2}$$

Solve the equation. (You may find the solution to Example 8.6(b) helpful.)

Part 14.3

(8) Obtain a graphical solution to the equation in Question 4 using isoclines.
(9) Use isoclines to obtain a solution of the equation $\mathrm{d}y/\mathrm{d}x = y$.
(10) Solve the differential equation $\mathrm{d}y/\mathrm{d}x = y/x$ by a graphical method. Compare your solution with the family of curves in Question 6.
(11) Solve the equation $\mathrm{d}y/\mathrm{d}x = 6x$ by integrating. By taking various values for the arbitrary constant, draw the family of curves. Solve the equation using isoclines and compare results.

(12) If a parallel sided tank with a small aperture in its base is filled with water which slowly seeps away and if
   (a) the volume of water in the tank, V, is directly proportional to the depth of water, h,
   (b) the pressure of water on the aperture, P, is directly proportional to h,
   (c) the rate of water loss is directly proportional to P,

   deduce that $dh/dt = -kh$, k being a positive constant. Show that if the tank is originally filled to a height h_0, this condition and the differential equation are consistent with the solution $h = h_0e^{-kt}$. Which is the assumption made in setting up the model that is most likely to be invaild?

(13) Show that if a freely falling body is subject to air resistance directly proportional to its velocity, v, then with the usual notation, $dv/dt = g - kv$, where k is a constant. Show also, that if the body is dropped, the equation and implied condition are consistent with the solution $v = g(1 - e^{-kt})/k$. What is the maximum velocity achievable?

(14) The rate of radioactive decay of radium is directly proportional to the amount, R, present. Show that this statement can be expressed mathematically as the differential equation $dR/dt = -kR$, k being a positive constant. If there is an amount R_0 originally present deduce that the quantity left at time t is R_0e^{-kt} using the result of Question 12. If it takes 1500 years for half of the radium to decompose, show that 63% remains after 1000 years.

(15) Newton's law of cooling states that the rate of cooling of a hot body is directly proportional to the temperature difference between that of the body, T_b, and its surroundings T_s. With the usual notation show that this can be written in the form $dT_b/dt = -k(T_b - T_s)$.

(16) The potential energy of the freely falling body described in Part 14.5 is equal to mgs and its kinetic energy is $\frac{1}{2}m(ds/dt)^2$. Use the principle of energy conservation to arrive at the equation $ds/dt = -\sqrt{(2g(s_0 - s))}$. Verify that the solution in the text, $s = s_0 - \frac{1}{2}gt^2$, satisfies this equation.

(17) Figure C5 shows a loaded spring which has been displaced a distance y below its equilibrium position. Assuming that the restoring force in the

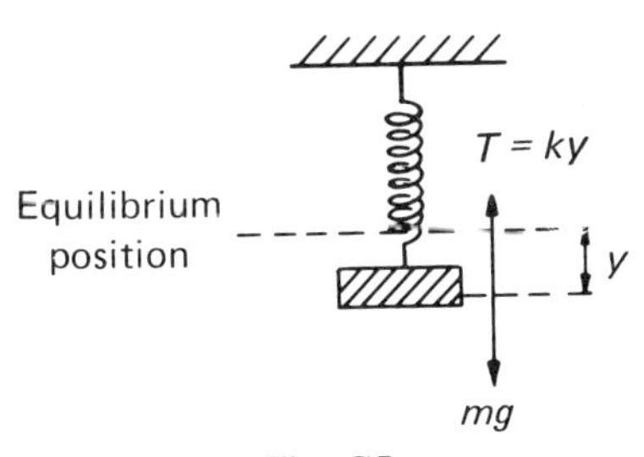

Fig. C5

spring, T, is proportional to its displacement, show that $T = ky$ and hence, that if air resistance can be neglected, $d^2y/dt^2 = -ky/m$.

(18) A substance X is changed chemically at a rate proportional to the amount present. Show that the quantity of X present at time t is described mathematically by the differential equation $dX/dt = -kX$, where k is a positive constant.

PART 15: SOLUTION OF FIRST ORDER DIFFERENTIAL EQUATIONS

15.1 Solution by Separation of the Variables

In Part 14, an indication was given of how some differential equations might arise. The rest of Section C will be concerned with how such equations can be solved.

As with integration, there are several methods of solution and, in order to determine the appropriate one, the type of equation must be recognised. If an equation can be expressed in the form $dy/dx = f(y)g(x)$, where $f(y)$ is a function of y only and $g(x)$ is a function of x only, it is permitted to separate the variables to the form

$$\frac{dy}{f(y)} = g(x)\,dx$$

and to obtain by integration

$$\int \frac{dy}{f(y)} = \int g(x)\,dx$$

These two steps are usually carried out simultaneously. Only one constant of integration need be inserted when the integration is performed. (Why?)

Example 15.1(a)
Solve the differential equation $y\,dy/dx = x - 3$, if $y = 2$ when $x = 0$.

$$y\,dy/dx = x - 3$$

$$\int y\,dy = \int (x - 3)\,dx$$

$$\tfrac{1}{2}y^2 = \tfrac{1}{2}x^2 - 3x + A$$

$$y^2 = x^2 - 6x + 2A$$

From the conditions, $2A = 4$, so that $y^2 = x^2 - 6x + 4$. If the general solution is required, the final line is omitted.

An alternative method of obtaining the particular solution is to incorporate the conditions into the integral limits. The integration step then becomes

$$\int_2^y y\,\mathrm{d}y = \int_0^x (x - 3)\,\mathrm{d}x$$

$$[\tfrac{1}{2}y^2]_2^y = [\tfrac{1}{2}x^2 - 3x]_0^x$$

$$\tfrac{1}{2}y^2 - 2 = \tfrac{1}{2}x^2 - 3x$$

$$y^2 = x^2 - 6x + 4 \quad \text{(as before)}$$

Example 15.1(b)
Solve the differential equation $\mathrm{d}y/\mathrm{d}x = y/(x + 2)$ given that $y = 12$ when $x = 0$.

$$\mathrm{d}y/\mathrm{d}x = y/(x + 2)$$

$$\int \frac{\mathrm{d}y}{y} = \int \frac{\mathrm{d}x}{x + 2}$$

$$\ln y = \ln (x + 2) + \ln A$$

$$= \ln A(x + 2)$$

$$y = A(x + 2)$$

From the conditions, $A = 6$, and so $y = 6(x + 2)$. This is probably the neatest way of handling the arbitrary constant if a logarithmic integral is involved.

Example 15.1(c)
A standard formula met in thermodynamics is

$$C_P \frac{dV}{V} + C_V \frac{dP}{P} = 0$$

where C_P and C_V are constants. Show that this is equivalent to the formula PV^γ is constant, where $\gamma = C_P/C_V$.

$$C_P \frac{dV}{V} + C_V \frac{dP}{P} = 0$$

$$C_P \int \frac{dV}{V} = -C_V \int \frac{dP}{P}$$

$$C_P \ln V = -C_V \ln P + \ln A$$

$$C_V \ln P + C_P \ln V = \ln A$$

$$\ln P + \gamma \ln V = \ln B \text{ where } \ln B = (\ln A)/C_V$$

$$\ln PV^\gamma = \ln B$$

$$PV^\gamma = B \ (= \text{constant})$$

15.2 Solution by Substitution

Any differential equation which can be expressed in the form $dy/dx = f(y/x)$ can be converted into one in which the variables separate by the substitution $y = vx$, v being a variable. Using this, dy/dx is replaced by $v + x\,dv/dx$ and y/x by v. Equations for which this is a suitable substitution are said to be HOMOGENEOUS in y/x.

The use of the substitution can be illustrated with the differential equation $(2xy - x^2)\,dy/dx = y^2$. It is convenient to express all variables in terms of y/x and this can be done by dividing throughout by x^2 to give

$$\left(\frac{2y}{x} - 1\right)\frac{dy}{dx} = \frac{y^2}{x^2}$$

The substitution leads to

$$(2v - 1)(v + x\,dv/dx) = v^2$$

Rearranging, separating variables and integrating gives

$$\int \frac{(2v - 1)}{v^2 - v} = -\int \frac{dx}{x}$$

$$\ln(v^2 - v) = -\ln x + \ln A = \ln A/x$$

$$v^2 - v = A/x$$

$$y^2/x^2 - y/x = A/x$$

$$y^2 - xy = Ax$$

expressing the solution in terms of x and y the variables of the original equation.

Example 15.2(a)
Solve the differential equation $2xy\,dy/dx = x^2 + 2y^2$, $y = 0$ when $x = 1$.

$$2xy\,dy/dx = x^2 + 2y^2$$

$$\frac{2y}{x}\frac{dy}{dx} = 1 + \frac{2y^2}{x^2}$$

If $v = y/x$,

$$2v(v + x\,dv/dx) = 1 + 2v^2$$

Rearranging, separating variables and integrating,

$$\int 2v\,dv = \int \frac{dx}{x}$$

$$v^2 = \ln x + \ln A = \ln Ax$$

$$y^2 = x^2 \ln Ax$$

From the conditions, $A = 1$, making $y^2 = x^2 \ln x$.

Example 15.2(b)
Find the family of curves described by the differential equation $dy/dx = (y + x)/(y - x)$. The members of a family of curves which are perpendicular to the members of another family are called its ORTHOGONAL TRAJECTORIES. Find the orthogonal trajectories of the family described above and, in particular, find the members which intersect at the point (1, 1).

$$\frac{dy}{dx} = \frac{y + x}{y - x}$$

$$= \frac{y/x + 1}{y/x - 1}$$

Substituting $v = y/x$, rearranging and integrating,

$$\int \frac{v - 1}{v^2 - 2v - 1}\, dv = -\int \frac{dx}{x}$$

$$\tfrac{1}{2} \ln (v^2 - 2v - 1) = -\ln x + \ln A$$

$$\sqrt{(v^2 - 2v - 1)} = A/x$$

Substituting back for v and simplifying,

$$y^2 - 2xy - x^2 = B$$

where B is a constant. Since orthogonal trajectories are perpendicular to the members of this family, their gradients are given by the differential equation $dy/dx = -(y - x)/(y + x)$. This is also homogeneous and is solved in the same way. The usual process leads to

$$\int \frac{v + 1}{v^2 + 2v - 1}\, dv = -\int \frac{dx}{x}$$

This is almost identical to the first family and can be solved in the same way giving $y^2 + 2xy - x^2 = C$, C being a further constant. At the point (1, 1), $B = -2$ and $C = 2$.

15.3 Solution by Use of an Integrating Factor

Very frequently there is a need to solve first order differential equations of the form $dy/dx + Py = Q$, where P and Q are either constants or functions of x.

In order to understand the method used, consider the derivative

$$\frac{d}{dx}\left(ye^{\int P\, dx}\right)$$

in which the power to which e is raised, a function of x, is written as an integral.

$$\frac{d}{dx}(ye^{\int P\,dx}) = e^{\int P\,dx}\cdot\frac{dy}{dx} + y\frac{d}{dx}(e^{\int P\,dx})$$

$$= e^{\int P\,dx}\cdot\frac{dy}{dx} + ye^{\int P\,dx}\cdot\frac{d}{dx}\left(\int P\,dx\right)$$

$$= e^{\int P\,dx}\left(\frac{dy}{dx} + Py\right)$$

In the first line the product rule has been used and in the second line the function of a function rule. In the third line the result

$$\frac{d}{dx}\left(\int P\,dx\right) = P$$

is used. It follows that if the differential equation $dy/dx + Py = Q$ is multiplied throughout by $e^{\int P\,dx}$, integrating the resulting left-hand side gives $ye^{\int P\,dx}$.

The term $e^{\int P\,dx}$ is called an INTEGRATING FACTOR of the differential equation, often abbreviated to IF. This factor may be simplified if $\int P\,dx$ is logarithmic as $e^{\ln x} = x$, $e^{\ln x^2} = x^2$, etc.

It should be noted that when the left-hand side of the differential equation has been handled in this way, the effect of the integration is to leave the product of y and the IF.

Example 15.3(a)
Solve the differential equation $dy/dx - 2y = 6x - 1$ if $y = 2$ when $x = 0$.

IF is $e^{\int -2\,dx} = e^{-2x}$.

$$e^{-2x}\left(\frac{dy}{dx} - 2y\right) = (6x-1)e^{-2x}$$

$$u = 6x - 1$$
$$dv = e^{-2x}\,dx$$
$$v = -\tfrac{1}{2}e^{-2x}$$
$$du = 6\,dx$$

$$ye^{-2x} = \int (6x-1)e^{-2x}\,dx$$

$$= -\tfrac{1}{2}e^{-2x}(6x-1) + 3\int e^{-2x}\,dx$$

$$= -\tfrac{1}{2}e^{-2x}(6x-1) - \tfrac{3}{2}e^{-2x} + A$$

$$= -3xe^{-2x} - e^{-2x} + A$$

$$y = -3x - 1 + Ae^{2x}$$

From the conditions $A = 3$ so that $y = 3e^{2x} - 3x - 1$.

In this case the integration of $\int(6x - 1)e^{-2x}$ is carried out using parts. Notice that in order to obtain y, it was necessary to multiply all terms by e^{2x} including A.

Example 15.3(b)
Solve the differential equation $x\,dy/dx - y = 2x^3$ if $y = 1$ when $x = 1$.

$$x\,dy/dx - y = 2x^3$$

$$\frac{dy}{dx} - \frac{y}{x} = 2x^2$$

IF is

$$e^{\int -dx/x} = e^{-\ln x} = e^{\ln 1/x} = 1/x$$

$$\frac{1}{x}\frac{dy}{dx} - \frac{y}{x^2} = 2x$$

$$y/x = \int 2x\,dx$$

$$= x^2 + A$$

$$y = x^3 + Ax$$

From the conditions $A = 0$ so that $y = x^3$.

In this example it was necessary to rearrange the differential equation (by dividing by x) in order to obtain it in the form $dy/dx + Py = Q$, before the IF could be determined.

Example 15.3(c)
Verify that $y = Ae^{-x} + 3$ is the general solution of the differential equation $dy/dx + y = 3$. Obtain this solution by (a) separating the variables, (b) using an integrating factor.

As $y = Ae^{-x} + 3$, $dy/dx = -Ae^{-x}$ and $dy/dx + y = 3$.

(a) If $dy/dx + y = 3$, $dy/dx = -(y - 3)$,

$$\int dy/(y - 3) = -\int dx$$

$$\ln(y - 3) = -x + \ln A$$

$$\ln(y - 3)/A = -x$$

$$(y - 3)/A = e^{-x}$$

$$y - 3 = Ae^{-x}$$

$$y = Ae^{-x} + 3$$

(b) If $dy/dx + y = 3$, IF is e^x

$$ye^x = \int 3e^x \, dx$$

$$= 3e^x + A$$

$$y = 3 + Ae^{-x}$$

This example illustrates the fact that there may be more than one method of solution.

Example 15.3(d)

Verify the solution of Question 13 Unworked Examples 1 using an integrating factor.

$$dv/dt = g - kv$$

$$dv/dt + kv = g$$

IF is $e^{\int k \, dt} = e^{kt}$

$$ve^{kt} = g \int e^{kt} \, dt = ge^{kt/k} + A$$

The implied conditions in the question are that $v = 0$ when $t = 0$, making $A = -g/k$.

$$ve^{kt} = ge^{kt}/k - g/k$$

$$v = g/k - ge^{-kt}/k$$

$$= g(1 - e^{-kt})/k$$

Unworked Examples 2

Part 15.1

(1) Solve the differential equation $dy/dx = (3x^2 + x)/y$, if $y = 1$ when $x = 1$.
(2) Find a general solution of the differential equation $dy/dx = y$.
(3) Find y if $dy/dx = 3x^2/2y$.
(4) If P(x,y) is a point on a circle centre O show that the gradient of the tangent at P is $-x/y$. Hence prove that the equation of the circle is $x^2 + y^2 = k^2$ where k is constant. What is a physical meaning for k?
(5) Solve the equation $\sec t \, dy/dt = y$ given that $y = 2$ when $t = 0$.
(6) By differentiating $y = Ae^x/x$ and eliminating A, show that this is the solution of $dy/dx = (xy - y)/x$. Verify the solution by separating the variables.

(7) Solve the differential equation $dh/dt = -kh$, $h = h_0$ when $t = 0$. *After solving*, compare your result with that given for Question 12 of Unworked Examples 1.

(8) A motor boat of mass m is moving with velocity v when the engine is shut off. If the resistance of the water is proportional to u^2 where u is instantaneous velocity of the boat after a further time t, show that $du/dt = -ku^2/m$ where k is a constant. Solve this differential equation to obtain the solution $1/u - 1/v = kt/m$.

Part 15.2

(9) Solve the differential equation $x\, dy/dx = 2x + y$ if $y = 0$ when $x = 1$.

(10) Obtain a general solution of the differential equation $dy/dx = y^2/(xy - x^2)$.

(11) Show that the substitution $t = x + 4$, $u = y - 2$ converts the differential equation $dy/dx = (2x + 2y + 4)/(3x + y + 10)$ into the homogeneous form $du/dt = (2t + 2u)/(3t + u)$. Use the substitution $u = vt$ to solve this modified equation using partial fractions to evaluate the integral in v. Find the solution of the original equation subject to the conditions $y = 6$ when $x = 6$.

(12) Show that the family of curves $x^2 + y^2 = A$ are orthogonal trajectories of the family $dy/dx = y/x$. Interpret the result geometrically.

(13) Solve the equation $x\, dy/dx = y + x \tan(y/x)$ if $y = \pi/2$ when $x = 1$.

Part 15.3

(14) Find a general solution of the differential equation $dy/dx + 2y/x = e^x/x^2$.

(15) If $dy/dx - 3y = 3x - 1$, and $y = 2$ when $x = 0$, express y in terms of x.

(16) A 500 litre tank is filled with water containing 2 kg of salt, thoroughly mixed and dissolved. A brine solution containing 10 g l^{-1} enters the tank at the rate of 2 l min^{-1} and leaves at the same rate. If the tank contains W kg of salt after t minutes, show that $dW/dt = 0.02 - 0.004\,W$. Solve this differential equation and hence find the quantity of salt ultimately remaining in the tank.

(17) A body is projected upwards through the air. After time t its velocity v is described mathematically by the differential equation

$$dv/dt = -(g + kv)$$

where g is the gravitational constant and k is a further constant. If u is the initial projection velocity, express v in terms of u, t, g and k and show that eventually v has a value of $-g/k$.

(18) Solve the differential equation $dy/dx + 2y/(x + 1) = 4(x + 1)$ subject to the conditions $y = 0$ when $x = 1$.

(19) Show that if

$$y = Ae^{x} + e^{-x} \tag{1}$$

then

$$dy/dx = Ae^{x} - e^{-x}$$

and, by eliminating A between these two equations, that Equation (1) is the general solution of the differential equation $dy/dx = y - 2e^{-x}$. Verify that by multiplying this differential equation by a suitable integrating factor, the general solution Equation (1) could be obtained in the conventional manner.

PART 16: SOLUTION OF SECOND ORDER HOMOGENEOUS EQUATIONS

16.1 Principle of Superposition

In Parts 16 and 17, we will be concerned with the solution of differential equations of the form $d^2y/dx^2 + a\, dy/dx + by = f(x)$ where a and b are constants, possibly zero, and $f(x)$ is any function of x, also possibly zero. These equations of the first degree are sometimes described as linear. It is convenient to begin with equations in which $f(x)$ actually is zero. Equations of this type are described as being HOMOGENEOUS second order. (This is a different meaning of the same word used in Part 15.2 in connection with first order equations.)

If $y = g_1(x)$ and $y = g_2(x)$ are two linearly independent solution of the homogeneous differential equation then

$$\frac{d^2}{dx^2}(g_1(x)) + a\frac{d}{dx}(g_1(x)) + bg_1(x) = 0 \tag{1}$$

and

$$\frac{d^2}{dx_2}(g_2(x)) + a\frac{d}{dx}(g_2(x)) + bg_2(x) = 0 \tag{2}$$

Adding (1) and (2),

$$\frac{d^2}{dx^2}(g_1(x) + g_2(x)) + a\frac{d}{dx}(g_1(x) + g_2(x)) + b(g_1(x) + g_2(x)) = 0$$

making $y = g_1(x) + g_2(x)$ another solution. This method of obtaining a third solution from two given solution uses the PRINCIPLE OF SUPERPOSITION.

16.2 Auxiliary Equation with Distinct Roots

The solution of the differential equation $dy/dx - my = 0$, where m is a constant, is found to be $y = Ae^{mx}$, either by the method of separating the variables or by using the integrating factor e^{-mx}.

This general solution is seen to satisfy the differential equation because $dy/dx = my = mAe^{mx}$. Since the derivatives of any expression involving e^{mx} themselves involve e^{mx}, this suggests that a similar expression might satisfy the differential

$$d^2y/dx^2 + a\,dy/dx + by = 0 \tag{1}$$

Substituting Ae^{mx} for y gives $m^2Ae^{mx} + amAe^{mx} + bAe^{mx} = 0$ which simplifies to

$$Ae^{mx}(m^2 + am + b) = 0 \tag{2}$$

Although y (or Ae^{mx}) $= 0$ would satisfy Equation (1), this is clearly a trivial solution as we are interested in y as a function of x. If we assume that $Ae^{mx} \neq 0$, Equation (2) is only satisfied if $m^2 + am + b = 0$. This is called the AUXILIARY EQUATION of Equation (1). If this quadratic equation has solutions m_1 and m_2, then both $y = Ae^{m_1x}$ and $y = Be^{m_2x}$ are solutions of Equation (1). (There is no reason why the arbitrary constants should be equal.) Using the principle of superposition, another solution of Equation (1) is $y = Ae^{m_1x} + Be^{m_2x}$. Since the general solution of a second order differential equation must involve two arbitrary constants, this is clearly the form of solution required.

Example 16.2(a)
Solve the differential equation $d^2y/dx^2 - 3\,dy/dx + 2y = 0$, given that $y = 3$ and $dy/dx = 5$ when $x = 0$. Check your solution by substitution.

The auxiliary equation is

$$m^2 - 3m + 2 = 0$$
$$(m - 1)(m - 2) = 0$$
$$m = 1 \quad \text{or} \quad m = 2$$
$$y = Ae^x + Be^{2x}$$
$$dy/dx = Ae^x + 2Be^2x$$

From the conditions $A + B = 3$, $A + 2B = 5$, making $A = 1$, $B = 2$,

$$y = e^x + 2e^{2x}$$

Check:

$$y = e^x + 2e^{2x}$$
$$dy/dx = e^x + 4e^{2x}$$
$$d^2y/dx^2 = e^x + 8e^{2x}$$
$$d^2y/dx^2 - 3\,dy/dx + 2y = e^x + 8e^{2x} - 3(e^x + 4e^{2x}) + 2(e^x + 2e^{2x}) = 0$$

When $x = 0$, $y = 3$ and $dy/\,dx = 5$.

It is normal to write down the auxiliary equation directly from the differential equation without further explanation. In the same way, it is acceptable to state the solution of the differential equation directly from the roots of the auxiliary equation.

If a solution is to be checked as shown above, it must satisfy the conditions as well as the differential equation itself.

Example 16.2(b)
Find the general solution of $d^2y/dx^2 - 2\,dy/dx = 0$.

Auxiliary equation

$$m^2 - 2m = 0$$
$$m(m - 2) = 0$$
$$m = 0 \quad \text{or} \quad m = 2$$
$$y = A + Be^{2x}$$

When $m = 0$, Ae^{mx} simplifies to the constant term A.

16.3 Auxiliary Equation with Repeated Roots

In the examples in Part 16.2, all the auxiliary equations had two distinct roots. Taking the equation

$$d^2y/dx^2 + 6\,dy/dx + 9y = 0 \tag{1}$$

the auxiliary equation is $m^2 + 6m + 9 = 0$ which factorises to $(m + 3)^2 = 0$. The auxiliary equation therefore has the repeated roots $m = -3$. Using the method of Part 16.2 and taking y as $Ae^{-3x} + Be^{-3x}$ is unhelpful as this is equivalent to $(A + B)e^{-3x}$ and effectively contains just one arbitrary constant $(A + B)$.

The general result can be obtained by letting $z = dy/dx + 3y$. Then $d^2y/dx^2 + 3\,dy/dx = dz/dx$ and $3\,dy/dx + 9y = 3z$. With this substitution, the original differential equation (1) becomes $dz/dx + 3z = 0$, which, as we have seen, has the solution $z = Ae^{-3x}$. This, in turn, means that $dy/dx + 3y = Ae^{-3x}$, an equation which is solved using the integrating factor e^{3x}. The solution of this and therefore of Equation (1) is $y = (Ax + B)e^{-3x}$. As the Part 16.2 it is standard practice to obtain the solution of the differential equation in this form directly from the repeated roots of the auxiliary equation. The value of the repeated root indicates the power to which e^x is raised.

Example 16.3(a)
Solve the differential equation $d^2y/dx^2 - 4\,dy/dx + 4y = 0$, given that $y = 3$, $dy/dx = 7$ when $x = 0$.

Auxiliary equation

$$m^2 - 4m + 4 = 0$$
$$(m - 2)^2 = 0$$
$$m = 2 \text{ (repeated)}$$

Hence $y = (Ax + B)e^{2x}$ and, using the product rule for differentiation, $dy/dx = e^{2x}(2Ax + A + 2B)$. From the conditions $B + 3$, $A + 2B = 7$. It follows that $A = 1$ and that $y = (x + 3)e^{2x}$.

Example 16.3(b)
If $y = ve^{2x}$ where v is a function of x, express dy/dx and d^2y/dx^2 in terms of v and x. Hence use the substitution $y = ve^{2x}$ to verify the solution of the differential equation of Example 16.3(a).

If $y = ve^{2x}$, $dy/dx = e^{2x}(dv/dx + 2v)$ and $d^2y/dx^2 = e^{2x}(d^2v/dx^2 + 4\,dv/dx + 4v)$. Substituting these values in the differential equation,

$$e^{2x}(d^2v/dx^2 + 4\,dv/dx + 4v - 4\,dv/dx - 8v + 4v) = 0$$
$$e^{2x}\,d^2v/dx^2 = 0$$

Since $e^{2x} \neq 0$, $d^2v/dx^2 = 0$. This is the differential equation obtained as the result of the substitution. Treating the conditions in the same way, we obtain $v = 3$, $dv/dx = 1$ when $x = 0$.

The solution of the differential equation $d^2v/dx^2 = 0$ is $v = Ax + B$ and the conditions lead to values $A = 1$, $B = 3$. Consequently $v = x + 3$ and $y = (x + 3)e^{2x}$.

16.4 Auxiliary Equation with Complex Roots

If the auxiliary equation has complex roots, they must be conjugate because they are the roots of an equation. In addition, the arbitrary constants will also be complex and these will only involve two unknowns if they are also conjugate.

Calling the roots of the auxiliary equation $m \pm jn$ and the arbitrary constants $\frac{1}{2}(A \pm jB)$—the reason for the factor $\frac{1}{2}$ will soon be clear—we obtain

$$\begin{aligned}
y &= \tfrac{1}{2}(A - jB)e^{(m+jn)x} + \tfrac{1}{2}(A + jB)e^{(m-jn)x} \\
&= \tfrac{1}{2}(A - jB)e^{mx} \cdot e^{jnx} + \tfrac{1}{2}(A + jB)e^{mx}e^{-jnx} \\
&= \tfrac{1}{2}e^{mx}((A - jB)e^{jnx} + (A + jB)e^{-jnx}) \\
&= \tfrac{1}{2}e^{mx}((A - jB)(\cos nx + j\sin nx) + (A + jB)(\cos nx - j\sin nx)) \\
&= \tfrac{1}{2}e^{mx}(2A\cos nx - 2j^2B\sin nx) \\
&= e^{mx}(A\cos nx + B\sin nx)
\end{aligned}$$

In the fourth line, the exponential form of a complex number was used (see Part 2.7). The final line shows the neatest form taken by the solution of the differential equation when the roots of the auxiliary equation are $m \pm jn$.

Table 16.1 summarises the various forms that the roots of the auxiliary equation may take and the associated solutions of the differential equation.

Table 16.1

Type of roots of the auxiliary equation	Value of the roots of the auxiliary equation	Solution of the differential equation
Real and distinct	m_1, m_2	$Ae^{m_1x} + Be^{m_2x}$
Repeated	m	$(Ax + B)e^{mx}$
Complex	$m \pm jn$	$e^{mx}(A \cos nx + B \sin nx)$

Example 16.4(a)
Solve the differential equation $d^2y/dt^2 - 4\,dy/dt + 5y = 0$ given that $y = 1$, $dy/dt = 2$ when $t = 0$

Auxiliary equation

$$m^2 - 4m + 5 = 0$$
$$m = \tfrac{1}{2}(4 \pm \sqrt{((-4)^2 - 4 \times 1 \times 5)}$$
$$= 2 \pm j$$
$$y = e^{2t}(A \cos t + B \sin t)$$
$$dy/dt = e^{2t}((2A + B) \cos t + (2B - A) \sin t)$$

From the conditions $A = 1$, $2A + B = 2$ making $B = 0$,

$$y = e^{2t} \cos t$$

Example 16.4(b)
Simple harmonic motion is defined as the motion of a body under the action of a force, directed towards a fixed point and proportional to the body's displacement, x, from it. Show that the motion may be described mathematically by the differential equation $d^2x/dt^2 + \omega^2 x = 0$, where ω is a constant and t is time.

Solve this equation taking conditions $x = a$, $dx/dt = 0$ when $t = 0$. Show that the motion is periodic and that the time of one oscillation is $2\pi/\omega$.

The motion is described by using the Newtonian expression $M\,d^2x/dt$ for force, where M is the mass of the body. We can therefore write

$$M\,d^2x/dt^2 \propto -x$$

the minus sign being inserted because force and displacement are measured in opposite directions. In order to obtain a differential *equation*, the proportionality symbol must be replaced by an equality symbol in the usual manner. It is conventional to have a positive constant of proportionality, so the differential equation is

$$M\,d^2x/dt^2 = -kx$$

which simplifies to

$$d^2x/dt^2 + \omega^2 x = 0$$

if ω^2 is taken as k/M.

The auxiliary equation is $m^2 + \omega^2 = 0$, making $m = \pm j\omega$. These are conjugate imaginary numbers, the real part being equal to zero. It follows that

$$x = A \cos \omega t + B \sin \omega t$$

$$dx/dt = -A\omega \sin \omega t + B\omega \cos \omega t$$

From the conditions $A = a$, $B\omega = 0$ making $B = 0$,

$$x = a \cos \omega t$$

The displacement is sinusoidal and consequently periodic. The time, T, taken for one oscillation is such that $\omega T = 2\pi$, so that $T = 2\pi/\omega$.

Unworked Examples 3

Parts 16.1 and 16.2

(1) Show that both $y = e^x$ and $y = xe^x$ satisfy the differential equation $d^2y/dx^2 - 2\,dy/dx + y = 0$. Deduce a third solution.
(2) If $y = g_1(x)$ and $y = g_2(x)$ are both solutions of a differential equation, explain why $y = g_1(x) + g_2(x)$ is another solution. If p and q are positive integers, deduce that $y = pg_1(x) + qg_2(x)$ is also a solution.
(3) Solve the differential equation $d^2y/dx^2 + 6\,dy/dx + 5y = 0$, giving a general solution.
(4) Find the solution of the differential equation $d^2z/dt^2 - 3\,dz/dt = 0$, with conditions $z = 2$, $dz/dt = 3$ when $t = 0$ using
    (a) the method of Part 16.2
    (b) the substitution $x = dx/dt$.
(5) Find the particular solution of the equation $d^2y/dt^2 + 6\,dy/dt + 8y = 0$, $y = 4$, $dy/dt = 0$ when $t = 0$.
(6) Show that the auxiliary equation of $d^2y/dx^2 = 4y$ has equal and opposite roots. Hence obtain a general solution of the equation and verify it by substitution.
(7) Show that the differential equation $d^2y/dx^2 + dy/dx - 6y = 0$ taken with conditions $y = 3$, $dy/dx = -4$ has a solution $y = e^{2x} + 2e^{-3x}$.

Part 16.3

(8) Solve the differential equation $d^2y/dt^2 = 0$
(9) Verify by direct substitution that $y = (Ax + B)e^{4x}$ is the general solution of the differential equation $d^2y/dx^2 - 8\,dy/dx + 16x = 0$.

(10) Show that the solution of the equation $d^2y/dx^2 + 4\,dy/dx + 4y = 0$ with conditions $y = dy/dx = 1$ when $x = 0$ is $y = (3x + 1)e^{-2x}$.

(11) Solve the differential equation $d^2x/dt^2 + 8\,dx/dt + 16x = 0$ if $x = 0$ and $dx/dt = 1$ when $t = 0$.

(12) Show that if v is a function of t, the substitution $x = ve^{-4t}$ converts the differential equation $d^2x/dt^2 + 8\,dx/dt + 16x = 0$ into the form $d^2v/dt^2 = 0$. Use this to verify the solution of Question 11.

(13) Form a second order differential equation with a general solution $y = (Ax + B)e^{2x}$.

Part 16.4

(14) Find y if $d^2y/dx^2 - 2\,dy/dx + 2y = 0$ and $y = 0$, $dy/dx = 1$ when $x = 0$.

(15) Find a differential equation with a general solution $y = e^{2t}(A \sin 3t + B \cos 3t)$.

(16) The differential equation representing damped harmonic motion can be taken as $d^2x/dt^2 + 2k\,dx/dt + \omega^2 x = 0$ where k and ω are constants. Solve this equation for cases where the damping is light ($k < \omega$).

(17) Obtain the general solution of the following differential equations
(a) $d^2x/dt^2 - 25\,dx/dt = 0$,
(b) $d^2x/dt^2 - 25x = 0$,
(c) $d^2x/dt^2 + 25x = 0$.

(18) By reference to tables or otherwise, deduce that if θ is measured in radians, $\sin\theta \approx \theta$ when θ is small. Figure C6 shows a simple pendulum consisting of a string OQ, of length l, to which is attached a light weight of mass m. By considering forces perpendicular to the string, show that the net restoring force tending to move m to the equilibrium position P is $mg \sin\theta$ and that the distance $QP = l\theta$. Deduce that $l\,d^2\theta/dt^2 = -g \sin\theta$. Show that if θ is small, the motion is approximately simply harmonic with a period $2\pi\sqrt{(l/g)}$.

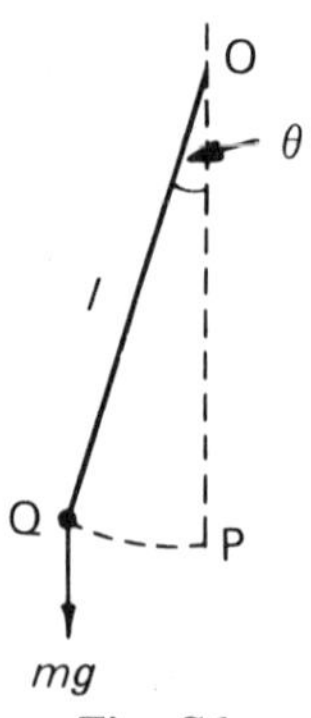

Fig. C6

(19) A wooden cube of side l and density ρ floats in water. If it is de
small distance x keeping its top horizontal, use Archimedes' P
find the net upthrust and hence deduce that $\mathrm{d}^2x/\mathrm{d}t^2 = -gx/\rho l$
the gravitational constant, when the pressure is released.
period of the resulting simple harmonic motion.

PART 17: SOLUTION OF SECOND ORDER INHOMOGENEOUS EQUATIONS

17.1 Complementary Function and Particular Integral

The second order differential equation $\mathrm{d}^2y/\mathrm{d}x^2 + a\mathrm{d}y/\mathrm{d}x + by = f(x)$, where $f(x)$ is non zero, is described as being INHOMOGENEOUS in contrast to the term used in Part 16.1.

If u is the complete general solution to the corresponding *homogeneous* differential equation then

$$\mathrm{d}^2u/\mathrm{d}x^2 + a\,\mathrm{d}u/\mathrm{d}x + bu = 0 \tag{1}$$

If v is *any* solution to the *inhomogeneous* differential equation then

$$\mathrm{d}^2v/\mathrm{d}x^2 + a\,\mathrm{d}v/\mathrm{d}x + bv = f(x) \tag{2}$$

Adding Equations (1) and (2) gives $\mathrm{d}^2(u + v)/\mathrm{d}x^2 + a\mathrm{d}(u + v)/\mathrm{d}x + b(u + v) = f(x)$ so that $(u + v)$ must be a solution to the inhomogeneous equation. Furthermore, as u is a complete general solution, it must involve two arbitrary constants and, consequently, so does $(u + v)$. This means that $(u + v)$ must be the complete general solution to the required equation.

As an example, consider the equation

$$\mathrm{d}^2y/\mathrm{d}x^2 - 3\,\mathrm{d}y/\mathrm{d}x + 2y = 2 \tag{3}$$

The solution of the corresponding homogeneous equation, $\mathrm{d}^2y/\mathrm{d}x^2 - 3\,\mathrm{d}y/\mathrm{d}x + 2y = 0$, is obtained as $u = A\mathrm{e}^x + B\mathrm{e}^{2x}$ using the method of Part 16.2. The constant value $v = 1$ satisfies Equation (3), since if $y = v = 1$, $\mathrm{d}y/\mathrm{d}x$ and $\mathrm{d}^2y/\mathrm{d}x^2 = 0$. This means that the complete general solution of Equation (3) is $y = u + v = A\mathrm{e}^x + B\mathrm{e}^{2x} + 1$. The term used for u is the COMPLEMENTARY FUNCTION of Equation (3) while v is called its PARTICULAR INTEGRAL. These are abbreviated to CF and PI respectively.

Example 17.1(a)
Solve the differential equation $\mathrm{d}^2y/\mathrm{d}t^2 + 6\,\mathrm{d}y/\mathrm{d}t + 5y = 10$, given that $y = 3$, $\mathrm{d}y/\mathrm{d}t = -3$ when $t = 0$.

CF: Auxiliary equation

$$m^2 + 6m + 5 = 0$$
$$(m + 1)(m + 5) = 0$$
$$m = -1 \quad \text{or} \quad m = -5$$
$$u = Ae^{-t} + Be^{-5t}$$

PI: By inspection, $v = 2$,

$$y = u + v = Ae^{-t} + Be^{-5t} + 2$$
$$dy/dt = -Ae^{-t} - 5Be^{-5t}$$

From the conditions $A + B + 2 = 3$, $A + 5B = 3$ so that $A = B = \frac{1}{2}$,

$$y = \tfrac{1}{2}e^{-t} + \tfrac{1}{2}e^{-5t} + 2$$

Notice that y and dy/dt (*not* u and du/dt) are the expressions involved in the conditions.

Example 17.1(b)
Verify that $\frac{1}{2}e^{3x}$ is a particular integral of the differential equation $d^2y/dx^2 + 3\,dy/dx - 10y = 4e^{3x}$. Hence find a complete general solution.
If $v = \frac{1}{2}e^{3x}$,

$$dv/dx = \tfrac{3}{2}e^{3x}, \qquad d^2v/dx^2 = \tfrac{9}{2}e^{3x}$$

and so

$$d^2v/dx^2 + 3\,dv/dx - 10v = 4e^{3x}$$

It follows that v is a particular integral.

CF: Auxiliary equation

$$m^2 + 3m - 10 = 0$$
$$(m - 2)(m + 5) = 0$$
$$m = 2 \quad \text{or} \quad m = -5$$
$$u = Ae^{2x} + Be^{-5x}$$

The complete general solution is $y = u + v = Ae^{2x} + Be^{-5x} + \frac{1}{2}e^{3x}$.

17.2 Trial Solutions

In the differential equations solved in the text and worked example section of Part 17.1, the expression on the right-hand side has either been a constant, which made the particular integral easy to find, or else the question was worded so that the particular integral was given. Obviously a more systematic approach will normally be needed and this may be carried out using a TRIAL SOLUTION described below or by the D OPERATOR METHOD described in Part 18.

Table 17.1

Right-hand side of the inhomogeneous differential equation	Trial solution
pe^{kx}	ae^{kx}
$p \cos kx + q \sin kx$	$a \cos kx + b \sin kx$
$px^n + qx^{n-1} + rx^{n-2} + \cdots$	$ax^n + bx^{n-1} + cx^{n-2} + \cdots$
$e^{kx}(p \cos lx + q \sin lx)$	$e^{kx}(a \cos lx + b \sin lx)$
$e^{kx}(px^n + qx^{n-1} + rx^{n-2} + \cdots)$	$e^{kx}(ax^n + bx^{n-1} + cx^{n-2} + \cdots)$
k, l, n, p, q and r are given but some may be zero	a, b and c are to be determined by equating coefficients

Table 17.1 shows the types of trial solution associated with various right-hand sides of an inhomogenous differential equation. Basically the functions are the same with only the coefficients needing to be determined, although a trial solution must always contain both sine and cosine terms whether or not both appear in the equation.

To obtain a particular integral of the differential equation $d^2y/dx^2 - 4\,dy/dx + 4y = 8e^{4x}$, the first line of Table 17.1 is used with $p = 8$, $k = 4$. This indicates a trial solution $v = ae^{4x}$ with a to be determined. Taking derivatives, $dv/dx = 4ae^{4x}$ and $d^2v/dx^2 = 16ae^{4x}$. Substituting these values in the differential equation and identifying the result with the right-hand side gives

$$16ae^{4x} - 16ae^{4x} + 4ae^{4x} \equiv 8e^{4x}$$
$$4ae^{4x} \equiv 8e^{4x}$$

Since this is an identity, coefficients can be equated, making $4a = 8$ and thus $a = 2$. It is easy to verify that $2e^{4x}$ is a particular integral of the equation in question by direct substitution.

Example 17.2(a)
Use the trial solution method to determine a particular integral of the differential equation $d^2y/dx^2 + 3\,dy/dx + 4y = 6 \cos x$.

From Table 17.1, $v = a \cos x + b \sin x$. Thus $dv/dx = -a \sin x + b \cos x$ and $d^2v/dx^2 = -a \cos x - b \sin x$.

$$(-a \cos x - b \sin x) + 3(-a \sin x + b \cos x) + 4(a \cos x + b \sin x)$$
$$\equiv 6 \cos x$$

Equating coefficients,

$$\cos x: \quad 3a + 3b = 6$$

$$\sin x: \quad 3b - 3a = 0$$

This leads to $a = b = 1$ and $v = \cos x + \sin x$.

Since there is no sine term on the right-hand side of the differential equation, it is necessary to equate $3b - 3a$ to zero.

Example 17.2(b)
Find a particular integral of the differential equation $\mathrm{d}^2y/\mathrm{d}x^2 + y = 2\mathrm{e}^x - 3 \sin 2x$.

Let $v = a\mathrm{e}^x + b \sin 2x + c \cos 2x$.

$$\mathrm{d}v/\mathrm{d}x = a\mathrm{e}^x + 2b \cos 2x - 2c \sin 2x$$

$$\mathrm{d}^2v/\mathrm{d}x^2 = a\mathrm{e}^x - 4b \sin 2x - 4c \cos 2x$$

Substituting in the differential equation,

$$(a\mathrm{e}^x - 4b \sin 2x - 4c \cos 2x) + (a\mathrm{e}^x + b \sin 2x + c \cos 2x) \equiv 2\mathrm{e}^x - 3 \sin 2x$$

Equating coefficients,

$$\mathrm{e}^x: \quad 2a = 2$$

$$\sin 2x: \quad -3b = -3$$

$$\cos 2x: \quad -3c = 0$$

Thus $a = 1$, $b = 1$, $c = 0$ and $v = \mathrm{e}^x + \sin 2x$

Whenever the right-hand side of the differential equation consists of two or more entries from Table 17.1, the trial solution required is the corresponding sum of trial solutions from the table with the coefficients suitably extended.

Example 17.2(c)
Solve the equation $\mathrm{d}^2z/\mathrm{d}t^2 - 4\,\mathrm{d}z/\mathrm{d}t + 5z = 5t^2 + 2t - 1$, if $z = 3$, $\mathrm{d}z/\mathrm{d}t = 4$ when $t = 0$.

CF: Auxiliary equation

$$m^2 - 4m + 5 = 0$$

$$m = (4 \pm \sqrt{(-4)})/2$$

$$= 2 \pm j$$

$$u = \mathrm{e}^{2t}(A \cos t + B \sin t)$$

PI:

$$\text{Let} \quad v = at^2 + bt + c$$

$$\mathrm{d}v/\mathrm{d}t = 2at + b$$

$$\mathrm{d}^2v/\mathrm{d}t^2 = 2a$$

$$2a - 4(2at + b) + 5(at^2 + bt + c) \equiv 5t^2 + 2t - 1$$

Equating coefficients,

$$t^2: \quad 5a = 5$$
$$t: \quad -8a + 5b = 2$$
$$t^0: \quad 2a - 4b + 5c = -1$$

making $a = 1$, $b = 2$, $c = 1$ and $v = t^2 + 2t + 1$.

$$z = u + v = e^{2t}(A \cos t + B \sin t) + t^2 + 2t + 1$$
$$dz/dt = e^{2t}((2A + B) \cos t) + (2B - A) \sin t) + 2t + 2$$

From the conditions $A + 1 = 3$, $2A + B + 2 = 4$, making $A = 2$, $B = -2$,

$$z = 2e^{2t}(\cos t - \sin t) + t^2 + 2t + 1$$

17.3 Breakdown of the Trial Solution Method

Consider the particular integral of the differential equation

$$d^2x/dt^2 - 3\ dx/dt + 2x = e^{2t} \qquad (1)$$

The 'obvious' trial solution is $v = ae^{2t}$, making $dv/dt = 2ae^{2t}$ and $d^2v/dt^2 = 4ae^{2t}$. Substituting in the differential equation gives

$$4ae^{2t} - 6ae^{2t} + 2ae^{2t} \equiv ae^{2t}$$

and the left-hand side vanishes!

If the complementary function had been obtained, it would have found to be $Ae^{2t} + Be^{t}$ by the usual method. Adding a particular integral of the form ae^{2t} is unhelpful since, whatever the value of a, it will be 'lost' when added to the term Ae^{2t} as $(A + a)$, the new coefficient is also arbitrary.

This type of breakdown occurs whenever the trial solution contains a term which also appears in the complementary function. If the unsatisfactory term in the trial solution is multiplied by t, the problem no longe arises. Referring back to Equation (1), if $v = ate^{2t}$, $dv/dt = 2ate^{2t} + ae^{2t}$ and $d^2v/dt^2 = 4ate^{2t} + 4ae^{2t}$. If these values are substituted in the equation it leads to

$$4ate^{2t} + 4ae^{2t} - 3(2ate^{2t} + ae^{2t}) + 2ate^{2t} \equiv e^{2t}$$

This simplifies to $ae^{2t} \equiv e^{2t}$, making $a = 1$ and yielding a particular integral of te^{2t}. (The term in te^{2t} introduced into the trial solution disappears.)

It may be necessary to multiply by t^2 rather than by t in certain cases. The complementary function of the equation $d^2x/dt^2 + 4\ dx/dt + 4x = e^{-2t}$ is $(At + B)e^{-2t}$. This means that neither ae^{-2t} nor ate^{-2t} is a satisfactory trial solution as terms in both of these appear in this complementary function. Taking $v = at^2e^{-2t}$ allows the particular integral to be found and, on substitution in the differential equation, terms in e^{-2t} and in te^{-2t} disappear.

Example 17.3(a)
Find a particular integral of the differential equation $d^2y/dt^2 + y = 2 \cos t$ given that the complementary function is $A \cos t + B \sin t$.

The trial solution must be $v = at \cos t + bt \sin t$. This makes dv/dt equal to $-at \sin t + bt \cos t + a \cos t + b \sin t$ and d^2v/dt^2 equal to $-at \cos t - 2a \sin t - bt \sin t + 2b \cos t$. Substituting in the differential equation,

$$(-at \cos t - 2a \sin t - bt \sin t + 2b \cos t) + (at \cos t + bt \sin t) \equiv 2 \cos t$$

Equating coefficients,

$$\cos t: \quad 2b = 2$$
$$\sin t: \quad -2a = 0$$

Hence $b = 1$, $a = 0$ and $v = t \sin t$.

Example 17.3(b)
Find the complete general solution of the differential equation $d^2z/dx^2 + 3\, dz/dx = 6 - e^{-3x}$

CF: Auxiliary equation

$$m^2 + 3m = 0$$
$$m(m + 3) = 0$$
$$m = 0 \quad \text{or} \quad m = -3$$
$$u = A + Be^{-3x}$$

PI: Let $v = ax + bxe^{-3x}$

$$dv/dx = a + be^{-3x} - 3bxe^{-3x}$$
$$d^2v/dx^2 = -6be^{-3x} + 9bxe^{-3x}$$

Substituting in the differential equation,

$$-6be^{-3x} + 9bxe^{-3x} + 3(a + be^{-3x} - 3bxe^{-3x}) \equiv 6 - e^{-3x}$$

Equating coefficients,

$$e^{-3x}: \quad -3b = -1$$
$$x^0: \quad 3a = 6$$

Hence $a = 2$, $b = \frac{1}{3}$ and $v = 2x + \frac{1}{3}xe^{-3x}$,

$$z = u + v = A + Be^{-3x} + 2x + \tfrac{1}{3}xe^{-3x}$$

The trial solution would 'normally' be $a + be^{-3x}$. As both a constant term and a term in e^{-3x} appear in the complementary function, it is taken as $ax + bxe^{-3x}$.

Unworked Examples 4

Part 17.1

(1) Find a particular integral of the equation $d^2y/dx^2 - 5\,dy/dx + 4y = 8$.
(2) Obtain a general solution of the differential equations (a) $d^2z/dt^2 - 4\,dz/dt + 3z = 60$ (b) $d^2z/dt^2 - 4\,dz/dt + 4z = 60$ (c) $d^2z/dt^2 - 4\,dz/dt + 5z = 60$
(3) Verify that $\sin 2x$ is a particular integral of the differential equation

$$d^2y/dx^2 - 3\,dy/dx - 2y = -6(\sin 2x + \cos 2x)$$

(4) Show that $v = 2x$ is a particular integral of the differential equation $d^2y/dx^2 + 2\,dy/dx = 4$. Hence obtain a general solution.
(5) Obtain the complete solution of the equation $d^2y/dx^2 - 6\,dy/dy/dx + 8y = 12$, given that $y = 4$, $dy/dx = 5$ when $x = 0$.

Part 17.2

(6) Use the trial solution method to obtain a particular integral of the equation $d^2y/dx^2 - 3\,dy/dx + y = e^{2x}(\cos x - 2\sin x)$:
(7) A body of mass m hangs from a string of stiffness s. Show that the extension of the string x satisfies the equation $m\,d^2x/dt^2 = mg - sx$, where t is time and g is the gravitational constant. Obtain a solution if $x = a$, $dx/dt = 0$ initially.
(8) Find a particular integral of $d^2x/dt^2 + 3\,dx/dt - 5x = f(t)$ where $f(t)$ is (a) $3e^t$ (b) $3\cos 2t - 24\sin 2t$ (c) $5t^2 - t$ (d) $e^t(5\cos t - 2\sin t)$.
(9) Solve the differential equation $dy/dx = 1/(e^y - x)$ if $y = 0$ when $x = 1$ by using the substitution $y = t$ and solving for x. (N.B. Although this is only a first order equation, the methods of Part 17 and still applicable.)
(10) Solve the equation $d^2y/dx^2 + 5\,dy/dx = 4\cos x - 6\sin x$ if $y = dy/dx = 11$ when $x = 0$.
(11) Find a particular integral of $d^2y/dx^2 + dy/dx + y = e^x(3x^2 + 9x + 8)$.

Part 17.3

(12) Solve the equation $d^2y/dx^2 = 3$ given that $y = 1$, $dy/dx = 1.5$ when $x = 0$.
(13) Solve $d^2x/dt^2 + 4\,dx/dt = e^{-4t}$ subject to conditions $x = 0$, $dx/dt = -\frac{1}{4}$ at $t = 0$.
(14) Find a particular integral of the equation $d^2x/dt^2 - 2\,dx/dt + x = e^t$.
(15) Show that the trial solution $v = ae^{3x} + be^x$ breaks down when used with the equation $d^2y/dx^2 - 4\,dy/dx + 3y = 4e^{3x} - 4e^x$. Use, instead, the trial solution $v = axe^{3x} + bxe^x$ and obtain values of a and b.

(16) Solve $d^2y/dt^2 - dy/dt - 2y = e^{-t}$ if $y = 1$, $dy/dt = 8/3$ when $t = 0$.

(17) Obtain the complementary function of $d^2y/dx^2 + y = 2 \sin x$ and hence suggest a trial solution which would lead to a particular integral. Find it.

PART 18: D OPERATOR RULES

18.1 Concept of an Operator

It is convenient to replace the derivative symbol d/dx by D, so that Dy means dy/dx and $D(Dy)$, generally written D^2y, means d^2y/dx^2. However the notation is more than just an abbreviation. Although D represents the process of differentiation, it can be manipulated as if it were an ordinary algebraic variable. A quantity which behaves in this way is called an OPERATOR. The expression on which the operator acts is called its OPERAND.

Consider the expression $d^2y/dx^2 - 8\, dy/dx + 15y$ when $y = e^{2x}$. This means that $dy/dx = 2e^{2x}$ and $d^2y/dx^2 = 4e^{2x}$ so that $d^2y/dx^2 - 8\, dy/dx + 15y = 3e^{2x}$. Using a D operator, the expression can be written $D^2y - 8Dy + 15y$. Since D is being treated as an algebraic variable, it is possible to write

$$\begin{aligned} D^2y - 8Dy + 15y &= (D^2 - 8D + 15)y \\ &= (D - 3)(D - 5)y \quad \text{or} \quad (D - 5)(D - 3)y \end{aligned}$$

Taking each of these equivalent expressions in turn with y equated to e^{2x},

$$\begin{aligned} (D - 3)(D - 5)e^{2x} &= (D - 3)(2e^{2x} - 5e^{2x}) \\ &= (D - 3)(-3e^{2x}) \\ &= -6e^{2x} + 9e^{2x} \\ &= 3e^{2x} \\ (D - 5)(D - 3)e^{2x} &= (D - 5)(2e^{2x} - 3e^{2x}) \\ &= (D - 5)(-e^{2x}) \\ &= -2e^{2x} + 5e^{2x} \\ &= 3e^{2x} \end{aligned}$$

The same result is obtained whatever the form that the operator takes. In all cases, D may be taken either as an algebraic variable or as the differential coefficient d/dx as required. The operand must always be written to the right of the operator.

Example 18.1(a)
Evaluate $(D^2 + D + 1) \sin x$.

$(D^2 + D + 1) \sin x = -\sin x + \cos x + \sin x = \cos x$.

18.2 D Operator Rules

Consider the result of $D^n e^{ax}$ where a is a constant and n takes the values 1,2,3 . . .

$$
\begin{aligned}
De^{ax} &= ae^{ax} \\
D^2e^{ax} &= DDe^{ax} \\
&= D(ae^{ax}) \\
&= a^2e^{ax}
\end{aligned}
$$

In the same way, $D^n e^{ax} = a^n e^{ax}$. If $f(D)$ is equal to $k_1D^n + k_2D^{n+1} + k_3D^{n-2} + \cdots$ then we can say

$$f(D)e^{ax} = f(a)e^{ax} \tag{1}$$

Next, consider the effect of D^n operating on the product ve^{ax} where v is a variable function of x.

$$
\begin{aligned}
D(ve^{ax}) &= e^{ax}Dv + vDe^{ax} \text{ (using the product rule)} \\
&= e^{ax}Dv + ave^{ax} \\
&= e^{ax}(D + a)v
\end{aligned}
$$

Similarly, $D^2(ve^{ax}) = e^{ax}(D + a)^2v$ and by extension $D^n(ve^{ax}) = e^{ax}(D + a)^n v$. Hence we can write

$$f(D)ve^{ax} = e^{ax}f(D + a)v \tag{2}$$

Finally, since $D^2 \sin ax = -a^2 \sin ax$ and $D^2 \cos ax = -a^2 \cos ax$, the same process can be used to arrive at the results

$$f(D^2) \sin ax = f(-a^2) \sin ax \tag{3}$$

$$f(D^2) \cos ax = f(-a^2) \cos ax \tag{4}$$

Rules 1, 2, 3 and 4 are important when using D operators and can save a considerable amount of work. Rule 2 is sometimes referred to as the SHIFT RULE. For convenience these four rules are shown in the first lines of Table 18.1 with $f(D)$ replaced by $1/g(D)$, the form in which it will be used to obtain particular integrals. Notice that in all cases the operands remain unchanged when the rule is applied.

Example 18.2(a)
Use a suitable rule, as far as possible, to verify the result of Example 18.1(a).

$$
\begin{aligned}
(D^2 + D + 1) \sin x &= (-1 + D + 1) \sin x \\
&= D \sin x \\
&= \cos x
\end{aligned}
$$

There is no rule for D operating on $\sin x$ (as opposed to D^2 operating on $\sin x$) and so this operator is interpreted as d/dx.

Table 18.1

Nature of the operand $F(x)$	Method of obtaining $1/g(\mathrm{D})$ when operating on $F(x)$
Exponential power, e^{ax}	Rule: $1/g(\mathrm{D})e^{ax} = e^{ax}/g(a)$
Product, one of whose terms is an exponential power, ve^{ax}	Shift rule: $1/g(\mathrm{D})ve^{ax} = e^{ax} \cdot 1/g(\mathrm{D} + a)v$
Sine or cosine	Rule: $1/g(\mathrm{D}^2) \sin ax = 1/g(-a^2) \sin ax$ $1/g(\mathrm{D}^2) \cos ax = 1/g(-a^2) \cos ax$
Polynomial in x	Express $1/g(\mathrm{D})$ suitably and use the binomial theorem

Example 18.2(b)
Use the rule for operating on an exponential power to verify the result for $(\mathrm{D}^2 - 8\mathrm{D} + 15)y$ when $y = e^{2x}$. Show that the shift rule could also have been used to obtain the result taking v as 1.

$$(\mathrm{D}^2 - 8\mathrm{D} + 15)e^{2x} = (4 - 16 + 15)e^{2x} = 3e^{2x}$$

Using the shift rule,

$$\begin{aligned}(\mathrm{D}^2 - 8\mathrm{D} + 15)1 \cdot e^{2x} &= e^{2x}((\mathrm{D} + 2)^2 - 8(\mathrm{D} + 2) + 15)1 \\ &= e^{2x}(\mathrm{D}^2 - 4\mathrm{D} + 3)1 \\ &= 3e^{2x}\end{aligned}$$

The final line is the result of D^2 and $-4\mathrm{D}$ operating on a constant 1 to give the value zero.

Example 18.2(c)
Find $\mathrm{D}(\mathrm{D} - 1)^2(\mathrm{D} + 1)^2 \cos 2t$.

$$\begin{aligned}\mathrm{D}(\mathrm{D} - 1)^2(\mathrm{D} + 1)^2 \cos 2t &= \mathrm{D}(\mathrm{D}^2 - 1)^2 \cos 2t \\ &= \mathrm{D}(-4 - 1)^2 \cos 2t \\ &= \mathrm{D}(25 \cos 2t) \\ &= -50 \sin 2t\end{aligned}$$

In this example D represents $\mathrm{d}/\mathrm{d}t$ rather than $\mathrm{d}/\mathrm{d}x$ but the method is the same. The final result could have been obtained in various ways but, in this example, it is best to manipulate the operators as shown. If the operator $\mathrm{D}(\mathrm{D} - 1)^2(\mathrm{D} + 1)^2$ had been expanded, it would contain terms in D^5 and D^3. The operation of these on $\cos 2t$ would then have been dealt with as follows

$$D^5 \cos 2t = D \cdot D^4 \cos 2t = D(16 \cos 2t) = -32 \sin 2t$$
$$D^3 \cos 2t = D \cdot D^2 \cos 2t = D(-4 \cos 2t) = 8 \sin 2t$$

18.3 Use of D Operator in Finding a Particular Integral

The complementary function of the equation $d^2y/dx^2 - 3\ dy/dx + 2y = 12e^{3x}$ is found to be $Ae^{2x} + Be^x$ using the method of Part 16. The particular integral could be obtained by the trial solution technique, taking $v = ae^{3x}$ since no term in e^{3x} appears in the complementary function.

Using this trial solution, $dv/dx = 3ae^{3x}$ and $d^2v/dx^2 = 9ae^{3x}$. Substituting in the differential equation gives

$$ae^{3x}(9 - 9 + 2) = 12e^{3x}$$
$$2a = 12$$
$$a = 6$$

so that the particular integral is $6e^{3x}$.

Writing the differential equation using D operator notation we have $(D^2 - 3D + 2)y = 12e^{3x}$ and, since this is satisfied by the particular integral v, we may write $(D^2 - 3D + 2)v = 12e^{3x}$. Treating the operator algebraically, v can be taken as

$$v = \frac{1}{D^2 - 3D + 2} 12e^{3x}$$

It may not be clear what is meant by the operator $1/(D^2 - 3D + 2)$ but there is a rule for operating on an exponential expression found on the first line of Table 18.1.

$$\text{Here} \quad v = \frac{1}{D^2 - 3D + 2} 12e^{3x}$$
$$= \frac{1}{9 - 9 + 2} 12e^{3x}$$
$$= 6e^{3x}$$

since the factor 12 is unaffected by differentiation or integration.

This is a useful method of obtaining particular integrals of equations whose right-hand side is a function $F(x)$ from Table 18.1.

Sometimes the operator must be algebraically manipulated. To evaluate

$$\frac{1}{D^2 - D + 5}(5 \cos 2x + 10 \sin 2x)$$

D^2 as usual is replaced by -4 giving

$$\frac{1}{1 - D}(5 \cos 2x + 10 \sin 2x)$$

As the rules in Table 18.1 allow D^2 to be replaced but not D, it is necessary to multiply the numerator and denominator of the operator by $1 + D$—corresponding to the rationalisation process for surds—to give

$$\frac{1 + D}{1 - D^2}(5 \cos 2x + 10 \sin 2x)$$

Replacing D^2 by -4 makes $1 - D^2$ equal to 5. This factor cancels with the operand leaving a result which simplifies to $(1 + D)(\cos 2x + 2 \sin 2x)$ and can be evaluated directly. The result is equal to $5 \cos 2x$.

Example 18.3(a)

Evaluate $\dfrac{1}{D^2 - D - 2}(5e^t \cos t)$

$$\begin{aligned}
\frac{1}{D^2 - D - 2}(5e^t \cos t) &= 5e^t \frac{1}{(D + 1)^2 - (D + 1) - 2} \cos t \\
&= 5e^t \frac{1}{D^2 + D - 2} \cos t \\
&= 5e^t \frac{1}{D - 3} \cos t \\
&= 5e^t \frac{D + 3}{D^2 - 9} \cos t \\
&= -\tfrac{1}{2}e^t(D + 3) \cos t \\
&= -\tfrac{1}{2}e^t(3 \cos t - \sin t)
\end{aligned}$$

Example 18.3(b)

Deduce a meaning for the operation 1/D. Hence evaluate $\dfrac{1}{D - 1} te^t$.

1/D is the reverse process to that of differentiation, i.e. integration. Using the shift rule,

$$\frac{1}{D - 1} te^t = e^t \frac{1}{D} t = \tfrac{1}{2}t^2e^t$$

Since the result is generally used as a particular integral, any value of the constant of integration may be chosen. The simplest result is obtained when C is made equal to 0.

Example 18.3(c)
Solve the differential equation $(\mathrm{D}^2 + 4\mathrm{D} - 5)y = 8\mathrm{e}^{-x}$, given that $y = \mathrm{D}y = -1$ when $x = 0$.

$$(\mathrm{D}^2 + 4\mathrm{D} - 5)y = 8\mathrm{e}^{-x}$$

CF: Auxiliary equation

$$m^2 + 4m - 5 = 0$$
$$(m + 5)(m - 1) = 0$$
$$m = -5 \quad \text{or} \quad m = 1$$
$$u = A\mathrm{e}^{-5x} + B\mathrm{e}^{x}$$

PI:

$$v = \frac{1}{\mathrm{D}^2 + 4\mathrm{D} - 5} 8\mathrm{e}^{-x}$$
$$= -\mathrm{e}^{-x}$$
$$y = u + v = A\mathrm{e}^{-5x} + B\mathrm{e}^{x} - \mathrm{e}^{-x}$$
$$\mathrm{D}y = -5A\mathrm{e}^{-5x} + B\mathrm{e}^{x} + \mathrm{e}^{-x}$$

From the conditions $A + B - 1 = -1$ and $-5A + B + 1 = -1$. This makes $A = \frac{1}{3}$, $B = -\frac{1}{3}$.

$$y = \tfrac{1}{3}(\mathrm{e}^{-5x} - \mathrm{e}^{x} - 3\mathrm{e}^{-x})$$

Notice the similarity in format of the operator $\mathrm{D}^2 + 4\mathrm{D} - 5$ and the left-hand side of the auxiliary equation.

Example 18.3(d)

Obtain $\dfrac{1}{\mathrm{D} - 1}\mathrm{e}^{x}(2x + \mathrm{e}^{x})$.

Method 1:

$$\frac{1}{\mathrm{D} - 1}\mathrm{e}^{x}(2x + \mathrm{e}^{x})$$
$$= \mathrm{e}^{x}\frac{1}{\mathrm{D}}(2x + \mathrm{e}^{x})$$
$$= \mathrm{e}^{x}(x^2 + \mathrm{e}^{x})$$

Method 2:

$$\frac{1}{D-1} e^x(2x + e^x)$$

$$= \frac{1}{D-1}(2xe^x + e^{2x})$$

$$= \frac{1}{D-1}(2xe^x) + \frac{1}{D-1}(e^{2x})$$

$$= e^x \frac{1}{D}(2x) + e^{2x}$$

$$= e^x x^2 + e^{2x}$$

$$= e^x(x^2 + e^x)$$

Only one of these methods would normally be used and since method 1 is shorter this is the obvious choice. Method 2 illustrates the approach needed when the operand consists of the sum or difference of terms denoted by $F(x)$ in Table 18.1.

18.4 Handling Polynomial Operands

Consider the operation

$$\frac{1}{D^2 + D}(1 + x^3)$$

No rule has been established for operands of this type. Some progress can be made by expressing the operator as

$$\frac{1}{D+1} \cdot \frac{1}{D}$$

since

$$\frac{1}{D+1}\frac{1}{D}(1 + x^3) = \frac{1}{D+1}(x + \tfrac{1}{4}x^4)$$

but without some new rule or an interpretation of the meaning of the operator $1/(D + 1)$, further progress seems to be impossible. In examples of this type, however, the binomial expansion can be used. Here we require $1/(D + 1) = (1 + D)^{-1} = 1 - D + D^2 - D^3 + \cdots$ but sometimes its counterpart $1/(D - 1) = -(1 - D)^{-1} = -1 - D - D^2 - D^3 - \cdots$ may be needed.

Both of these expansions are infinite. However, if the operand is a polynomial, it can only be differentiated a finite number of times. It can be seen that

$D^n x = 0$ when $n > 1$ and that $D^n(\frac{1}{4}x^4) = 0$ when $n > 4$, so that $(1/[D + 1])(x + \frac{1}{4}x^4)$ is *exactly* equal to $(1 - D + D^2 - D^3 + D^4)(x + \frac{1}{4}x^2)$, all higher powers of D contributing nothing further to the result. The binomial expansion need only be taken as far as the term in D equal to the highest power of x.

Where the operator is the reciprocal of a quadratic expression, it is often helpful to use partial fractions (see Part 9.2) to simplify the working. To obtain

$$\frac{1}{D^2 - 3D + 2} x^2$$

the procedure is as follows

$$\begin{aligned}\frac{1}{D^2 - 3D + 2} x^2 &= \frac{1}{(D - 2)(D - 1)} x^2 \\ &= \left(\frac{1}{D - 2} - \frac{1}{D - 1}\right) x^2 \\ &= (-\tfrac{1}{2}(1 - \tfrac{1}{2}D)^{-1} + (1 - D)^{-1})\, x^2 \\ &= (-\tfrac{1}{2}(1 + \tfrac{1}{2}D + \tfrac{1}{4}D^2) + (1 + D + D^2))\, x^2 \\ &= (\tfrac{1}{2} + \tfrac{3}{4}D + \tfrac{7}{8}D^2) x^2 \\ &= \tfrac{1}{2}x^2 + \tfrac{3}{2}x + \tfrac{7}{4}\end{aligned}$$

In order to use the binomial expansion, each part of the operator must be expressed as $(1 + kD)^{-1}$ where k is a positive or negative constant.

Example 18.4(a)

Obtain $\dfrac{1}{D^2 - 6D + 8}(16x + 4)$.

$$\begin{aligned}\frac{1}{D^2 - 6D + 8}(16x + 4) &= \frac{1}{(D - 2)(D - 4)}(16x + 4) \\ &= \left(\frac{-\frac{1}{2}}{D - 2} + \frac{\frac{1}{2}}{D - 4}\right)(16x + 4) \\ &= (\tfrac{1}{4}(1 - \tfrac{1}{2}D)^{-1} - \tfrac{1}{8}(1 - \tfrac{1}{4}D)^{-1})\,(16x + 4) \\ &= \tfrac{1}{4}(1 + \tfrac{1}{2}D) - \tfrac{1}{8}(1 + \tfrac{1}{4}D)(16x + 4) \\ &= (\tfrac{1}{8} + \tfrac{3}{32}D)\,(16x + 4) \\ &= 2x + \tfrac{3}{2} + \tfrac{1}{2} \\ &= 2x + 2\end{aligned}$$

Example 18.4(b)

Solve the differential equation $(D^2 - 4D + 3)y = 9x^2 + 1$, given $y = 9$, $Dy = 8$ when $x = 0$.

CF: Auxiliary equation

$$m^2 - 4m + 3 = 0$$
$$(m - 1)(m - 3) = 0$$
$$m = 1 \text{ or } m = 3$$
$$u = Ae^x + Be^{3x}$$

PI:

$$v = \frac{1}{D^2 - 4D + 3}(9x^2 + 1)$$
$$= \frac{1}{(D - 1)(D - 3)}(9x^2 + 1)$$
$$= \left(\frac{-\frac{1}{2}}{D - 1} + \frac{\frac{1}{2}}{D - 3}\right)(9x^2 + 1)$$
$$= (\tfrac{1}{2}(1 - D)^{-1} - \tfrac{1}{6}(1 - \tfrac{1}{3}D)^{-1})(9x^2 + 1)$$
$$= \tfrac{1}{2}(1 + D + D^2) - \tfrac{1}{6}(1 + \tfrac{1}{3}D + \tfrac{1}{9}D^2)(9x^2 + 1)$$
$$= (\tfrac{1}{3} + \tfrac{4}{9}D + \tfrac{13}{27}D^2)(9x^2 + 1)$$
$$= 3x^2 + 8x + 9$$
$$y = Ae^x + Be^{3x} + 3x^2 + 8x + 9$$
$$Dy = Ae^x + 3Be^{3x} + 6x + 8$$

From the conditions, $A + B + 9 = 9$ and $A + 3B + 8 = 8$, making $A = B = 0$.

$$y = 3x^2 + 8x + 9$$

Example 18.4(c)
Show that if $g(D) = k_1D^n + k_2D^{n-1} + \cdots k_{n+1}$, the result of $1/g(D)$ operating on a constant a is equal to a/k_{n+1}.

$$\frac{1}{g(D)}a = \frac{a}{k_{n+1}}\left(\frac{1}{1 + \cdots \frac{k_2}{k_{n+1}}D^{n-1} + \frac{k_1}{k_{n+1}}D^n}\right) \cdot 1$$
$$= \frac{a}{k_{n+1}}\left(1 + \cdots \frac{k_2}{k_{n+1}}D^{n-1} + \frac{k_1}{k_{n+1}}D^n\right)^{-1} \cdot 1$$
$$= a/k_{n+1}$$

The final line occurs because the binomial expansion is only required as far as the first term, i.e. 1. The same result could also have been obtained by considering the operand a as ae^{px} with $p = 0$ and using the first rule in Table 18.1

18.5 Special Operator Techniques

Sometimes special techniques are needed when using D operators. For example the operation

$$\frac{1}{D - 1} e^x$$

breaks down when the 'normal' rule is applied. However, if the operand is considered as the product of e^x and 1 and the shift rule is applied, a result is obtained

$$\frac{1}{D - 1}(e^x \cdot 1) = e^x \frac{1}{D} 1 = xe^x$$

As usual, *any* integral will suffice in the final answer so that $e^x(x + 1)$ or $e^x(x - 5)$ would be equally acceptable as particular integrals but it is most convenient to choose the simplest form in which the constant of integration is taken as zero. Nevertheless this may still cause ambiguity in certain cases. In calculating

$$\frac{1}{D^2 + D}(x + 4)$$

the operator may be treated as either

$$\frac{1}{D} \cdot \frac{1}{D + 1} \quad \text{or} \quad \frac{1}{D + 1} \cdot \frac{1}{D}$$

This leads to apparently different results.

$$\begin{aligned} 1/D\ (1 + D)^{-1}(x + 4) &= 1/D\ (1 - D)(x + 4) \\ &= 1/D\ (x + 3) \\ &= \tfrac{1}{2}x^2 + 3x \end{aligned}$$

$$\begin{aligned} \frac{1}{D + 1}\frac{1}{D}(x + 4) &= (1 + D)^{-1}(\tfrac{1}{2}x^2 + 4x) \\ &= (1 - D + D^2)(\tfrac{1}{2}x^2 + 4x) \\ &= \tfrac{1}{2}x^2 + 3x - 3 \end{aligned}$$

In both cases the constant of integration has been taken as zero at the integration step but nevertheless an extra constant term appears in the second result. Both results are equally acceptable as particular integrals because the constant term is actually arbitrary but as the first method is slightly shorter, we may as well use it. The lesson to learn from this is that whenever an expression 1/D exists as one factor of an operator, this part of the operation should be the last to be handled.

A further example where a special technique is required is

$$\frac{1}{D^2 + 1} \cos x$$

where replacing D^2 by -1 will give a meaningless result. The method used this time is to work with the real part of e^{jx}—written Re (e^{jx})—since e^{jx} is the exponential form of the complex number $\cos x + j \sin x$ (see Part 2.7).

$$\frac{1}{D^2 + 1} \cos x = \text{Re}\left(\frac{1}{D^2 + 1} e^{jx}\right)$$

$$= \text{Re}\left(e^{jx} \frac{1}{(D + j)^2 + 1} \cdot 1\right) \qquad \text{(shift rule)}$$

$$= \text{Re}\left(e^{jx} \frac{1}{D^2 + 2jD} \cdot 1\right)$$

$$= \text{Re}\left(e^{jx} \frac{1}{D} \frac{1}{D + 2j} \cdot 1\right)$$

$$= \text{Re}\left(e^{jx} \frac{1}{D} \cdot \frac{1}{2j}\right)$$

(using the result of Example 18.4(c))

$$= \text{Re}\,(e^{jx} \cdot x/2j)$$

$$= \text{Re}\,(-\tfrac{1}{2}jx(\cos x + j \sin x))$$

$$= \tfrac{1}{2}x \sin x$$

This uses many of the results of Part 18. At all stages j is treated in the same way as any other constant.

Example 18.5(a)

Show that $\sin x = \text{Im}\,(e^{jx})$ where Im means 'the imaginary part of'. Hence deduce that

$$\frac{1}{D^2 + 1} (\sin x) = -\tfrac{1}{2}x \cos x$$

Since $e^{jx} = \cos x + j \sin x$, $\text{Im}\,(e^{jx}) = \sin x$

$$\frac{1}{D^2 + 1} (\sin x) = \text{Im}\left(\frac{1}{D^2 + 1} e^{jx}\right)$$

$$= \text{Im}\,(-\tfrac{1}{2}jx(\cos x + j \sin x))$$

$$= -\tfrac{1}{2}x \cos x$$

The penultimate line is taken from the text above.

Example 18.5(b)

Obtain $\dfrac{1}{D^2 + 4D}(e^{-4x} + 2)$.

$$\frac{1}{D^2 + 4D}(e^{-4x} + 2) = \frac{1}{D^2 + 4D} e^{-4x} + \frac{1}{D^2 + 4D} 2$$

$$= e^{-4x} \frac{1}{(D - 4)^2 + 4(D - 4)} \cdot 1 + \frac{1}{D} \frac{1}{D + 4} 2$$

$$= e^{-4x} \frac{1}{D^2 - 4D} 1 + \frac{1}{D} \frac{1}{2}$$

$$= e^{-4x} \frac{1}{D} \frac{1}{D - 4} 1 + \frac{1}{2} x$$

$$= e^{-4x} \frac{1}{D} \left(-\frac{1}{4}\right) + \frac{1}{2} x$$

$$= -\frac{1}{4} x e^{-4x} + \frac{1}{2} x$$

Example 18.5(c)
Find a general solution of the differential equation $(D + 1)^2 y = e^{-x}$.

CF: Auxiliary equation

$$(m + 1)^2 = 0$$
$$m = -1 \text{ (repeated)}$$
$$u = (Ax + B)e^{-x}$$

PI:

$$v = \frac{1}{(D + 1)^2} e^{-x}$$

$$= e^{-x} \frac{1}{D^2} 1$$

$$= \tfrac{1}{2} x^2 e^{-x}$$

$$y = u + v = (Ax + B)e^{-x} + \tfrac{1}{2} x^2 e^{-x}$$

$1/D^2$ means integrate twice w.r.t. x.

18.6 Evaluation of Integrals

The evaluation of an indefinite integral such as $\int e^{2x} \sin x \, dx$ is usually carried out by parts. In fact, this very integral was evaluated by the method in Example 9.3(c). The D operator offers an alternative method.

$$\begin{aligned}
\int e^{2x} \sin x \, dx &= \frac{1}{D}(e^{2x} \sin x) + C \\
&= e^{2x} \frac{1}{D+2} \sin x + C \\
&= e^{2x} \frac{D-2}{D^2-4} \sin x + C \\
&= -\frac{1}{5} e^{2x} (D-2) \sin x + C \\
&= -\frac{1}{5} e^{2x} (\cos x - 2 \sin x) + C \\
&= \frac{1}{5} e^{2x} (2 \sin x - \cos x) + C
\end{aligned}$$

As this is an indefinite integral rather than a particular integral, it is necessary to include a constant of integration. The method allows the result to be reached in one step instead of the two needed when using parts. With an integral of the type $\int x^3 e^{-x} \, dx$, which requires three sets of integrations using parts, the D operator method is even more time saving.

Example 18.6(a)
Integrate $\int x^4 e^{-x} \, dx$.

$$\begin{aligned}
\int x^4 e^{-x} \, dx &= \frac{1}{D}(x^4 e^{-x}) + C \\
&= e^{-x} \frac{1}{D-1} x^4 + C \\
&= -e^{-x}(1 \quad D)^{-1} x^4 + C \\
&= -e^{-x}(1 + D + D^2 + D^3 + D^4)x^4 + C \\
&= -e^{-x}(x^4 + 4x^3 + 12x^2 + 24x + 24) + C
\end{aligned}$$

The time and space saved in using this method rather than parts is considerable.

Example 18.6(b)
Evaluate $\int_0^\infty e^{-2x} \cos 3x \, dx$.

$$\int_0^\infty e^{-2x} \cos 3x \, dx = \left[\frac{1}{D}(e^{-2x} \cos 3x)\right]_0^\infty$$

$$= \left[e^{-2x} \frac{1}{D-2} \cos 3x\right]_0^\infty$$

$$= \left[e^{-2x} \frac{D+2}{D^2-4} \cos 3x\right]_0^\infty$$

$$= \left[-\frac{1}{13} e^{-2x}(D+2) \cos 3x\right]_0^\infty$$

$$= \left[-\frac{1}{13} e^{-2x}(2 \cos 3x - 3 \sin 3x)\right]_0^\infty$$

$$= \frac{2}{13}$$

Since both $\cos 3x$ and $\sin 3x$ lie in the range -1 to 1 whatever the value of x $e^{-2x} \cos 3x$ and $e^{-2x} \sin 3x$ both tend to zero as $x \to \infty$.

Unworked Examples 5

Parts 18.1 and 18.2

(1) Obtain $(D + 1)xe^{2x}$.
(2) Show that $D^4 y = y$ when y is equal to (i) $\sin x$ (ii) $\cos x$ (iii) e^x.
(3) Find $(D^2 - 7D + 4)y$ when $y = e^{2x} - 2x$.
(4) Show that $(D^2 + 4) \cos 2x = 0$.
(5) Use suitable rules to verify the results of Questions 1 and 4.
(6) Obtain $(D + 1)^4 e^x$.
(7) Prove that $(D^2 + 9)^8(\sin 3t - 5 \cos 3t) = 0$.
(8) Obtain $(D + 1)x^4 e^{-x}$.

Part 18.3

(9) Carry out the operation

$$\frac{1}{D+2} e^{2x} \sin 2x$$

(10) Obtain

$$\frac{1}{D^2 - 2D + 1} 24te^t$$

(11) Solve the differential equation $(D^2 + 2D - 3)y = 5e^{2x}$ subject to the conditions $y = 3$, $Dy = 0$ when $x = 0$.

(12) Find the general solution of the equation $(D^2 + 2D + 2)y = \sin x + 2 \cos x$.

(13) Obtain

$$\frac{1}{D + 2}(3e^x - \sin x)$$

(14) Find a particular integral of the differential equation $d^2y/dx^2 + 3\,dy/dx + y = 5e^x$.

(15) Solve the differential equation $(D^2 - D - 2)z = 10\,e^t \sin t$ given that $z = 2$, $Dz = -4$ when $t = 0$.

Part 18.4

(16) Obtain $(D^2 + 4D + 3)(x - 2)$ and

$$\frac{1}{D^2 + 4D + 3}(x - 2)$$

(17) Show that

$$\frac{1}{D^2 - 5D + 4}8 = 2$$

(18) Solve the differential equation $(D^2 - 5D + 4)y = 8x - 6$ given the conditions $y = -6$ and $Dy = 4$ when $x = 0$.

(19) Find

$$\frac{1}{D^2 - 2D - 2}(4 - 6t^2)$$

(20) Verify that

$$\frac{1}{D^2 + 5D + 2}(e^x + 4) = \tfrac{1}{8}e^x + 2$$

(21) Find

$$\frac{1}{D^2 - 4}(-8x^3)$$

Part 18.5

(22) Obtain

$$\frac{1}{D^2 + 4}(4 \cos 2t + 12 \sin 2t)$$

(23) Solve the following differential equations:
 (a) $(D^2 - 4D + 4)y = 2e^{2t}$ if $y = 1$, $Dy = 5$ when $t = 0$
 (b) $(D^2 + \omega^2)y = \sin \omega t$ (no conditions).

(24) Find a general solution of the differential equation $(D^2 - 1)z = 2e^t$.

(25) Show that

$$\frac{1}{D^2 + 4} \sin 2t = -\tfrac{1}{4}t \cos 2t$$

Part 18.6

(26) Prove that $\int e^{-2x} \sin 3x \, dx = -\frac{1}{13}e^{-2x}(2 \sin 3x + 3 \cos 3x) + C$.

(27) Integrate $\int 17e^{-x} \cos 4x \, dx$.

(28) Prove that $\int_0^\infty t^2 e^{-2t} \, dt = \frac{1}{4}$ and that $\int_0^\infty t^3 e^{-2t} \, dt = \frac{3}{16}$.

SECTION D
Numerical Methods

PART 19: ITERATIVE METHODS

19.1 The Idea of an Iterative Method

For the solution of some algebraic equations, such as simultaneous linear equations, there are standard methods. For others, such as quadratic equations, a formula may be used. If the equation is such that neither a standard method nor a formula is applicable, another possibility is a graphical solution, but this will be slow and of limited accuracy.

A completely different approach is to use an ITERATIVE method. In this an approximate root of the equation x_1 is used as a starting value. An ITERATIVE FORMULA (sometimes called an ITERATIVE EQUATION) is used to obtain a better approximation, x_2, of this root, from x_1. The same formula is used to obtain an even better aproximation x_3 from x_2 and this, in turn, is used to obtain a more accurate root x_4. The process is repeated until a root of the required degree of accuracy is obtained. The notation for x_n is the nth iterate.

The method can be illustrated using the quadratic equation $x^2 + 10x + 8 = 0$, the roots of which can be obtained by formula as -9.1231 and -0.8769, both values being correct to four places of decimals.

Iterative methods only yield one root at a time and so the starting value x_1 might be taken as -9 if the smaller value (i.e. the more negative value) is to be obtained.

There are many iterative formulae that can be used, some good, some poor and even some useless. It is possible to obtain one by rearranging the quadratic equation in the form $x = (-10x - 8)/x$ and expressing x_2 in terms of x_1 as the iterative formula $x_2 = (-10x_1 - 8)/x_1$. Since the same formula is used to obtain x_3 from x_2 it will be written $x_3 = (-10x_2 - 8)/x_2$ and subsequently the formula $x_4 = (-10x_3 - 8)/x_3$ is used. The general form of this iterative equation is expressed as $x_{n+1} = (-10x_n - 8)/x_n$. With the aid of a calculator, preferably one with a memory, it is easy to obtain x_2, x_3, x_4, etc. The results are quoted correct to four places of decimals

$$x_1 = -9.0000$$
$$x_2 = -9.1111$$
$$x_3 = -9.1220$$
$$x_4 = -9.1230$$
$$x_5 = -9.1231$$

Subsequent values are unchanged as far as the first four places of decimals are concerned. The value -9.1231 will therefore be one of the roots of the equation corrected to this degree of accuracy.

As this example illustrates there are three steps involved in solving equations by an iterative method:

(a) obtain a starting value
(b) carry out the iterative process repeatedly
(c) establish a terminating condition.

If for some value of n, $x_n = x_{n+1} = x_{n+2}$, etc. to a given degree of accuracy, the value of x_n can be taken as the root of the equation. Steps (a) and (b) will be considered in more detail in Parts 19.2 and 19.3.

19.2 Obtaining a Starting Value

In the example in Part 19.1, the starting value $x_1 = -9$ was chosen. In many cases an approximate solution to the equation is given or implied and this can be used as the starting value. In other cases an approximate root may be obtained graphically. With no clues whatsoever, a starting value can be obtained from a table of values of the type used for plotting a graph. With $f(x) = x^2 + 10x + 8$, this might be laid out as in Table 19.1. In particular let us concentrate on the results

$$f(0) = 8 \qquad\qquad f(-9) = -1$$
$$\text{and}$$
$$f(-1) = -1 \qquad\qquad f(-10) = 8$$

In drawing the graph of $y = f(x)$ the points corresponding to $x = 0$ and $x = 1$ would be plotted on either side of the x-axis and when these were

Table 19.1

Table of values for $f(x) = x^2 + 10x + 8$

x	0	−1	−2	−3	−4	−5	−6	−7	−8	−9	−10
$f(x)$	8	−1	−8	−13	−16	−17	−16	−13	−8	−1	8

joined, the graph would intersect the x-axis at a point lying somewhere between these values. Since each point of intersection of the graph with the x-axis is a root of the equation, we can deduce that a root of the equation $f(x) = 0$ lies between $x = 0$ and $x = 1$. As $f(0)$ is further from the x-axis than is $f(-1)$, it is likely, but not certain, that the root is nearer to -1 than to zero. This suggests -1 as a starting value.

In the same way, a study of $f(-9)$ and $f(-10)$ suggests that $x_1 = -9$ is an approximation to another root. Basically we are looking for an integer a, such that $f(a)$ and $f(a+1)$ have opposite signs. Then $x_1 = a$ or $x_1 = a + 1$ as appropriate.

It should be noticed, however, that the correct root may be obtained even if the starting value is not particularly accurate. In Part 19.1, if a starting value of -8 were taken for x_1, the first iteration makes $x_2 = -9$ and subsequent steps would be the same as those illustrated except that the suffix attached to the iterates would be increased by 1. The effect of choosing a poor starting value is merely to extend the process.

Example 19.2(a)
Show that choosing a starting value $x_1 = -6$ but using the iterative formula of Part 19.1 leads to the same eventual root.

If $x_{n+1} = (-10x_n - 8)/x_n$ the iterative process leads to

$$x_1 = -6.0000$$
$$x_2 = -8.6667$$
$$x_3 = -9.0769$$
$$x_4 = -9.1186$$
$$x_5 = -9.1227$$
$$x_6 = -9.1231$$

All results have been corrected to four places of decimals. It should be checked that even with a very negative starting value such as $x_1 = -50$, the iterative formula will eventually lead to the root -9.1231.

Example 19.2(b)
Show that if $x^3 + x - 9 = 0$ then $x \approx 2$. Use the iterative formula $x_{n+1} = (9 - x_n)^{1/3}$ to find the root correct to three places of decimals.

When $x = 2$, $x^3 + x = 10$. Hence if $x \approx 2$ then $x^3 + x - 9 \approx 0$. We take $x_1 = 2$ as a starting value. Using this together with the iterative formula gives

$$x_1 = 2.000$$
$$x_2 = 1.913$$
$$x_3 = 1.921$$
$$x_4 = 1.920$$

All subsequent iterations lead to the same value corrected to three places of decimals. This must be the required root correct to the level of accuracy specified.

19.3 Convergence and Divergence of the Process

The iterative formula $x_{n+1} = (-10_n - 8)/x_n$ used in Part 19.1 to solve the quadratic equation $x^2 + 10x + 8 = 0$, was based on a rearrangement of this equation. However there are quite a few ways of rearranging the equation. Consider an alternative iterative formula $x_{n+1} = -\sqrt{(-10x_n - 8)}$. Using the same starting value the iterates are

$$
\begin{aligned}
x_1 &= -9.0000\\
x_2 &= -9.0554\\
x_3 &= -9.0859\\
x_4 &= -9.1027\\
x_5 &= -9.1119\\
x_6 &= -9.1170\\
x_7 &= -9.1198\\
x_8 &= -9.1213\\
x_9 &= -9.1221\\
x_{10} &= -9.1225\\
x_{11} &= -9.1228\\
x_{12} &= -9.1229\\
x_{13} &= -9.1230\\
x_{14} &= -9.1231
\end{aligned}
$$

and although the same root of the equation is eventually found the method is obviously not so efficient.

But worse is to come! If we had used the iterative formula $x_{n+1} = (-x_n{}^2 - 8)/10$, the first few iterates would have been

$$
\begin{aligned}
x_1 &= -9.0000\\
x_2 &= -8.9000\\
x_3 &= -8.7210\\
x_4 &= -8.4056\\
x_5 &= -7.8654, \text{ etc.}
\end{aligned}
$$

Instead of approaching the expected root -9.1231, each iteration leads to a value further from it.

An iterative formula whose use causes iterates to approach a particular root is said to be CONVERGENT to that root. Convergence may be rapid or slow. When the formula causes iterates to move away from a particular root, then it is said to be DIVERGENT from that root. Essentially the concepts of convergence and divergence are the same as the ideas met in Part 12.

Even a formula convergent to one root is not necessarily convergent to another. The iterative equation $x_{n+1} = (-10x_n - 8)/x_n$ shown in Part 19.1 to converge to the root -9.1231 of the quadratic equation $x^2 + 10x + 8 = 0$, diverges from the root lying close to -1. Even using mental arithmetic, it is possible to find the first few iterates

$$x_1 = -1.0000$$

$$x_3 = -2.0000$$

$$x_3 = -6.0000$$

$$x_4 = -6.6667, \text{ etc.}$$

Example 19.3(a)

Obtain the first few iterates from the formula $x_{n+1} = (-x_n^2 - 8)/10$, $x_1 = -1$. Show that they appear to be converging to the less negative root of the quadratic equation $x^2 + 10x + 8 = 0$.

The values are

$$x_1 = -1.0000$$

$$x_2 = -0.9000$$

$$x_3 = -0.8810$$

$$x_4 = -0.8789$$

$$x_5 = -0.8772$$

Example 19.3(b)

Show that if an equation has an exact root a, the corresponding iterative form $x_{n+1} = f(x_n)$ converges if $f'(a)$ is numerically less than 1.

Suppose that the error in the nth iterate x_n is e_n. Then $x_n - e_n = a$ and similarly $x_{n+1} - e_{n+1} = a$. Hence

$$\begin{aligned} a + e_{n+1} &= f(a + e_n) \\ &\approx f(a) + e_n f'(a) \end{aligned}$$

But if a is an exact solution then $a = f(a)$ so that

$$e_{n+1} = e_n f'(a)$$

Therefore $e_{n+1} < e_n$ if $|f'(a)| < 1$ and this is the condition for convergence.

The expansion of $f(a + e_n)$ uses Taylor's theorem (see Part 10.3) and it is assumed that the error terms in e_n^2 and higher powers can be neglected.

19.4 Newton Raphson Formula

Since choosing an iterative formula by chance can lead to either a convergent or a divergent series of approximations (and even if convergent the rate of convergence may still be slow), a better method is required. It is true that it is possible to analyse the method to determine its suitability but it would clearly be more satisfactory if an iterative formula existed which, in the majority of cases, led to rapid convergence. Let us consider such a formula.

If the exact solution of the equation $f(x) = 0$ is $x = a$ and if x_n, the nth iterate, is in error by an amount e_n, then $x_n - e_n = a$ and so $f(x_n - e_n) = 0$. Using Taylor's theorem and ignoring terms in e_n^2 and higher powers we have $f(x_n - e_n) = f(x_n) - e_n f'(x_n)$, so that $f(x_n) - e_n f'(x_n) = 0$ and hence that $e_n = f(x_n)/f'(x_n)$. An improvement on x_n will therefore be

$$\begin{aligned} x_{n+1} &= x_n - e_n \\ &= x_n - f(x_n)/f'(x_n) \end{aligned}$$

This is the NEWTON RAPHSON iterative formula, which, with certain exceptions, leads to rapid convergence of each root of the equation $f(x) = 0$.

This formula can also be verified graphically. In Figure D1, if P is the point on the curve $y = f(x)$ where $x = x_n$ and PR is a tangent to the curve, its gradient is $f'(x_n)$. This means that $PQ/QR = f'(x_n)$ but as $PQ = f(x_n)$, $QR = f(x_n)/f'(x_n)$ and the x co-ordinate of R is $x_n - f(x_n)/f'(x_n)$.

The exact solution of the equation $f(x) = 0$ is, of course, given by the x co-ordinate of S. As R is nearer to S than is Q, an improved approximation to x_n will be $x_n - f(x_n)/f'(x_n)$. Hence $x_{n+1} = x_n - f(x_n)/f'(x_n)$.

Example 19.4(a)
Use the Newton Raphson formula to obtain the smaller root of the equation $x^2 + 10x + 8 = 0$ taking $x_1 = -9$.

Taking $f(x) = x^2 + 10x + 8$, $f'(x) = 2x + 10$. The required formula is

$$\begin{aligned} x_{n+1} &= x_n - (x_n^2 + 10x_n + 8)/(2x_n + 10) \\ &= (x_n^2 - 8)/(2x_n + 10) \end{aligned}$$

and the iterates are

$$x_1 = -9.0000$$
$$x_2 = -9.1250$$
$$x_3 = -9.1231$$

As a rough rule of thumb, the number of accurate decimal figures at each stage doubles with every iteration.

Either of the forms of the iterative equation may be used. The equation $x_{n+1} = (x_n^2 - 8)/(2x_n + 10)$ is simpler but the form containing $(x_n^2 + 10x_n + 8)$ allows continual monitoring of progress because this expression should approach zero with each iteration. If preferred, since $2x_n + 10$ is approximately equal to -8 whatever the value of n, the formula can be taken as $x_{n+1} = x_n + (x_n^2 + 10x_n + 8)/8$. This slows the convergence somewhat but simplifies each step.

Example 19.4(b)
Show that $x = 2$ is an approximate solution of the equation $e^x = 4x$. Find this root correct to four places of decimals.

Let $f(x) = e^x - 4x = 0$, $f'(x) = e^x - 4$. The iterative equation is $x_{n+1} = x_n - (e^{x_n} - 4x_n)/(e^x n - 4)$.

$$x_1 = 2.0000$$
$$x_2 = 2.1803$$
$$x_3 = 2.1540$$
$$x_4 = 2.1533$$

Further iterations lead to the same value corrected to four places of decimals. As this example illustrates the Newton Raphson formula can be used to solve non-polynomial algebraic equations.

Example 19.4(c)
Use the Newton Raphson formula to derive an iterative formula for $\sqrt{N}$. Use it to obtain $\sqrt{37}$ correct to six places of decimals.

We require N as the root of an equation. This equation can be expressed in various ways but the simplest is $f(x) \equiv x^2 - N = 0$, making $f'(x) = 2x$.

$$x_{n+1} = (x_n^2 - N)/2x_n$$
$$= \tfrac{1}{2}(x_n + N/x_n)$$

Taking N as 37 and x_1 as 6, the iterates are

$$x_1 = 6.000\,000$$
$$x_2 = 6.083\,333$$
$$x_3 = 6.082\,763$$

This is the standard iterative formula for finding square roots but is not the only one.

19.5 Iterative Solution of Simultaneous Linear Equations

Iterative methods can also be used to solve simultaneous linear equations. To solve the equations

$$8x - 2y + z = 12 \quad (1)$$
$$x + 7y + 4z = 4 \quad (2)$$
$$x - 3y + 9z = 4 \quad (3)$$

which have the exact solution $x = 1\frac{1}{2}$, $y = \frac{1}{6}$ and $z = \frac{1}{3}$, we rearrange the equations making x the subject of Equation (1), y the subject of Equation (2)

Table 19.2

n	x_n	y_n	z_n
1	0	0	0
2	1.500	0.357	0.397
3	1.540	0.125	0.315
4	1.492	0.178	0.338
5	1.502	0.164	0.332
6	1.500	0.167	0.333

and z the subject of Equation (3). These can then be expressed iteratively as

$$x_{n+1} = (12 + 2y_n - z_n)/8$$
$$y_{n+1} = (4 - x_{n+1} - 4z_n)/7$$
$$z_{n+1} = (4 - x_{n+1} + 3y_{n+1})/9$$

In the absence of any starting values it is possible to use $x_1 = y_1 = z_1 = 0$. The results are obtained in the usual way and are shown in Table 19.2, corrected to three places of decimals. This is the GAUSS SEIDEL method.

If the iterative equations are arranged so that x is made the subject of Equation (2), y the subject of Equation (3) and z the subject of Equation (1), we would have

$$x_{n+1} = (4 - 7y_n - 4z_n)$$
$$y_{n+1} = -(4 - x_{n+1} - 9z_n)/3$$
$$z_{n+1} = (12 - 8x_{n+1} + 2y_{n+1})$$

Starting with $x_1 = y_1 = z_1 = 0$ and using these equations, the values obtained are shown in Table 19.3. It is obvious that the equations are wildly divergent.

The reason for this difference in the two sets of iterates lies with the coefficients of the variables made the subject of the equations in each case. In the original set of equations, these coefficients were larger than all the others totalled together and are called DOMINANT coefficients. It is important that the variable prefixed with a dominant coefficient is made the subject of the equation from which it is derived. It will be seen in Equation (2) that the

Table 19.3

n	x_n	y_n	z_n
1	0	0	0
2	4	0	−20
3	84	33.333	−593.333
4	2144	−1066.667	−19 273.333

coefficient of y is dominant but not so pronouncedly so as the other two coefficients. This is reflected in Table 19.2 where y_n is seen to be less rapidly convergent than x_n or z_n.

Where sets of equations do not possess dominant coefficients they must be suitably rearranged before the Gauss Seidel method is used. Given the equations

$$2x + 3y + 3z = 1 \quad (4)$$

$$5x + 6y - 5z = 4 \quad (5)$$

$$3x - 7y - 8z = 2 \quad (6)$$

which certainly do not possess dominant coefficients, the set

$$9x + 2y + z = 5 \quad (3 \times (4) + (6))$$

$$2x + 13y + 3z = 2 \quad ((5) - (6))$$

$$x + 3y + 14z = -1 \quad (3 \times (4) - (5))$$

could be obtained.

Example 19.5(a)

Rearrange the equations

$$6x + 5y + 5z = 2$$

$$5x - 6y + 4z = 3$$

$$5x + 4y + 7z = 5$$

suitably and hence solve them using the Gauss Seidel method, giving values correct to two places of decimals.

It is reasonably straightforward to obtain equations containing a dominant coefficient of y is dominant but not so pronouncedly as the other two steps. If

$$6x + 5y + 5z = 2 \quad (1)$$

$$5x - 6y + 4z = 3 \quad (2)$$

$$5x + 4y + 7z = 5 \quad (3)$$

(1) − (3) gives

$$x + y - 2z = -3 \quad (4)$$

(1) + (2) + 4 × (4) gives

$$15x + 3y + z = -7 \quad (5)$$

(1) − (2) gives

$$x + 11y + z = -1 \quad (6)$$

4 × (3) − 3 × (1) gives

$$2x + y + 13z = 14 \quad (7)$$

Table 19.4

n	x_n	y_n	z_n
1	0	0	0
2	−0.47	−0.05	1.15
3	−0.52	−0.15	1.17
4	−0.51	−0.15	1.17

Equations (5), (6) and (7) are a suitable set for solution using the Gauss Seidel method. The associated iterative equations are

$$x_{n+1} = (-7 - 3y_n - z_n)/15$$
$$y_{n+1} = (-1 - x_{n+1} - z_n)/11$$
$$z_{n+1} = (14 - 2x_{n+1} - y_{n+1})/13$$

and the iterates are shown in Table 19.4.

To check the solution, we can substitute the values in the original equations

$$6x + 5y + 5z = 2.04$$
$$5x - 6y + 4z = 3.03$$
$$5x + 4y + 7z = 5.04$$

These values check out if allowance is made for rounding errors.

19.6 Gaussian Elimination

The Gauss Seidel method is particularly suitable for use with a computer or a programmable calculator but it can be a little tedious without these aids. If this is the case the standard methods described in Part 6.6 are more suitable and, indeed, Example 19.5(b) is easily solved using row equivalence techniques, giving the exact solutions $x = -89/173$, $y = -26/173$ and $z = 202/173$.

When the coefficients of the set of equations are not integral, however, although the method is still applicable, it is wise to lay the calculation out so that a continuous running check can be made. This is the GAUSSIAN - ELIMINATION METHOD. It is not an iterative method.

Consider the set of equations

$$4.5x - 0.7y + 1.2z = 8.59 \quad (1)$$
$$-0.8x + 5.2y - 1.1z = -16.39 \quad (2)$$
$$1.2x + 1.3y + 5.8z = -17.56 \quad (3)$$

These are laid out as in Table 19.5. The column headings are line number (for reference), operation (explaining how each new line is formed), x, y, z and k

Table 19.5

Line	Operation	x	y	z	k	Check
(1)		4.5	−0.7	1.2	8.59	13.59
(2)		−0.8	5.2	−1.1	−16.39	−13.09
(3)		1.2	1.3	5.8	−17.56	−9.26
(4)	(3)/5.8	0.207	0.224	1.000	−3.027	−1.596
(5)	(2) + 1.1 × (4)	−0.572	5.446		−19.720	−14.846
(6)	(1) − 1.2 × (4)	4.252	−0.969		12.222	15.505
(7)	(5)/5.466	−0.105	1.000		−3.621	−2.726
(8)	(6) + 0.969 × (7)	4.150			8.713	12.864

(containing the coefficients of the variables and the right side constant) and, most important, a check figure. This check figure consists of the sum of the values in the preceding four columns. As each line of figures is obtained from the previous lines, as described in the operation column, the same arithmetic process should affect the figures in the x, y, z, k and check columns. Apart from a small rounding error, the check figure should remain the sum of all the other values on each line. If a discrepancy arises, the line must be re-checked and the error corrected.

Lines (1), (2) and (3) are the original set of equations with the check figure inserted. Line (4) consists of one of the previous three lines divided by a constant figure known as a PIVOT. This should be the numerically largest of all coefficients in order to reduce rounding errors. Using the pivot, the associated variables are eliminated in lines (5) and (6) and the process is repeated for the remaining two variables in lines (7) and (8).

When the table is complete, Equations (4), (7) and (8) form a triangular set from which the unknowns can be evaluated. This evaluation process is called BACK SUBSTITUTION and is carried from Table 19.5 as follows

From (8): $4.150x = 8.713$

$$x = 2.10$$

From (7): $-0.105 \times 2.10 + y = -3.621$

$$y = -3.40$$

From (4): $0.207 \times 2.10 + 0.224 \times (-3.40) + z = -3.027$

$$z = -2.70$$

The results have been corrected to two places of decimals.

It is important to choose the correct pivot at lines 4 and 7 because in this method, unlike iterative methods where the solution is being continually refined, rounding errors tend to build up and, in less well-conditioned equations could seriously affect the results. Normally, however, if the figures in the table are corrected to a given degree of accuracy, the final results can be taken as being accurate to one place of decimals below with a fair degree of confidence.

It cannot be over-stressed that the figures in the check column should not be omitted. It is time-consuming to insert them and, as they do not contribute directly to the final result, it is tempting to omit them. On balance, however, it will be found that the time 'wasted' in calculating them is likely to be less than the time lost in correcting a table subsequently found to contain invalid figures.

When the results have been obtained, a check should be carried out on at least one of the original equations.

Example 19.6(a)

In a pulley system there are three tensions in the strings T_1, T_2 and T_3. Applying the laws of statics, the results can be reduced to the simultaneous equations

$$7.17T_1 - 2.53T_2 - 1.29T_3 = 2.62$$
$$1.85T_1 + 8.78T_2 + 1.19T_3 = -8.74$$
$$3.55T_1 - 4.16T_2 + 5.90T_3 = -2.59$$

Obtain values of T_1, T_2 and T_3 each correct to two places of decimals.

Using the Gaussian elimination method, the values are shown in Table 19.6. The back substitution gives

From (8): $7.011T_3 = -6.787$

$T_3 = -0.97$

Table 19.6

Line	Operation	T_1	T_2	T_3	k	Check
(1)		7.17	−2.53	−1.29	2.62	5.97
(2)		1.85	8.78	1.19	−8.74	3.08
(3)		3.55	−4.16	5.90	−2.59	2.70
(4)	(2)/8.78	0.211	1.000	0.136	−0.995	0.351
(5)	(1) + 2.53 × (4)	7.704		−0.946	0.103	6.858
(6)	(3) + 4.16 × (4)	4.428		6.466	−6.729	4.160
(7)	(5)/7.704	1.000		−0.123	0.013	0.890
(8)	(6) − 4.428 × (7)			7.011	−6.787	0.129

From (7): $T_1 - 0.123 \times (-0.97) = 0.013$
$T_1 = -0.11$

From (4): $T_2 + 0.211 \times (-0.11) + 0.136 \times (-0.97) = -0.995$
$T_2 = -0.84$

The check is carried out in the usual way.

Unworked Examples 1

Parts 19.1–19.3

(1) What is meant by an iterative method?
(2) Obtain three approximate roots of the cubic equation $x^3 - 12x - 15 = 0$.
(3) Show that -0.7 is an approximate root of the equation $x^3 = -(x + 1)$. By considering the first few iterates show that the iterative equation
 (a) the iterative equation $x_{n+1} = -(x_n + 1)/x_n^2$ diverges from this root,
 (b) the iterative equation $x_{n+1} = -(x_n + 1)^{1/3}$ slowly converges towards this root.
(4) Show that the equation $2x^3 - 2x^2 + 11 = 0$ has a root lying between -1 and -2.
(5) Prove that the equation $4x^5 + 8x^4 - 1 = 0$ has a root in the region of -2. Show that the equation could be rearranged in the form $x = 1/4x^4 - 2$ and obtain an iterative formula based on this. Use it to find the root correct to five places of decimals.

Part 19.4

(6) Obtain the root of the equation $x^3 + 5x - 7 = 0$, close to 1 using the Newton Raphson method. Give your answer correct to four places of decimals.

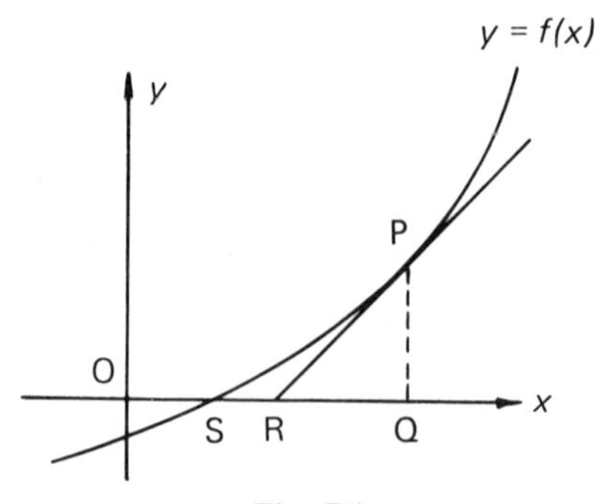

Fig. D1

(7) Show that $x^3 - x^2 - 5x - 2 = 0$ has a root lying close to -1. What happens if the Newton Raphson method is used in an attempt to refine the root?

(8) Show that the Newton Raphson formula can still be obtained graphically when the tangent to $y = f(x)$ lies above the curve rather than below it as shown in Figure D1.

(9) In Figure D2, if the arc of the circle PQR is of length 12 cm and the chord length PR is 10 cm, show that the angle x, measured in radians, is given by the equation $6 \sin x = 5x$. Hence find x in degrees and minutes.

(10) Solve the equation $\ln x + x = 2$ giving the root lying in the region of 1.5 correct to five places of decimals.

(11) Use the Newton Raphson formula to derive an iterative equation for the reciprocal of a number N which does not involve division. Hence obtain the reciprocal of 48 correct to five places of decimals. Take $f(x) \equiv N - 1/x = 0$.

(12) Prove that the pth root of a number N can be obtained using the iterative formula $x_{n+1} = x_n(1 - 1/p) + N/px_n^{p-1}$. Use it to obtain the 10th root of 100 correct to three places of decimals.

(13) Find the root of $e^x = 2 - x$ correct to three places of decimals.

Parts 19.5 and 19.6

(14) Solve the equations

$$13x + 4y + 2z = 5$$
$$5x + 16y + 2z = 9$$
$$2x + 5y + 12z = 11$$

by an iterative method giving all values correct to three places of decimals.

(15) Rearrange the equations

$$9x + 9y + 11z = 19 \quad (1)$$
$$8x + 10y + 3z = 14 \quad (2)$$
$$5x + y - 2z = 5 \quad (3)$$

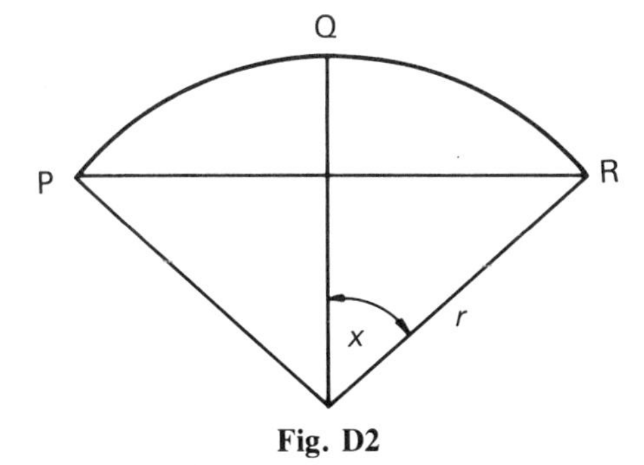

Fig. D2

so that they form a convergent set when the Gauss Seidel method is used. Find the value of x, y and z correct to two places of decimals.

(16) Solve the equations

$$\begin{aligned} 97l + 66m - 55n &= 7.94 \\ 45l - 33m - 23n &= 8.35 \\ 37l - 67m + 95n &= 2.02 \end{aligned}$$

by any method giving solutions correct to three places of decimals.

(17) Use Gaussian elimination to obtain a, b and c if

$$\begin{aligned} 7.2a + 2.1b - 0.8c &= 4.34 \\ 1.7a - 9.3b + 2.1c &= 2.27 \\ 1.5a - 2.4b + 6.8c &= 10.09 \end{aligned}$$

(18) Solve the following equations using the Gauss Seidel method.

$$\begin{aligned} 36x + 7y - 9z &= 48 \\ 6x - 28y - 9z &= 178 \\ 3x - 7y - 45z &= -215 \end{aligned}$$

Give answers correct to three places of decimals.

(19) Evaluate x, y and z correct to two places of decimals if

$$\begin{aligned} 3x + 5y + 9z &= -3.79 \\ x - 2y - 8z &= 0.90 \\ 6x + 2y + z &= 4.49 \end{aligned}$$

PART 20: FINITE DIFFERENCE TABLES AND THEIR USE

20.1 Finite Difference Tables

If $f(x) = x^2 + 4x + 1$, it is not too difficult to evaluate $f(0)$, $f(1)$, $f(2)$, $f(3)$ and $f(4)$ without any calculating aid. The values are shown in Table 20.1, laid out in the form of a table of values as used to plot a graph. It is convenient to use the notation $f(x)$ for $x = 0(1)4$ for this, where 0(1)4 means 'from 0 to 4 in steps of 1'. The value 1 in this example is called the STEP LENGTH.

If the values of the polynomial $a_0 + a_1x + a_2x^2 + \cdots$ are to be found using a calculator, it will be best to obtain the expression in NESTED FORM. This consists of a series of linked bracketed expressions in the form $a_0 + x(a_1 + x(a_2 + x(\ldots)))$. In the example used above $x^2 + 4x + 1$ would be nested as $1 + x(4 + x)$. By setting the value of x into the calculator and working from the innermost bracket outwards, the value of the expression can be obtained directly.

Table 20.1

Values of $f(x) = x^2 + 4x + 1$					
x	0	1	2	3	4
$f(x)$	1	6	13	22	33

If successive values of $f(x)$ in Table 20.1 are studied, a pattern in the values should be noticed. The difference between $f(0)$ and $f(1)$ is 5, between $f(1)$ and $f(2)$ is 7, between $f(2)$ and $f(3)$ is 9 and between $f(3)$ and $f(4)$ is 11.

In order to show the pattern more clearly, it is convenient to set up values of x and $f(x)$ in columns and to insert the differences in a position half way between the horizontal lines to which they apply. The process can then be extended taking the difference between differences called SECOND DIFFERENCES and, given enough values, third, fourth and higher differences. This is a FINITE DIFFERENCE TABLE and is laid out for $x^2 + 4x + 1$ in Table 20.2. Values in the even difference columns will be aligned with the values of x and $f(x)$ while those in odd difference columns will be shifted half a line. In this case as second differences are constant third and subsequent differences are zero.

It is found that when the difference table of a polynomial with a highest power n such that the coefficient of x^n is a, is laid out

(i) the nth difference values are constant (and subsequent differences zero)

(ii) the value of the constant $n!ah^n$ where h is the step length.

In Table 20.2, $n = 2$, $h = 1$ and $a = 1$ making the constant difference 2 as shown.

Example 20.1(a)
Express $f(x) \equiv 3x^3 - 7x^2 + 2x^2 + 2x + 5$ in nested form. Use the result to find $f(1.7)$.

Table 20.2 Finite Difference Table for $f(x) = x^2 + 4x + 1$

x	$f(x)$	1st. diff.	2nd. diff.	3rd. diff.
0	1			
		5		
1	6		2	
		7		0
2	13		2	
		9		0
3	22		2	
		11		
4	33			

$f(x) = 5 + x(2 + x(-7 + x(3)))$. Working outwards from the innermost brackets,

$$f(1.7) = 2.909$$

Example 20.1(b)
Obtain $f(x) \equiv x^3 + 4x^2 + 3x - 1$ for $x = 0(0.1)0.3$. Use differences to extend the table as far as $x = 1$.

$$f(x) \equiv x^3 + 4x^2 + 3x - 1 = -1 + x(3 + x(4 + x))$$

Hence

$$f(0) = -1.000$$
$$f(0.1) = -0.659$$
$$f(0.2) = -0.232$$
$$f(0.3) = 0.287$$

These four values can be put in a finite difference table and are sufficient to allow the constant difference to be found and checked. Using this constant

Table 20.3 $f(x) = x^3 + 4x^2 + 3x - 1$

x	$f(x)$	δ	δ^2	δ^3
0	−1.000			
		341		
0.1	−0.659		86	
		427		6
0.2	−0.232		92	
		519		6
0.3	0.287		98	
		617		6
0.4	0.904		104	
		721		6
0.5	1.625		110	
		831		6
0.6	2.456		116	
		947		6
0.7	3.403		122	
		1069		6
0.8	4.472		128	
		1197		6
0.9	5.669		134	
		1331		
1.0	7.000			

difference, further differences and therefore further values of $f(x)$, can be built up until the table is complete, the value of $f(1)$ giving a final check. Here $n!ah^n = 3! \times 1 \times (0.1)^3 = 0.006$.

It is conventional to omit non-significant zeros and the decimal point in the differences since, in an example like this, all values of $f(x)$ and therefore all differences involve three decimal figures. The complete finite difference table is shown as Table 20.3, the dotted line indicating the point from which values were calculated by differences.

The notation δ, δ^2, δ^3, etc. for first, second, third and higher differences will be used from this point on.

20.2 Rounded Functions

The comments on finite difference tables made in Part 20.1 apply only if values of $f(x)$ are exact. For polynomial functions this will always be so provided that enough decimal places are included. With other functions, however, this is not the case. Table 20.4 shows a finite difference table for $\sin x°$, $x = 0(5)40$, with values corrected to three places of decimals.

The first three columns of differences are 'normal', increasing or decreasing in a regular pattern. The fifth, sixth and subsequent differences alternate

Table 20.4 Finite Difference Table for $f(x) = \sin x°$

x	$f(x)$	δ	δ^2	δ^3	δ^4	δ^5	δ^6
0	0.000 00						
		8716					
5	0.087 16		−67				
		8649		−65			
10	0.173 65		−132		0		
		8517		−65		2	
15	0.258 82		−197		2		−3
		8320		−63		−1	
20	0.342 02		−260		1		4
		8060		−62		3	
25	0.422 62		−322		4		−6
		7738		−58		−1	
30	0.500 00		−380		1		
		7358		−57			
35	0.573 58		−437				
		6921					
40	0.642 79						
		Regular Columns			Transitional Column	Irregular Columns	

in sign while the fourth differences are irregular in size but not in sign. The first set of columns are termed REGULAR, the next TRANSITIONAL and subsequent ones IRREGULAR. This pattern occurs in all finite difference tables where values are rounded but there may be more or fewer regular columns depending on the accuracy of $f(x)$.

The analogy between finite differences and differential calculus, which deals with infinitesimal differences is clear from the examples chosen so far, because when $f(x) = ax^n + \ldots$, $f^n(x)$ is constant and higher derivatives are zero and a similar relationship occurs with differences. Again for values of x in the first quadrant when $f(x) = \sin x$, $f(x) > 0$, $f'(x) < 0$ and $f''(x) > 0$. Similar results for differences can be seen in the regular part of Table 20.4.

Example 20.2(a)
Produce a finite difference table for $f(x) = e^x$, $x = 0(0.1)1$ giving values correct to four places of decimals. Which is the first irregular column?

The results are shown in Table 20.5. Clearly δ^4 is the first irregular column.

Table 20.5 Finite Difference Table for $f(x) = e^x$

x	$f(x)$	δ	δ^2	δ^3	δ^4	δ^5	δ^6
0	1.0000						
		1052					
0.1	1.1052		110				
		1162		13			
0.2	1.2214		123		−2		
		1285		11		7	
0.3	1.3499		134		5		−13
		1419		16		−6	
0.4	1.4918		150		−1		10
		1569		15		4	
0.5	1.6487		165		3		−8
		1734		18		−4	
0.6	1.8221		183		−1		12
		1917		17		8	
0.7	2.0138		200		7		−18
		2117		24		−10	
0.8	2.2255		224		−3		
		2341		21			
0.9	2.4596		245				
		2587					
1	2.7183						

Table 20.6

Error in $f(x)$	Error in δ	Error in δ^2	Error in δ^3	Error in δ^4
				e
			e	
		e		$-4e$
	e		$-3e$	
e		$-2e$		$6e$
	$-e$		$3e$	
		e		$-4e$
			$-3e$	
				e

20.3 Error Correction

If a polynomial function is calculated exactly but contains a single error e in one value, this will affect the two values on either side of it in the first difference column. These errors will, in turn, affect three values in the next difference column, four in the next and so on. The pattern of errors is shown in Table 20.6 but as this pattern is superimposed on the values that will occur in these columns anyway, it will be difficult to decipher until the first zero difference column is reached. If $f(x)$ is a quadratic polynomial with exactly calculated values given in Table 20.7 but there is reason to suspect that one of the values is incorrect, both this value and the size of the error can be determined using Table 20.8.

The values in δ^3 should all be zero. By comparison with the error pattern $e = 9$ (i.e. 0.009) and occurs in $f(4)$ which should therefore be 34.456. (This can be checked directly since $f(x) = 1.4x^2 + 3.009x + 0.02$.) In practice, however, $f(x)$ is unlikely to be a polynomial of known degree and, if it is rounded, all that can be done is to compare the error pattern with the values in the first irregular column. This will only give an indication of the error size.

Table 20.9 contains differences of a rounded function $f(x)$ containing one error. Obviously the first difference column is regular but the unusual pattern of the second difference column makes it difficult to be sure of its nature. Bearing in mind that one error in $f(x)$ gives rise to three incorrect figures in this column, the table suggests that second differences are regular and that the error occurs in $f(6)$. Taking δ^4 as the first irregular column, this can be compared with the error pattern. Unfortunately sporadic rounding errors

Table 20.7

x	0	1	2	3	4	5	6
$f(x)$	0.020	4.429	11.638	21.647	34.465	50.065	68.474

Table 20.8

x	$f(x)$	δ	δ^2	δ^3	Error pattern
0	0.020				
		4409			
1	4.429		2800		
		7209		0	
2	11.638		2800		
		10 009		9	e
3	21.647		2809		
		12 818		−27	$-3e$
4	34.465		2782		
		15 600		27	$3e$
5	50.065		2809		
		18 409			
6	68.474				

Table 20.9

x	$f(x)$	δ	δ^2	δ^3	δ^4	Error pattern
2	1.0488					
		466				
3	1.0954		−18			
		448		0		
4	1.1402		−18		5	e
		430		5		
5	1.1832		−13		−9	$-4e$
		417		−4		
6	1.2249		−17		10	$6e$
		400		6		
7	1.2649		−11		−6	$-4e$
		389		0		
8	1.3038		−11		−1	e
		378		−1		
9	1.3416		−10			
		368				
10	1.3784					

Table 20.10

x	0	1	2	3	4	5	6	7	8
$f(x)$	2.9151	3.0209	3.1215	3.2174	3.3091	3.3970	3.4807	3.5631	3.6418

Table 20.11

x	$f(x)$	δ	δ^2	δ^3	δ^4	Error pattern
0	2.9151					
		1058				
1	3.0209		−52			
		1006		5		
2	3.1215		−47		0	
		959		5		
3	3.2174		−42		−1	
		917		4		
4	3.3091		−38		−8	e
		879		−4		
5	3.3970		−42		33	$-4e$
		837		29		
6	3.4807		−13		−53	$6e$
		824		−24		
7	3.5631		−37			
		787				
8	3.6418					

interfere and it is difficult to be certain of a value for e. However it looks as if the most likely value is 2 and that $f(6) = 1.2247$.

Example 20.3(a)
Find and correct the error in Table 20.10.

The difference table and associated error pattern are laid out in Table 20.11. The values in the second difference column pinpoint the error to $f(6)$ and the error pattern suggests that $e = -8$, making $f(6) = 3.4815$, although it might not be unreasonable to take e as -9.

20.4 Difference Operators

It is convenient to represent sets of values of x and corresponding values of $f(x)$ using the suffix notation $x_0, x_1, x_2 \ldots$ and $f_0, f_1, f_2 \ldots$ so that corresponding values are linked. Using this notation the nth entry in the first

Table 20.12

x_0	f_0									
		Δf_0			∇f_1			$\delta f_{0.5}$		
x_1	f_1		$\Delta^2 f_0$			$\nabla^2 f_2$			$\delta^2 f_1$	
		Δf_1		$\Delta^3 f_0$	∇f_2		$\nabla^2 f_3$	$\delta f_{1.5}$		$\delta^3 f_{1.5}$
x_2	f_2		$\Delta^2 f_1$			$\nabla^2 f_3$			$\delta^2 f_2$	
		Δf_2		$\Delta^3 f_1$	∇f_3		$\nabla^2 f_4$	$\delta f_{2.5}$		$\delta^3 f_{2.5}$
x_3	f_3		$\Delta^2 f_2$			$\nabla^2 f_4$			$\delta^2 f_3$	
		Δf_3		$\Delta^3 f_2$	∇f_4		$\nabla^2 f_5$	$\delta f_{3.5}$		$\delta^3 f_{3.5}$
x_4	f_4		$\Delta^2 f_3$			$\nabla^2 f_5$			$\delta^2 f_4$	
		Δf_4			∇f_5			$\delta f_{4.5}$		
x_5	f_5									
			Forward differences			Backward differences			Central differences	

difference column is $f_n - f_{n-1}$ and there are three notations for describing this in terms of operators (see Part 18.1). One is the FORWARD DIFFERENCE OPERATOR Δ, such that $\Delta f_n = f_{n+1} - f_n$, another is the BACKWARD DIFFERENCE OPERATOR ∇, such that $\nabla f_{n+1} = f_{n+1} - f_n$. The third is the one that we have been using, δ, the CENTRAL DIFFERENCE OPERATOR, which, in many respects, is the most convenient notation but which requires fractional suffixes since, to be consistent with the notation adopted $f_{n+1} - f_n$ must be expressed as $\delta_{n+0.5}$. Higher differences are written using index notation as Δ^2 or ∇^3, etc. The three operators are laid out in Table 20.12 and it will be seen that, for example, Δf_1, Δf_2 and $\delta f_{1.5}$ are all different ways of representing the same difference. Notice also that in writing $\delta^m f_n$ if m is even, n must be integral.

As with the D operator, the difference operators may be used symbolically or algebraically as required.

20.5 Newton's Interpolation Formulae

It can be shown that the nth entry in a difference table f_n is related to f_0 by the two formulae:

$$\text{(a)}\ f_n = (1 + \Delta)^n f_0 = f_0 + n\Delta f_0 + n(n-1)\Delta^2 f_0/2! + \cdots$$

$$\text{(b)}\ f_0 = (1 - \nabla)^n f_n = f_n - n\nabla f_n + n(n-1)\nabla^2 f_n/2! - \cdots$$

These are called NEWTON'S FORWARD FORMULA and NEWTON'S BACKWARD FORMULA respectively. A reference to a difference table, say Table 20.3, will illustrate how they work. Using the forward formula with

$n = 4$ we have

$$f_4 = f_0 + 4\Delta f_0 + 6\Delta^2 f_0 + 4\Delta^3 f_0 + \Delta^4 f_0$$

and, from the table, $f_0 = -1.0000$, $\Delta f_0 = 0.341$, $\Delta^2 f_0 = 0.086$, $\Delta^3 f_0 = 0.006$ and $\Delta^4 f_0 = 0$ (not shown in the table). Substituting these values in the formula makes f_4 equal to 0.904. Any value could be taken as f_0, say $f(0.4)$, and then f_4 would give the value of $f(0.8)$. The backward difference formula works in the same way but gives values higher up the table.

There is really little point in working in this way when n is an integer as tabular values can be read directly or obtained using differences if the table is not long enough. If, however, we wish to INTERPOLATE, that is to obtain non-tabular values, the formulae apply using non-integral values of n.

If, for example, we require $f(0.35)$ from Table 20.3, assuming that $f(x)$ is not given explicitly, rather than merely taking the average of $f(0.3)$ and $f(0.4)$ which is a very crude method, one of the Newton formulae could be used.

Taking $f(3)$ as f_0, f_n with $n = 0.5$ is required. The formula is infinitely long but as $\Delta^4 f_0$ and higher differences are zero, no error is introduced in truncating the formula after the term in $\Delta^3 f_0$. This gives

$$f_{0.5} = f_0 + \tfrac{1}{2}\Delta f_0 - \tfrac{1}{8}\Delta^2 f_0 + \tfrac{1}{16}\Delta^3 f_0$$

when simplified. Working in the units of the difference columns the required value is

$$287 + 617/2 - 104/8 + 6/16 = 583 \text{ (nearest integer)}$$

Hence $f(3.5) = 0.583$. When dealing with rounded functions, higher differences do not disappear but generally become quite small, only building up as rounding errors increase. These formulae should therefore not include differences from the irregular part of the table.

Central difference interpolation formulae exist but are rather more involved.

Example 20.5(a)
Use a Newton formula to obtain $e^{0.82}$ from Table 20.5.

Being near the foot of the table, this interpolation requires the use of the backward formula. Taking $n = 0.8$ and using $f(0.9)$ as the base of the calculation, the value is given by

$$\begin{aligned}(1 - \nabla)^{0.8} f(0.9) &= f(0.9) - 0.8\nabla f(0.9) + 0.8 \times (-0.2) \times \nabla^2 f(0.9)/2! \\ &\quad - 0.8 \times (-0.2) \times (-1.2) \times \nabla^3 f(0.9)/3! \\ &= 24\,596 - 0.8 \times 2341 - \tfrac{1}{2} \times 0.8 \times 0.2 \times 224 \\ &\quad - 0.8 \times 0.2 \times 1.2 \times 24/6 \\ &= 22\,705 \text{ (nearest integer)}\end{aligned}$$

Hence $e^{0.82} = 2.2705$.

20.6 Numerical Differentiation

In examples where values of x and $f(x)$ are given in the form of a table it is possible to obtain values of $f'(x)$ and higher derivatives using differences. The most practical formulae involve central differences but as terms of the form $\delta^n f_0$ do not exist when n is odd, it is necessary to introduce a further operator μ, the AVERAGING OPERATOR, such that

$$\mu\delta f_n = \tfrac{1}{2}(\delta f_{n-1/2} + \delta f_{n+1/2})$$

Using this operator, the derivative formulae are

$$\text{(a)}\ f_n' = \frac{1}{h}(\mu\delta f_n - \tfrac{1}{6}\mu\delta^3 f_n + \tfrac{1}{30}\mu\delta^5 f_n) + \cdots)$$

$$\text{(b)}\ f_n'' = \frac{1}{h^2}(\delta^2 f_n - \tfrac{1}{12}\delta^4 f_n + \cdots)$$

$$\text{(c)}\ f_n''' = \frac{1}{h^3}(\mu\delta^3 f_n - \tfrac{1}{4}\mu\delta^5 f_n + \cdots)$$

where h is the step length in each case. All of these formulae are infinitely long but the coefficients of further terms decrease rapidly.

In Table 20.5, $f(x) = e^x$ so that all derivatives should be equal to $f(x)$. Using the first formula taking $f(0.5)$ as f_0 gives

$$f'(0.5) = \frac{1}{0.1}(\tfrac{1}{2} \times (1569 + 1734) - \tfrac{1}{12} \times (15 + 18) + \cdots)$$
$$= 16\,488$$

Hence $f'(0.5) = 1.6488$, the small difference between this and $f(0.5)$ being due to rounding errors. The values of higher derivatives tend to be less accurate still.

$$f''(0.5) = \frac{1}{0.01}(165 - \tfrac{1}{12} \times 3) = 16\,475 \text{ (nearest integer)}$$

so that $f''(0.5) = 1.6475$.

$$f'''(0.5) = \frac{1}{0.001}(\tfrac{1}{2} \times (15 + 18) - \tfrac{1}{8} \times (4 - 4))$$
$$= 16\,500 \text{ (nearest integer)}$$

so that $f'''(0.5) = 1.65$.

Example 20.6(a)
Use Table 20.13 to obtain (a) $f'(0.4)$ (b) $f''(5)$.

Table 20.13

x	$f(x)$	δ	δ^2	δ^3	δ^4	δ^5
1	1.1350					
		1532				
2	1.2882		207			
		1739		28		
3	1.4621		235		4	
		1974		32		−2
4	1.6595		267		2	
		2241		34		7
5	1.8836		301		9	
		2542		43		
6	2.1378		344			
		2886				
7	2.4264					

(a) $f_0' = (\frac{1}{2} \times (1974 + 2241) - \frac{1}{12} \times (32 + 34) + \frac{1}{60} \times (-2 + 7)$

$= 2103$ (nearest integer)

$f'(4) = 0.2103$

(b) $f_0'' = 301 - \frac{1}{12} \times 9 = 300$ (nearest integer)

$f(5) = 0.03$

20.7 Numerical Integration

Various difference formulae for integration exist but having a table of values, the trapezoidal rule or Simpson's rule can be used with rather less work. These and some useful integration formulae are listed. In each case, h is the step length and there are $(n + 1)$ ordinates $f_0, f_1, f_2 \ldots f_n$.

(a) *Trapezoidal rule*

$$\int_{x_0}^{x_n} f(x)\,\mathrm{d}x = h(\tfrac{1}{2}(f_0 + f_n) + f_1 + f_2 + \cdots f_{n-1})$$

(b) *Simpson's rule*

$$\int_{x_0}^{x_n} f(x)\,\mathrm{d}x = \tfrac{1}{3}h(f_0 + f_n + 4(f_1 + f_3 + \cdots f_{n-1}) + 2(f_2 + f_4 + \cdots f_{n-2})$$

This requires an odd number of tabular values.

Table 20.14

x	0	0.1	0.2	0.3	0.4	0.5	0.6
$f(x)$	0	0.1003	0.2027	0.3093	0.4228	0.5463	0.6841

(c) *Three-eighths rule*

$$\int_{x_0}^{x_n} f(x)\,dx = \tfrac{3}{8}h(f_0 + f_n + 3(f_1 + f_2 + f_3 + \cdots + f_{n-1})$$
$$-(f_3 + f_6 + f_9 + \cdots f_{n-3})$$

This requires $3k + 1$ values where k is an integer.

(d) *Weddle's rule*

$$\int_{x_0}^{x_n} f(x)\,dx = \tfrac{3}{10}h\,(f_0 + f_2 + f_4 + \cdots f_n + 5(f_1 + f_3 + f_5 + \cdots)$$
$$+(f_3 + f_6 + f_9 + \cdots)$$

This requires $6k + 1$ values where k is an integer.

Example 20.7(a)

Obtain $\int_0^{0.6} f(x)\,dx$ where $f(x)$ is the function in Table 20.14 using each of the rules above.

(a) Trapezoidal rule

$$\int_0^{0.6} f(x)\,dx = (\tfrac{1}{2} \times 6841 + 15\,814) \times 10^{-5} = 0.192\,345$$

(b) Simpson's rule

$$\int_0^{0.6} f(x)\,dx = \tfrac{1}{3}(6841 + 4 \times 9559 + 2 \times 6255) \times 10^{-5} = 0.191\,957$$

(c) Three-eighths rule

$$\int_0^{0.6} f(x)\,dx = \tfrac{3}{8}(6841 + 3 \times 15\,814 - 3093) \times 10^{-5} = 0.191\,963$$

(d) Weddle's rule

$$\int_0^{0.6} f(x)\,dx = \tfrac{3}{10}(13\,096 + 5 \times 9559 + 3093) \times 10^{-5} = 0.191\,952$$

Unworked Examples 2

Part 20.1

(1) If $f(x) = 4x^2 - 5x - 2$, evaluate $f(2)$ and $f(-1.72)$.

(2) Obtain $x^3 - 5x + 7$ in nested form. Evaluate this expression when $x = -0.6$.

Table 20.15

x	ln x	δ	δ^2	δ^3	δ^4	δ^5	δ^6
1	0						
		69					
2	0.69		−28				
		41		16			
3	1.10		−12		−11		
		29		5		9	
4	1.39		−7		−2		−9
		22		3		0	
5	1.61		−4		−2		3
		18		1		3	
6	1.79		−3		1		
		15		2			
7	1.94		−1				
		14					
8	2.08						

(3) Draw up a finite difference table for $2x^3 - 5x^2 + 7x - 1$ for $x = 0(0.2)1$. Do the constant differences
   (a) occur in the expected column?
   (b) have a value consistent with the formula $n!ah^n$?

(4) Obtain values of $f(x) = 3x^4 + 2x^2 - 1$ for $x = 0(0.1)0.4$. Use differences to extend the table up to $x = 1$. What is the value of $f(0.8)$?

(5) Verify the result of the final part of Question 4 by expressing $f(x)$ in nested form.

(6) If $f(x) = 2x^3 + 7x^2 - 5x - 3$, in which column of a finite difference table drawn for $x = 0(0.2)1$ would the constant occur if all values of $f(x)$ were exact? Obtain the expected constant by formula and verify it using differences.

(7) Express $3x^3 - 3x^2 - 3x - 3$ in nested form.

Parts 20.2 and 20.3

(8) By use of tables or a calculator verify values of ln x correct to two places of decimals for $x = 1(1)8$. Which are the regular differences and which is the transitional column? (See Table 20.15.)

(9) Find and correct the error in Table 20.16.

(10) What can you say about the sum of the errors in any column of a difference table due to an error e in a value of $f(x)$? Give a reason for your answer.

Table 20.16

x	0.16	0.17	0.18	0.19	0.20	0.21	0.22	0.23
$f(x)$	0.5398	0.5793	0.6179	0.6545	0.6915	0.7257	0.7580	0.7881

(11) Draw up an error table, similar to Table 20.6, showing the pattern arising from errors e_1 and e_2 in successive values of $f(x)$ in a difference table.

(12) If the values of $f(x)$ in Table 20.3 were corrected to two places of decimals how many regular columns would it contain?

Parts 20.4 and 20.5

(13) Interpolate Table 20.3 to obtain $f(0.37)$. Verify the result algebraically.

(14) Use Table 20.4 to find (i) sin 17° (b) sin 36°30′.

(15) Are there occasions when Newton's backward formula is preferable to his forward formula?

(16) Use Table 20.5 to find $e^{0.55}$. Verify your result by using the power series for e^x.

Parts 20.6 and 20.7

(17) Obtain $f'(0.4)$, $f''(0.4)$ and $\int_0^1 f(x)\,dx$ where $f(x) = x^3 + 4x^2 + 3x - 1$ using numerical calculus together with Table 20.3. The integration should be carried out using Simpson's rule. Verify the results using normal methods of calculus.

(18) Find $\int_0^{0.6} e^x\,dx$ using Table 20.5 and a suitable integration method.

(19) Use the differences in Table 20.17 to find $f'(0.4)$ if the step length is 0.1.

(20) Find $f'(3)$ and $\int_1^7 f(x)\,dx$ from Table 20.13.

(21) Use Table 20.2 and Simpson's rule to obtain $\int_0^4 (x^2 + 4x + 1)\,dx$. Verify the result by integrating directly.

Table 20.17

x	$f(x)$	δ	δ^2	δ^3	δ^4	δ^5
		1918		136		49
0.4	1.1918		445		96	
		2363		232		79

SECTION E

Statistics

PART 21: SAMPLING AND QUALITY CONTROL

21.1 The Main Distributions

A probability distribution is a set of probabilities, characteristic of a statistical population and obtained either by formula or from tables. It is assumed that the reader is familiar with the BINOMIAL, the POISSON and the NORMAL (or GAUSSIAN) distributions. This section contains a brief summary of their properties and some examples to illustrate their use.

The Binomial Distribution

This is a distribution of discrete probabilities, arising from a situation in which there are two possible outcomes, favourable or unfavourable. If the individual probability of a favourable outcome is p and of an unfavourable outcome is q (equal, of course, to 1-p), then in n cases, the probabilities of 0, 1, 2, 3 . . . n favourable outcomes are given by

$$P(0) = q^n$$

$$P(1) = \binom{n}{1} q^{n-1} p$$

$$P(2) = \binom{n}{2} q^{n-2} p^2$$

$$P(3) = \binom{n}{3} q^{n-3} p^3$$

$$\vdots \qquad \vdots$$

$$P(n) = p^n$$

As is the case with all probability distributions, the sum of the probabilities is unity. The probabilities themselves are the terms of the expansion of $(q + p)^n$ and the general probability $P(r)$ is equal to $\binom{n}{r}q^{n-r}p^r$. In all cases $\binom{n}{r}$ means the number of combinations of r items from n.

The mean number of favourable outcomes is equal to np and the standard deviation from that mean is $\sqrt{(npq)}$. The values of n, p and q are the PARAMETERS of the distribution.

The Poisson Distribution

This is a distribution of discrete probabilities arising from a series of random and comparatively rare events. If the mean number of successful outcomes is m, the probability of 0, 1, 2, 3, . . . successful outcomes can be expressed in terms of the single parameter m and is given by

$$P(0) = e^{-m}$$

$$P(1) = me^{-m}$$

$$P(2) = \frac{m^2}{2!}e^{-m}$$

$$P(3) = \frac{m^3}{3!}e^{-m}$$

$$\vdots \qquad \vdots$$

The individual probabilities are each the product of e^{-m} with the terms of the expansion of e^m and the total probability is once again unity. A general probability is given by

$$P(r) = \frac{m^r}{r!}e^{-m}$$

The mean number of successful outcomes is m and the standard deviation from that mean is $\sqrt{m}$.

The Normal Distribution

This is a distribution of continuous probabilities, which occurs when random fluctuations from a mean measurement are caused by many independent factors, no one of which is significant enough to dominate the others. It is the most commonly occurring distribution giving rise to a characteristic bell-shaped distribution curve. A standardised curve is arranged symmetrically about the y-axis, the x units measured in standard deviates (standard deviations from the mean) and the y-scale chosen to make the total area under the curve between $\pm\infty$ equal to unity. In this standardised form, its equation is $y = e^{-x^2/2}/\sqrt{(2\pi)}$. It is shown in Figure E1.

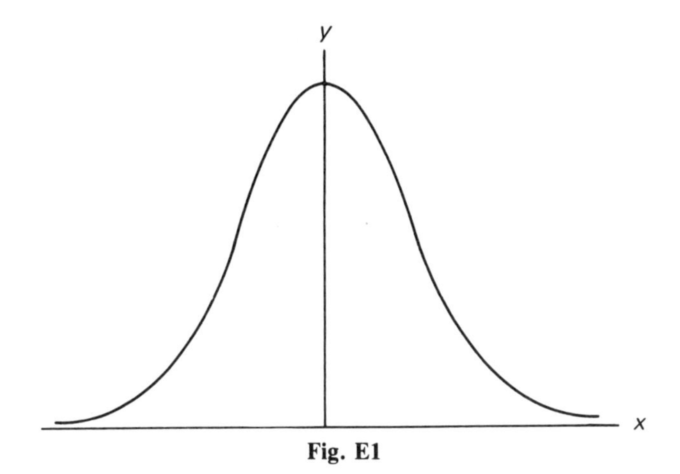

Fig. E1

Probabilities of normally distributed variable are determined from the areas under sections of the curve, which are given in Table 21.1. The notation $A(x)$ will be used to denote the area under a normal distribution curve between $-\infty$ and the ordinate at x measured in standard deviates. Two areas that are widely used are

(i) 95% of the area lies between the ordinates at $x = \pm 1.96$

(ii) 99.8% of the area lies between the ordinates at $x = \pm 3.09$.

Example 21.1(a)

A manufacturer claims that his mass-produced goods are no more than 2% substandard. A potential buyer agrees to place an order if a random sample of 100 of the units gives no more than 2 defective items upon thorough testing. What is the probability of the order being placed if the manufacturer's claim is valid?

The distribution of substandard items is binomial with paremeters $n = 100$, $p = 0.02$ and $q = 0.98$. If $P(r)$ is the probability of obtaining r substandard units in the sample of 100

$$P(0) = (0.98)^{100} = 0.133$$

$$P(1) = 100(0.98)^{99}(0.02) = 0.271$$

$$P(2) = \binom{100}{2}(0.98)^{98}(0.02)^2 = 0.273$$

Probability of order being placed $= 0.133 + 0.271 + 0.273 = 0.677$.

Example 21.1(b)

A small machine shop has three sets of welding equipment. On average, during any hourly period, one set is in use. Find the probability that (a) no sets are in use, (b) all sets are in use and there is a demand for a fourth set. If one set is unserviceable, what is the probability of the demand exceeding the supply of available sets?

Table 21.1 Differences in x

	0.00	0.01	0.02	0.03	0.04	0.05	0.06	0.07	0.08	0.09
0.0	.5000	.5040	.5080	.5120	.5160	.5199	.5239	.5279	.5319	.5359
0.1	.5398	.5438	.5478	.5517	..5557	.5596	.5636	.5675	.5714	.5753
0.2	.5793	.5832	.5871	.5910	.5948	.5987	.6026	.6064	.6103	.6141
0.3	.6179	.6217	.6255	.6293	.6331	.6368	.6406	.6443	.6480	.6517
0.4	.6554	.6591	.6628	.6664	.6700	.6736	.6772	.6808	.6844	.6879
0.5	.6915	.6950	.6985	.7019	.7054	.7088	.7123	.7157	.7190	.7224
0.6	.7257	.7291	.7324	.7357	.7389	.7422	.7454	.7486	.7517	.7549
0.7	.7580	.7611	.7642	.7673	.7703	.7734	.7764	.7793	.7823	.7852
0.8	.7881	.7910	.7939	.7967	.7995	.8023	.8051	.8078	.8106	.8133
0.9	.8159	.8186	.8212	.8238	.8264	.8289	.8315	.8340	.8365	.8389
1.0	.8413	.8438	.8461	.8485	.8508	.8531	.8554	.8577	.8599	.8621
1.1	.8643	.8665	.8686	.8708	.8729	.8749	.8770	.8790	.8810	.8830
1.2	.8849	.8869	.8888	.8906	.8925	.8943	.8962	.8980	.8997	.9015
1.3	.9032	.9049	.9066	.9082	.9099	.9115	.9131	.9147	.9162	.9177
1.4	.9192	.9207	.9222	.9236	.9251	.9265	.9279	.9292	.9306	.9319
1.5	.9332	.9345	.9357	.9370	.9382	.9394	.9406	.9418	.9429	.9441
1.6	.9452	.9463	.9474	.9484	.9495	.9505	.9515	.9525	.9535	.9545
1.7	.9554	.9564	.9573	.9582	.9591	.9599	.9608	.9616	.9625	.9633
1.8	.9641	.9648	.9656	.9664	.9671	.9678	.9686	.9693	.9699	.9706
1.9	.9713	.9719	.9726	.9732	.9738	.9744	.9750	.9756	.9761	.9767
2.0	.9772	.9778	.9783	.9788	.9793	.9798	.9803	.9808	.9812	.9817
2.1	.9821	.9826	.9830	.9834	.9838	.9842	.9846	.9850	.9854	.9857
2.2	.9861	.9864	.9868	.9871	.9875	.9878	.9881	.9884	.9887	.9890
2.3	.9893	.9896	.9898	.9901	.9904	.9906	.9909	.9911	.9913	.9916
2.4	.9918	.9920	.9922	.9924	.9927	.9929	.9930	.9932	.9934	.9936
2.5	.9938	.9940	.9941	.9943	.9945	.9946	.9948	.9949	.9951	.9952
2.6	.9953	.9955	.9956	.9957	.9959	.9960	.9961	.9962	.9963	.9964
2.7	.9965	.9966	.9967	.9968	.9969	.9970	.9971	.9972	.9973	.9974
2.8	.9974	.9975	.9976	.9977	.9977	.9978	.9979	.9979	.9980	.9981
2.9	.9981	.9982	.9982	.9983	.9984	.9984	.9985	.9985	.9986	.9986
3.0	.9986	.9987	.9987	.9988	.9988	.9989	.9989	.9989	.9990	.9990
3.1	.9990	.9991	.9991	.9991	.9992	.9992	.9992	.9992	.9993	.9993
3.2	.9993	.9993	.9994	.9994	.9994	.9994	.9994	.9995	.9995	.9995
3.3	.9995	.9995	.9995	.9996	.9996	.9996	.9996	.9996	.9996	.9996
3.4	.9997	.9997	.9997	.9997	.9997	.9997	.9997	.9997	.9997	.9998
3.5	.9998	.9998	.9998	.9998	.9998	.9998	.9998	.9998	.9998	.9998
3.6	.9998	.9998	.9998	.9999	.9999	.9999	.9999	.9999	.9999	.9999

Assuming that jobs requiring the use of welding equipment are programmed independently, the probabilities will follow a Poisson pattern with a parameter $m = 1$. With the notation of Example 21.1(a)

$$P(0) = e^{-1} = 0.368$$
$$P(1) = e^{-1} = 0.368$$
$$P(2) = e^{-1}/2! = 0.184$$
$$P(3) = e^{-1}/3! = 0.061$$
$$P(\text{exceeding } 3) = 1 - (0.368 + 0.368 + 0.184 + 0.061) = 0.019$$

(a) The probability of no sets being in use is 0.368.
(b) The probability of a need for a fourth set is 0.019.
If a set is unserviceable, probability of demand exceeding supply is $0.019 + 0.061$, i.e. 0.08.

Example 21.1(c)
A machining operation is timed by a work-study practitioner. It is found that 20% of operatives take more than 40 s while 10% can accomplish the task in under 30 s. Assuming that the distribution of operation times is normal, find the mean operation time and the standard deviation from that mean both correct to 0.1 s.

The 20% of slow operatives are represented by the shaded area on the right of Figure E2 while the speedy 10% are represented by the shaded area on the left. Using normal probability areas from Table 21.1, the values x_1 and x_2 are such that $A(x_1) = 0.8$ and $A(x_2) = 0.1$. The values of x_1 and x_2 are approximately 0.84 and -1.28 respectively. If the mean value is m s and the standard deviation s s then

$$0.84s = 40 - m$$
$$-1.28s = 30 - m$$

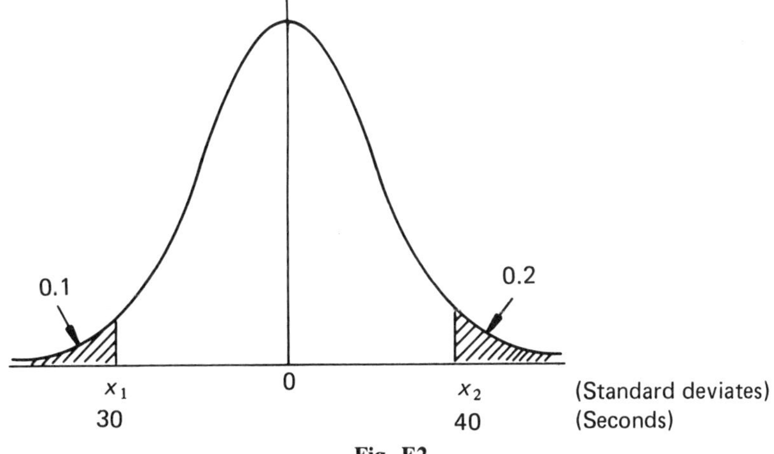

Fig. E2

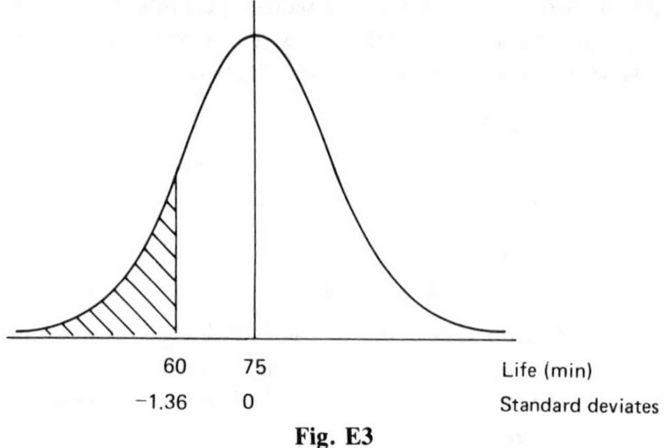

Fig. E3

Using any suitable method of solution, it will be found that $m = 36.0$ and $s = 4.7$. The mean operation time is 36.0 s and the standard deviation from that mean is 4.7 s.

Example 21.1(d)
The lives of cutting tools are normally distributed with a mean value of 75 min and a standard deviation of 11 min. What percentage of tools have a life of less than 1 h?

60 min $\equiv (60 - 75)/11 = -1.36$ standard deviates. Referring to Figure E3, $A(-1.36) = 1 - A(1.36) = 1 - 0.913 = 0.087$. Just under 9% of tools have a life of less than 1 h.

Example 21.1(e)
Bottles which are filled by machine have their exact contents normally distributed. When the nominal capacity is 1 l, the mean contents are actually

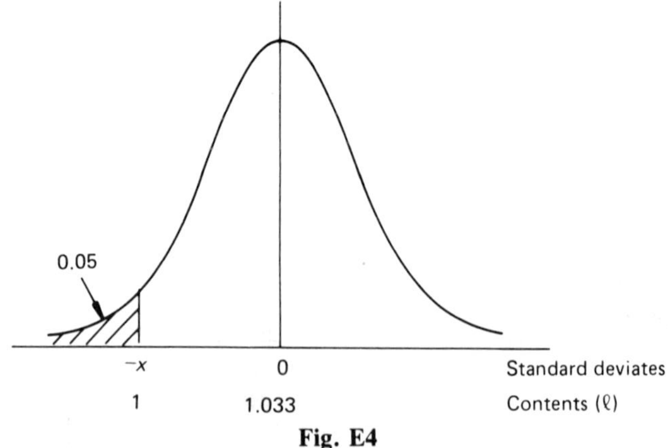

Fig. E4

1.033 l. If, by law, no more than 5% of the bottles must contain less than the nominal contents, find the standard deviation needed to meet this requirement.

Referring to Figure E4, $A(-x) = 0.05$. Hence $A(x) = 0.95$ and $x = 1.65$. If the required standard deviation is s, then $1.65s = 0.033$,

$$s = 0.02$$

The standard deviation is 0.02 l.

21.2 Approximating One Distribution by Another

Although for a particular application, only one of the distributions described may be strictly applicable, under certain conditions the necessary probability may be obtained, at least to a reasonable degree of accuracy, by using an alternative distribution. This can be very convenient because it is generally easier to obtain probabilities using a Poisson formula than using a binomial formula. Since normal probabilities are obtained from a table rather than by calculation, it may, in turn, be preferable to use a normal distribution in place of one distributed in Poisson fashion. This is especially true when the parameter m is large because its probability will be the product of terms m^r, $r!$ and e^{-m} all of which are very large or very small. When n is large, say greater than 10, STIRLING'S FORMULA $n! \approx (n/e)^n\sqrt{(2\pi n)}$ allows a reasonable approximation to $n!$

Table 21.2 shows the conditions under which the probabilities of one distribution approximate to those of another, giving agreement to about two places of decimals, although the exact degree of accuracy varies. In cases where a lesser degree of accuracy is needed, the conditions can be more flexible.

Where a normal distribution replaces a binomial or a Poisson distribution allowance must be made for the fact that the former is a continuous distribution but the others generate only discrete probabilities.

Suppose that an event occurs, at random, on average 16 times each hour. To obtain the probability of 18 occurrences in the following hour, strictly speaking a Poisson distribution should be used. The probability found is

Table 21.2

Distribution being approximated	Distribution replacing it	Conditions applying
Binomial	Poisson	$p < 0.1$
Binomial	Normal	$np > 5$, $nq > 5$
Poisson	Normal	$m > 10$

$16^{18}e^{-16}/18!$, which is equal to 0.083, a value which is not very convenient to find, even using Stirling's formula and aided by a calculator!

It will be seen from Table 21.2 that when $m = 16$, this probability can be approximated using a normal distribution. However, the nearest approach that can be made to obtaining a discrete probability of 18 using a normal curve is to calculate the area lying under it bounded by ordinates at 17.5 and 18.5, these being the values associated with the corresponding histogram of Poisson probabilities. Since $m = 16$, the values 16 and 4 can be taken for the mean and standard deviation of the distribution. The ordinates, expressed in standard deviates will therefore be 2.5/4 and 1.5/4 which can be approximated to 0.63 and 0.38 respectively, since the result is not expected to be too precise. These are shown in Figure E5.

From Table 21.1, taking areas correct to three places of decimals

$$\begin{aligned} A(0.63) &= 0.736 \\ A(0.38) &= \underline{0.648} \\ \text{Required probability} &= 0.088 \end{aligned}$$

Example 21.2(a)
Verify the results of Example 21.1(a) using a Poisson distribution.

Since $p < 0.1$, the probabilities may be obtained reasonably accurately using a Poisson distribution. The parameter is $m = np = 2$

$$\begin{aligned} P(0) &= e^{-2} &&= 0.135 \\ P(1) &= 2e^{-2} &&= 0.271 \\ P(2) &= 2^2e^{-2}/2! &&= 0.271 \end{aligned}$$

The probability of an order being placed is $0.135 + 0.271 + 0.271 = 0.677$.

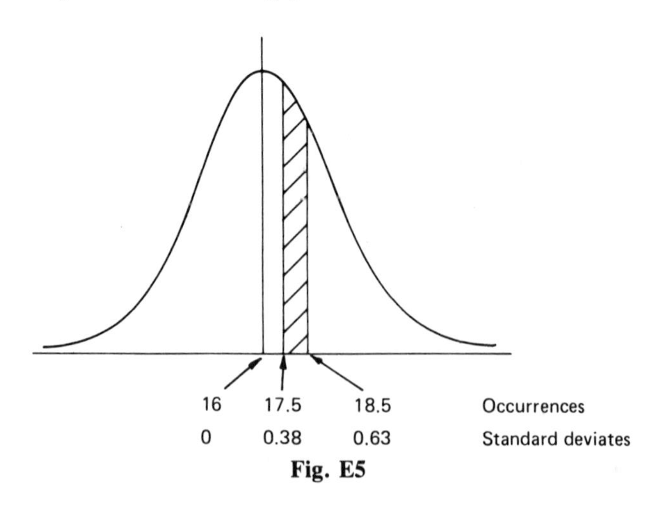

Fig. E5

Example 21.2(b)
A manufacturer uses components from two different sources. If 25% of his stock is from source A and the rest from source B and these are not separated at shop floor level find the probability that a manufactured unit requiring 80 components has exactly 20 originating from source A.

Method 1: Using a binomial distribution with parameters $n = 80, p = 0.25, q = 0.75$

$$P(20) = \binom{80}{20}(0.75)^{60}(0.25)^{20}$$

$$= \frac{80!(0.75)^{60}(0.25)^{20}}{20!\ 60!}$$

$$= 0.103 \text{ (using Stirling's formula)}$$

Method 2: Since np and nq are both larger than 5, a normal distribution may be used. Its parameters are $m = np = 20$, $s = \sqrt{(npq)} = 3.87$.

Referring to Figure E6, the area bounded by 19.5 and 20.5 will give the required probability. Using standard deviates

$$20.5 \equiv (20.5 - 20)/3.87 = 0.13$$
$$19.5 \equiv (19.5 - 20)/3.87 = -0.13$$

Required probability $= A(0.13) - A(-0.13) = 2A(0.13) - 1 \approx 0.104$.

The difference in the work needed in the two methods is enormous yet there is only a marginal difference in the results.

Example 21.2(c)
A company manufacturing in a number of factories receives on average 25 accident reports a month. If the figure drops to 21 in the first month after a

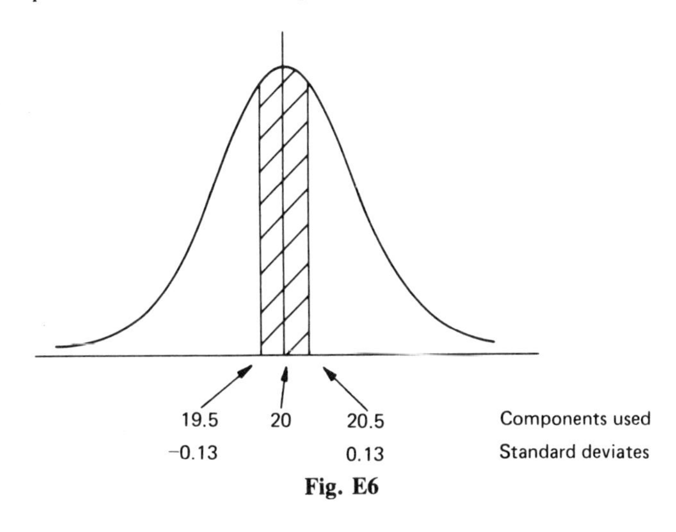

Fig. E6

safety campaign, find the probability that the number of accidents could fall to this figure (or less) by chance.

Method 1: Using a Poisson distribution with $m = 25$ and working to two places of decimals

$$
\begin{aligned}
P(21) &= 25^{21}e^{-25}/21! = 0.06 \\
P(20) &= 25^{20}e^{-25}/20! = 0.05 \\
P(19) &= 25^{19}e^{-25}/19! = 0.04 \\
P(18) &= 25^{18}e^{-25}/18! = 0.03 \\
P(17) &= 25^{17}e^{-25}/17! = 0.02 \\
P(16) &= 25^{16}e^{-25}/16! = 0.02 \\
P(15) &= 25^{15}e^{-25}/15! = 0.01 \\
P(14) &= 25^{14}e^{-25}/14! = 0.01 \\
&\text{Total} \quad 0.24
\end{aligned}
$$

All other probabilities are too small to register to two places of decimals.

Method 2: Using a normal distribution with $m = 25$, $s = \sqrt{m} = 5$, the required probability is equal to the shaded area in Figure E7,

$$21.5 \equiv (21.5 - 25)/5 = -0.7 \text{ standard deviates}$$

The required probability is $A(-0.7)$

$$= 1 - A(0.7)$$
$$= 0.24$$

This example illustrates another advantage of using a normal distribution in place of a Poisson or a binomial distribution. When the parameters are

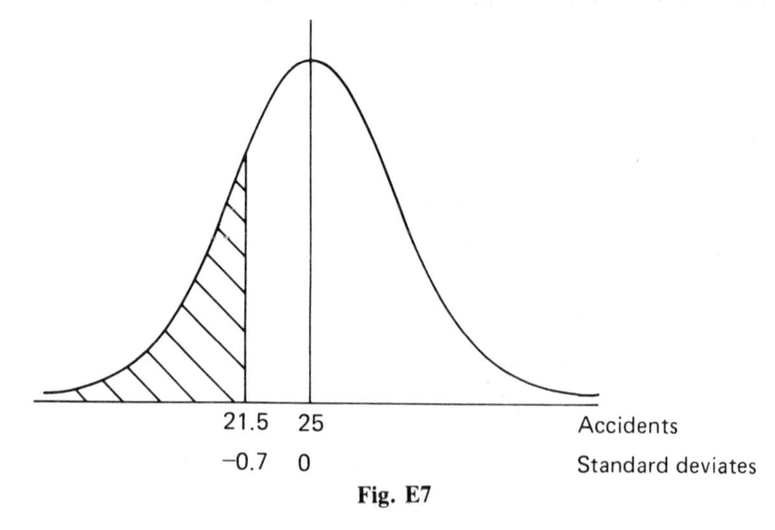

Fig. E7

large, successive probabilities alter slowly and it may be necessary to find several in order to solve a problem such as this. Using a normal distribution, the required probability can be found in one step.

Unworked Examples 1

Part 21.1

(1) If a batch of components contains 5% substandard items, find the probability of (a) exactly 1 (b) more than 1 substandard item in a sample of 20. Give your answer correct to three places of decimals.
(2) Over recent years the average number of fatal accidents in an industry has been 3.8. Find, correct to two significant figures, the probability that next year will be free of fatal accidents in the industry.
(3) Units of mean length 15 mm with a standard deviation of 0.04 mm are produced mechanically. If tolerance limits of 15 ± 0.14 mm are specified for length, find the percentage of unacceptable units produced, assuming that production is normally distributed.
(4) Light bulbs are normally distributed with a mean life of 1100 h and a standard deviation from that mean of 20 h. If a bulb is selected at random, find the probability that it has a life (a) exceeding 1050 hours (b) exceeding 1120 h.
(5) Sparking plugs are sold in sets of fours. One hundred sets were thoroughly examined and the distribution of faulty plugs is shown in Table 21.3. Assuming that the distribution is binomial, find the probability of an individual plug being faulty.
(6) An office has three telephone lines and at any one time an average of 1.2 lines are in use. Find the probability that there are no free lines when the need to make a call arises, giving your answer corrected to three places of decimals.
(7) A vending machine sells paraffin in nominal units of 4.5.1. If the mean delivery is 4.52 l find the standard deviation needed so that at least 90% of sales exceed the nominal delivery figure.
(8) In a factory, the output of operatives is normally distributed with a mean daily figure of 200 units and a standard deviation from that mean of 11 units. What percentage of operatives
    (a) exceed a daily output of 215 units
    (b) produce less than 180 units each day.
    If a bonus scheme were introduced for anyone whose output exceeded 225 units, what percentage of operatives would be expected to qualify?

Table 21.3

Number of faulty plugs in a packet	0	1	2	3	4
Frequency	82	17	1	0	0

(9) If 100 unbiased coins are spun, find the probability of obtaining more than 52 heads. Give your answer correct two significant figures.

(10) On average, 10% of the workforce of a large factory employing 2500 persons are not at work. What is the probability that this could drop to 9% tomorrow?

(11) On average, one item in every 25 fails to reach a manufacturer's standard. If a batch of 60 items is examined in detail, use a Poisson distribution to find the probabilities of 0, 1, 2 and 3 defective items in the sample.

(12) If the average number of people failing to report to work in a factory is 25 each day, find the probability of this absentee figure falling to 20 or less on a given day, giving the answer correct to two significant figures.

(13) A product is in great demand to such an extent that the average waiting list is 100 customers. Find the probability that demand could fluctuate so that the waiting list reached 120.

(14) For a binomial distribution if $n = 40$, $p = 0.1$ and $q = 0.9$, find $P(0)$, $P(1)$ and $P(2)$ correct to three places of decimals. Compare the results with the corresponding Poisson probabilities.

(15) A large company has an average weekly turnover of staff of 20. What is the probability that this figure will exceed 22 next week?

(16) For a Poisson distribution with a parameter $m = 36$, obtain $P(36)$ correct to two significant figures using Stirling's formula. Compare this value with the corresponding probability obtained using a normal distribution.

21.3 Sampling Schemes

It is a standard practice for customers to test samples of goods for quality level. The scheme outlined in Example 21.1(a) is a simple example. As the solution to that question illustrates, it is possible for a customer to reject a consignment of acceptable standard—called PRODUCER'S RISK—or, in fact, to accept a consignment which is below specification—called the CONSUMER'S RISK.

The nominal quality level, agreed by producer and customer, is called the ACCEPTABLE QUALITY LEVEL (AQL). A lower standard, called the LOT TOLERANCE PERCENTAGE DEFECTIVE (LTPD) is nominated by the consumer as the minimum acceptable standard. These standards may be expressed as probabilities although it is more usual to handle them in percentage form.

Suppose that in a sampling scheme that AQL is 2% and the LTPD is 10%. A consumer agrees to accept a consignment if a sample of 30 items contains no more than one which is below standard.

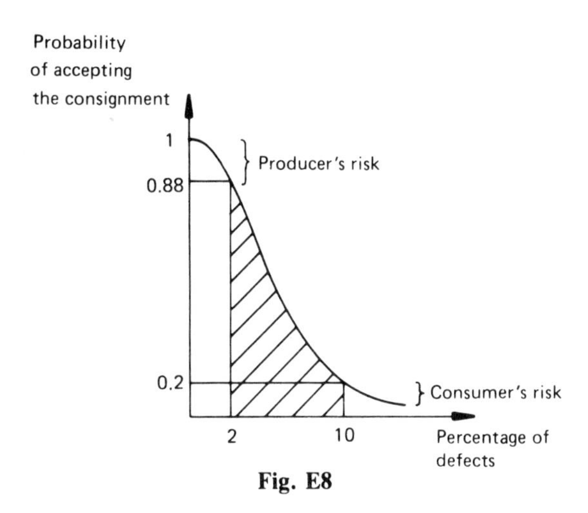

Fig. E8

The probabilities of 0 or 1 defective item in a sample can be found using a binomial distribution but with a 2% AQL, the Poisson distribution may be used. Taking $m = 0.6$ (i.e. 2% of 30), $P(0) = 0.55$, $P(1) = 0.33$ and the probability of not accepting a consignment is 0.12. This is the producer's risk.

On the other hand, taking the LTPD of 10%, the corresponding probabilities are $P(0) = 0.05$, $P(1) = 0.15$ and there is a probability of 0.2 of accepting an unsuitable consignment. This is the consumer's risk. Short of 100% inspection, these risks cannot be eliminated but, by a suitable choice of sampling scheme, they can be minimised. Generally, however, the degree of risk depends on the intensity of the sampling.

Figure E8 is the graph of probability of consignment acceptance against percentage of substandard goods, using this particular sampling scheme. It is known as an OPERATING CHARACTERISTIC CURVE (OCC) and, although for a sample with other statistical characteristics, the probabilities would differ, the curve would be of the same general shape.

Figure E9 shows the ideal OCC. This can only be achieved using 100% inspection of course, but the merit of any sampling scheme is judged by how

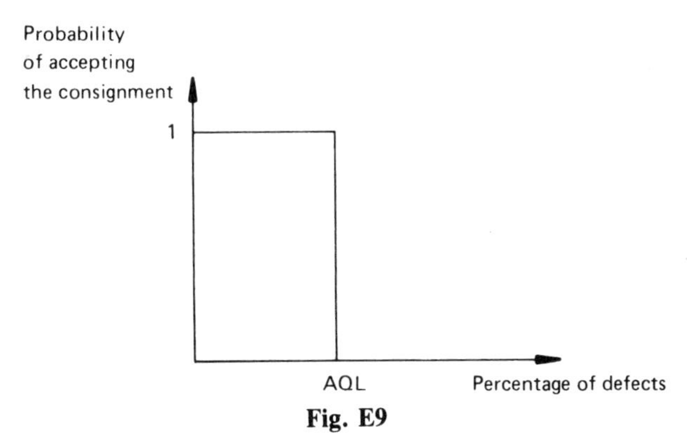

Fig. E9

Table 21.4

Number of substandard items in a sample of 30	Overall quality level		
	2% defects	6% defects	10% defects
0	0.55	0.17	0.05
1	0.33	0.30	0.15
Probability of acceptance	0.88	0.47	0.20

Table 21.5

Number of substandard items in a sample of 90	Overall quality level		
	2% defects	6% defects	10% defects
0	0.17	0	0
1	0.30	0.02	0
2	0.27	0.07	0
3	0.16	0.12	0.02
Probability of acceptance	0.90	0.21	0.02

closely its OCC approaches this ideal for a given cost. Tables 21.4 and 21.5 show the effect of increasing the sample size. All the probabilities have been corrected to two places of decimals. It will be seen that while the producer's risk is only affected marginally the consumer's risk is reduced considerably. The price to be paid is the extra time and expense involved in handling larger samples. As sampling is being used to keep down these cost, although the improvement is desirable, it may not be economically justified. A compromise might be to inspect samples of 60 but it can be shown that a DOUBLE SAMPLING scheme involving the inspection of two samples of 30 is more cost-effective.

A double sampling scheme might be

(a) inspect a sample of 30 items

(b) if the sample contains no defective items, accept the consignment

(c) if the sample contains two or more defective items, reject the consignment

(d) if the sample contains one defective item, examine a second sample. If this contains no defective items, accept the consignment.

This is not as reliable as the single sampling of 60 items but has the merit of only requiring 30 inspections in the majority of cases. It might therefore be compared with, say, the single sampling of 35 items and, on that basis, gives more reliable results.

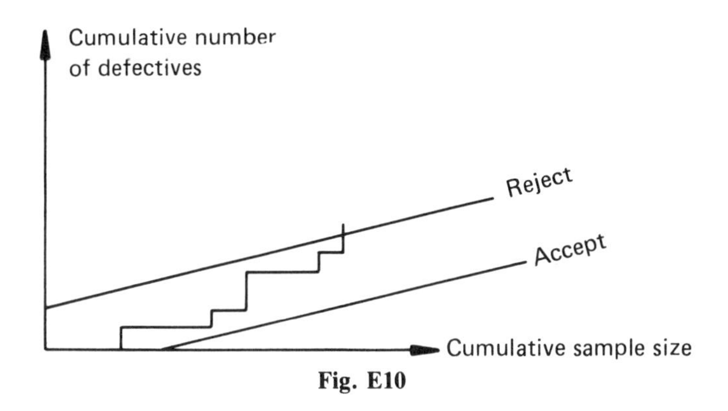

Fig. E10

The process can be extended to include any number of stages, but although these MULTIPLE SAMPLING schemes give an increase in reliability, a point is eventually reached when the complexity of the scheme outweighs its reliability.

SEQUENTIAL SAMPLING is a completely different process. In it, items are inspected singly and the decision to accept or reject the consignment is made as the examination proceeds. A sequential sampling chart is shown in Figure E10. It consists of a graph of the cumulative number of defective items against cumulative sample size. 'Accept' and 'reject' lines, whose positions are calculated from the statistical criteria involved, are added. The graph is drawn as a series of steps until it crosses one of the lines.

It is possible for the test to last for a very long time in exceptional circumstances, although it can be proved mathematically that it will ultimately terminate. On average, however, sequential sampling involves less inspection than other methods carrying the same risk.

Example 21.3(a)

A single sampling scheme operates as follows. Inspect a sample of size 20. Accept the batch if it contains no more than two defective items otherwise reject the batch. Find the producer's and consumer's risks with an AQL of 4% and an LTPD of 25%.

The probabilities can be obtained using a binomial distribution (or one that approximates its results). These give the following probabilities:

(a) When $n = 20$, $p = 0.04$, $q = 0.96$ $P(0) = 0.44$, $P(1) = 0.37$ and $P(2) - 0.15$

(b) When $n = 20$, $p = 0.25$, $q = 0.75$ $P(0) = 0$, $P(1) = 0.02$ and $P(2) = 0.07$

The producer's risk is 0.04 and the consumer's risk is 0.09. All probabilities have been corrected to two places of decimals.

Example 21.3(b)
Suppose the sampling plan of Example 21.3(a) were replaced by the following double sampling scheme. Inspect a sample of size 10.

(a) Accept the consignment if it contains no substandard items.
(b) Reject the consignment if it contains two or more substandard items.
(c) In other cases, inspect a second sample of the same size. Accept the consignment if this contains no more than one substandard item.

Find the producer's and consumer's risks with an AQL of 4% and an LTPD of 25%. Find also the mean number of items inspected in each case.

(a) When $n = 10$, $p = 0.04$, $q = 0.96$ $P(0) = 0.66$, $P(1) = 0.28$ and so $P(x > 1) = 0.06$.
Producer's risk $= 0.06 + 0.28 \times 0.06 = 0.08$ (corr. 2 pl. dec.).
(b) When $n = 10$, $p = 0.25$, $q = 0.75$ $P(0) = 0.06$, $P(1) = 0.19$ and so $P(x > 1) = 0.75$.
Consumer's risk $= 0.06 + 0.19 \times 0.25 = 0.11$ (corr. 2 pl. dec.).

In the first case, the average number of items inspected is $10 + 0.28 \times 10 \approx 13$. In second case the figure is $10 + 0.19 \times 10 \approx 12$.

21.4 Sampling Distributions

The sampling plans discussed depend on taking representative samples from a statistical population with known characteristics. However, the existence of producer's and consumer's risks indicates that however fairly the samples are obtained, they will vary in content.

The same variation occurs when dealing with samples selected to check the progress of production. Although the 'satisfactory/unsatisfactory' tests are used, more sensitive results are obtained by measuring quantitive features. The measurements will be a range of values, which are normally distributed in the majority of cases.

If a series of samples of size n are taken from a statistical population with a mean value μ and a standard deviation from that mean of σ and if the means of the samples are calculated, it can be shown that

(a) the mean of the sample means is μ
(b) the standard deviation of the sample means, called the STANDARD ERROR, is $\sigma/\sqrt{n}$
(c) the distribution of sample means is normal. This applies even in cases where the statistical population is markedly non-normal, although in such cases n should be at least 25.

These points can be illustrated using the figures in Table 21.6, which contains a population of 50 normally distributed values with a mean value μ of 1000 and a standard deviation σ of 19. They could, for example, be the lives of electric light bulbs, measured in hours.

Table 21.6

964	980	982	1011	1010	981	993	1000	997	995
1039	1020	988	1004	1013	976	996	1004	1020	1037
981	1007	986	1051	988	993	1003	981	1019	1024
977	991	1001	980	947	994	1037	1016	1000	991
1003	1004	994	990	989	991	994	974	1019	1000

Dividing the 50 values into 10 samples of size 5, these have mean values of 992.8, 1000.4, 990.2, 1007.2, 989.4, 987.0, 1004.6, 995.0, 1011.0 and 1009.4. The mean of these 10 samples means is 998.7 and their standard deviation is 8.49. These values compare favourably with the theoretical values $\mu = 1000$ and $\sigma/\sqrt{n} = 8.50$.

The usual method of obtaining σ is to group the data and use one of the standard deviation formulae for grouped data. However, if the distribution is known to be normal and a number of random samples are extracted, there is a relationship between the mean range of the samples $\bar{w}$ and the standard deviation of the population. Multiplying $\bar{w}$ by the appropriate constant from Table 21.7 gives a good approximation to the standard deviation. The ranges of the 10 samples in Table 21.6 are 75, 40, 19, 71, 66, 18, 44, 42, 23 and 46. These ranges have a mean value of 44.4. Using Table 21.7, an estimate of σ is $44.4 \times 0.43 \approx 19.1$.

Table 21.7

Adapted from *Biometrika Tables for Statisticians*, Vol. 1, Table 22, by permission of Biometrika Trustees.

Sample size	2	3	4	5	6	7
Constant	0.89	0.59	0.49	0.43	0.39	0.37

21.5 Quality Control Charts for Means

As we have seen, measurable factors on manufactured items vary. When such items are begin mass-produced, it is necessary to monitor the manufacturing process to detect changes in it and, where necessary, to take corrective action so that the overall quality of production is maintained. The method used is to plot sample means on a QUALITY CONTROL CHART so that a continuous visible record of the state of production is made. By this means it is a relatively simple matter to maintain production standards. The skill in quality control work lies not in maintaining the standards but in determining the form of the control chart.

A typical quality control chart for means is shown in Figure E11. It consists of a central mean line flanked by a pair of INNER CONTROL LINES which are, themselves, flanked by a pair of OUTER CONTROL LINES. Their positions are chosen so that when production is UNDER CONTROL only 1

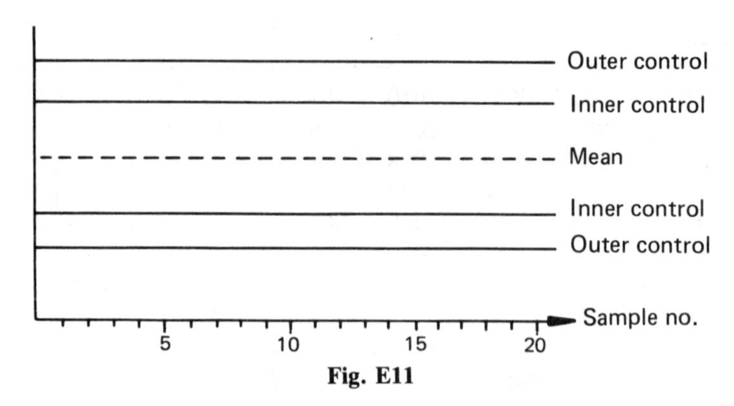

Fig. E11

sample mean in every 40 will fall outside either of the inner lines and only 1 in every 1000 outside either of the outer lines. It is considered that any event likely to occur 1 time in every 40 is sufficiently infrequent to constitute a warning that the process may be drifting out of control but that if an event which theoretically should occur on one occasion in every 1000 actually does take place, something has gone awry. Production is then considered to be OUT OF CONTROL. For these reasons the inner and outer lines are sometimes designated WARNING LINES and ACTION LINES respectively.

Since sample means are close to, if not actually, normally distributed, the positioning of these control lines is determined using Table 21.1. As $A(1.96) = 0.975$, it follows that $A(x > 1.96) = 0.025 = 1/40$ and $A(x < -1.96) = 1/40$. This is just another way of saying that the shaded area in Figure E12 is 95% of the whole area under the normal curve.

In exactly the same way $A(x > 3.09) = A(x < -3.09) = 1/1000$. The inner and outer lines on a control chart must therefore be at positions equivalent to 1.96 and 3.09 standard deviates from the mean position. Since the standard error of samples of size n is $\sigma/\sqrt{n}$, the position of the control lines are at $\mu \pm 1.96\,\sigma/\sqrt{n}$ and $\mu \pm 3.09\,\sigma/\sqrt{n}$ respectively.

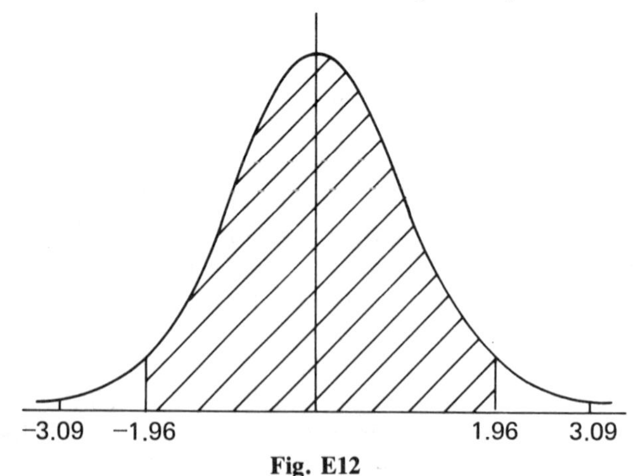

Fig. E12

Example 21.5(a)
Components are being mass-produced with their length the critical factor. The planned mean length is 15 mm with a standard deviation of 0.08 mm. Determine the positions of the lines on a quality control chart if samples of size 10 are taken.

The inner lines will be at $15 \pm 1.96 \times 0.08/\sqrt{10}$, i.e. 15.05 mm and 14.95 mm.

The outer lines will be at $15 \pm 3.09 \times 0.08/\sqrt{10}$, i.e. 15.08 mm and 14.92 mm.

Example 21.5(b)
In order to set up a quality control chart to check the breaking strength of rods, the 10 samples in Table 21.8 were taken once production was considered stable. All breaking loads were measured in newtons, each value being correct to three significant figures. Determine the mean breaking load and the standard deviation from that mean and hence establish the positions of the lines on the control chart.

The mean value and the range of each of the 10 samples are as follows.

Sample 1	Mean	501.75	Range	7
Sample 2	Mean	499	Range	9
Sample 3	Mean	500.25	Range	9
Sample 4	Mean	499.75	Range	1
Sample 5	Mean	499	Range	7
Sample 6	Mean	499.5	Range	5
Sample 7	Mean	500.25	Range	3
Sample 8	Mean	500	Range	6
Sample 9	Mean	499	Range	9
Sample 10	Mean	500	Range	9

The mean of the sample means is approximately 499.9. This can be taken as the population mean. The mean range $\bar{w}$ is 6.5. Using Table 21.7, this gives an estimate of the population standard deviation of $6.5 \times 0.49 \approx 3.2$.

Table 21.8

Sample	1	2	3	4	5	6	7	8	9	10
	501	503	497	500	504	501	502	503	502	505
	506	500	506	500	497	497	501	500	502	500
Loads	501	494	498	499	497	498	499	497	493	496
	499	499	500	500	498	502	499	500	499	499

The inner control lines are at $499.9 \pm 1.96 \times 3.2/\sqrt{4}$, i.e. 503 N and 496.8 N.

The outer control lines are at $499.9 \pm 3.09 \times 3.2/\sqrt{4}$, i.e. 504.8 N and 495.0 N.

21.6 Quality Control Charts for Range

Since the positions of the lines on a control chart for means depend partly on σ, it is necessary to monitor its value and ensure that it remains reasonably constant during the manufacturing process. This is effected by keeping a check on the ranges as samples are taken and plotting the results on a chart similar to that used for means. It is not strictly necessary to worry about lower limits (although some range charts include them) and the charts in this book will have two upper limits calculated at positions such that not more than 1 in 40 and not more than 1 in 1000 range values exceed them.

The positions of the lines are calculated from the formulae $D_{0.999}\bar{w}$ and $D_{0.975}\bar{w}$, where $D_{0.999}$ and $D_{0.975}$ are constants whose suffix indicates the probability of the range value falling below the line. The values of these constants are given in Table 21.9. They are, of course, dependent on the value of the sample size n.

Table 21.9

From O. Davis & P. Goldsmith: *Statistical Methods in Research and Production*, published by Longman Group Ltd., London, 1976, by permission of the authors and publisher.

Sample size	2	3	4	5	6	7
$D_{0.975}$	2.81	2.17	1.93	1.81	1.72	1.66
$D_{0.999}$	4.12	2.99	2.58	2.36	2.22	2.12

Example 21.6(a)

Use the values in Table 21.8 and the constants in Table 21.9 to verify that the ranges are all within the expected limit.

As $n = 4$ and $\bar{w} = 6.5$ the position of the inner line on a range chart is at $1.93 \times 6.5 = 12.54$. No range value exceeds this.

21.7 Choice of Sample Size

The choice of sample size may be influenced by practical factors such as the time and expenses involved in taking samples and overall production and quality levels. Other than this, the main effect of increasing the sample size is to increase the sensitivity of the testing.

Suppose that testing is planned around the production of units with a mean length of 6 cm and a standard deviation of 0.1 cm. If samples of size 4 are

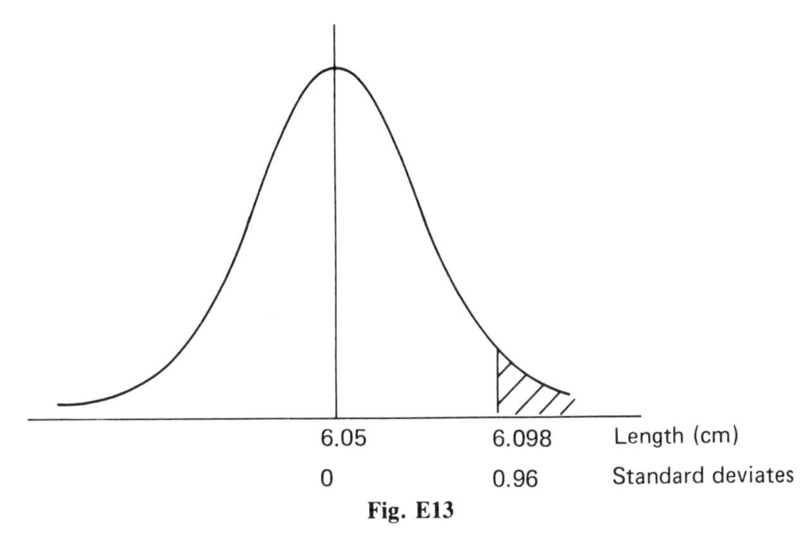

Fig. E13

taken, the inner control lines would be positioned at $6 \pm 1.96 \times 0.1/\sqrt{4} = 6.098$ cm and 5.902 cm. If the mean production length actually increased to 6.05 cm, leaving the standard deviation unchanged, the probability of a point falling outside the upper warning line can be obtained using a normal distribution. Referring to Figure E13, a length of 6.098 cm is at a position corresponding to $0.048/(0.1/\sqrt{4}) = 0.96$ standard deviates. The probability of a sample mean exceeding this can be obtained from Table 21.1 as $1 - A(0.96)$, i.e. 0.169. This means that about 1 point in 6 will fall above the line rather than 1 in 40. Using the same argument it can be shown that about 1 point in 55 will fall outside the action line.

If the sample size were doubled, leaving all other figures unchanged, the positions of the inner lines would be at $6 \pm 1.96 \times 0.1/\sqrt{8}$, i.e. 6.069 and 5.931. The length 6.069 cm. would correspond with 0.537 standard deviates and the probability of a point falling outside the upper line would increase to 0.3 so that about 1 point in 3 would warn of a process failure. Similarly about 1 point in 21 would fall outside the action line.

Example 21.7(a)
The exact contents of bottles containing nominally 1 litre of wine are actually normally distributed with a mean capacity of 1004 ml and a standard deviation of 1.2 ml. Find the positions of the 1 in 40 and 1 in 1000 control lines if samples of size 5 are used for inspection. If the mean capacity drifted up to 1004.5 ml, leaving the population standard deviation unchanged, find the probability of a point falling above (i) the upper 1 in 40 line (ii) the upper 1 in 1000 line.

The inner control lines are $1004 \pm 1.96 \times 1.2/\sqrt{5}$, i.e. 1005.05 and 1002.95.

The outer control lines are at $1004 \pm 3.09 \times 1.2/\sqrt{5}$, i.e. 1005.66 and 1002.34.

If the capacity increases to 1004.5 ml, the 1 in 40 line lies at a distance $0.55/(1.2/\sqrt{5}$, i.e. 1.02 standard deviates. The probability of a point falling above this line $= 1 - A(1.02) = 0.15$.

The corresponding figure for the 1 in 1000 line is $1.16/(1.2/\sqrt{5}) = 2.16$ standard deviates and the corresponding probability is $1 - A(2.16) = 0.015$.

Example 21.7(b)

If components are produced with a planned mean life of 2000 working hours and a standard deviation from that mean of 15h and are normally distributed, find the minimum sample size required such that more than 20% of the samples have a mean life that falls outside the inner control line of a quality control chart if the life of the components falls by 6 h.

Let the sample size be n. From Table 21.1 $A(0.84) \approx 0.8$. Since the position of the inner control line is $200 - 1.96 \times 15/\sqrt{n}$, this must correspond to 0.84 standard deviates from a mean life of 1994 hours. Hence

$$2000 - 1.96 \times 15/\sqrt{n} = 1994 - 0.84 \times 15/\sqrt{n}$$
$$6 = 1.12 \times 15/\sqrt{n}$$
$$n = 7.84$$

Samples of size 8 are required.

21.8 Quality Control Charts for Attributes

Where testing is qualitative rather than quantitative, a quality control chart must be set up for attributes. This can take several forms, but if samples of a fixed size are inspected and the number of unsatisfactory items in a sample is determined, these values can be plotted on a chart of the type shown in Figure E14.

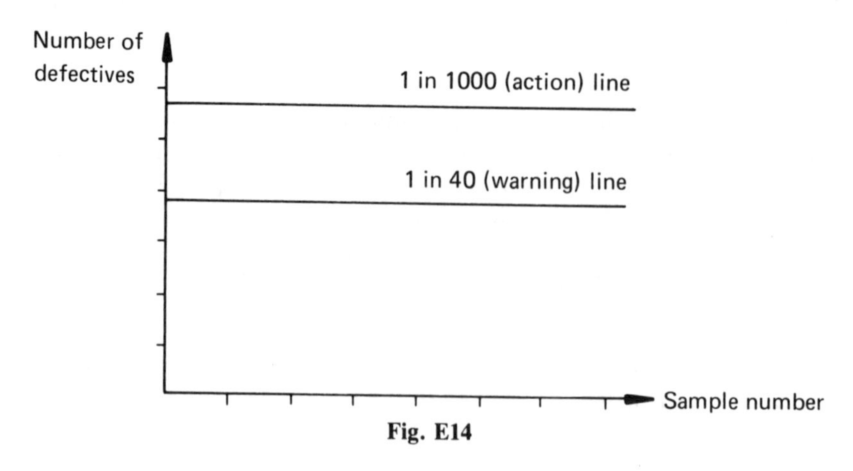

Fig. E14

If the probability of an item being unsatisfactory is p and of its being satisfactory is q, then with samples of size n, the distribution of defective items is binomial. However, qualitative testing is often visual and therefore fairly rapid and it is not unusual to have large samples with small numbers of defective items. This means that the distribution can be approximated by either a normal or a Poisson distribution and therefore the action line can be taken at $np + 3.09\sqrt{(np)}$ and the warning line at $np + 1.96\sqrt{(np)}$, i.e. $m + 3.09\sqrt{m}$ and $m + 1.96\sqrt{m}$.

Example 21.8(a)
In a testing process, the sample is destroyed and consequently the inspection is qualitative. Fifteen samples, each of size 50 are inspected and the numbers of defective items are respectively 0, 1, 0, 0, 1, 2, 1, 1, 0, 2, 0, 0, 1, 1, and 0. Find the positions of the inner and outer control lines on an attributes chart.

The mean number of defectives per sample is $10/15 = 0.67$. Taking this value as m, the position of
(a) the inner control line is at $0.67 + 1.96\sqrt{0.67} = 2.27$,
(b) the outer control line is at $0.67 + 3.09\sqrt{0.67} = 3.19$.

Example 21.8(b)
In a finished product, member is used for strengthening purposes. Batches of them are tested for breaking strength and the numbers in each batch failing to reach the required standard are 2, 1, 2, 1, 1, 0, 3, 2, 2, 4, 2, 1, 5, 2, 1, 3, 1, 2, 3 and 3. Plot these values on an attributes chart whose lines are at positions determined from an expected mean number of failures of 2.2 per sample. Is the process under control?

The 1 in 40 and 1 in 1000 lines are positioned at

$$2.2 + 1.96\sqrt{2.2} = 5.11$$
$$2.2 + 3.09\sqrt{2.2} = 6.78$$

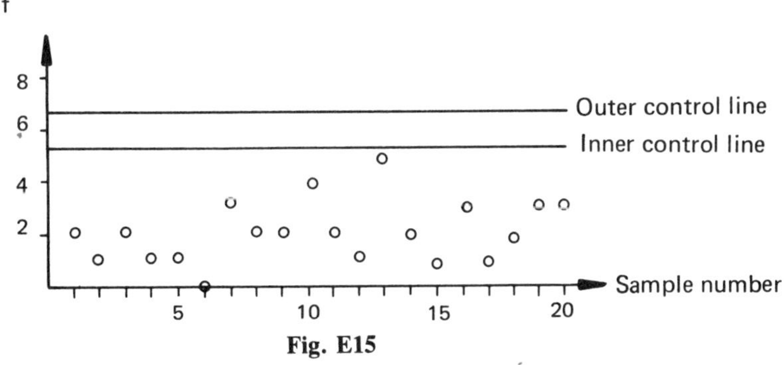

Fig. E15

The chart with the points plotted is shown in Figure E15. The process appears to comfortably under control.

Unworked Examples 2

Part 21.3

(1) A single sampling scheme operates as follows. Inspect a sample of size 40. Accept the batch if it contains three or fewer defectives otherwise reject it. Find the producer's and consumer's risks for an AQL of 2% and an LTPD of 20%. Give your answers correct to three places of decimals.

(2) Repeat Question 1 with
  (a) a sample size of 100, an acceptance condition of six or fewer failures but with the same AQL and LTPD.
  (b) the sample size and acceptance conditions of Question 1 but with an AQL of 3% and an LTPD of 25%.

(3) Which of the following schemes is
  (a) more likely to accept a batch containing 4% defectives?
  (b) more likely to reject a batch containing 20% defectives?
  Scheme A: Allow 0 or 1 defective in a sample of 10.
  Scheme B: Allow 0, 1 or 2 defectives in a sample of 20.

(4) In a single sampling scheme, a batch is accepted if it contains no more than one substandard item per sample. If the sample size is n and the AQL is 3%, show that the probability of accepting the batch is $e^{-0.03n}(1 + 0.03n)$, if a Poisson distribution is used. By drawing a graph of y against n for $y = e^{-0.03n}(1 + 0.03n)$ or by using a suitable method from Section D, deduce the value of n which gives a producer's risk of 3%.

(5) Find the producer's and consumer's risks using the following double sampling scheme with an AQL of 3% and an LTPD of 20%. Inspect a sample of size 8.
  (a) Reject the consignment if it contains two or more substandard items.
  (b) Accept the consignment if it contains no substandard items.
  (c) In other cases inspect a second sample and accept the consignment if this contains no substandard items.

(6) Fine the mean number of items inspected in the scheme described in Question 5.

Parts 21.4, 21.5 and 21.6

(7) Automatically produced components are known to be normally distributed with a mean length of 8.5 cm and a standard deviation of

0.12 cm. If samples of size 10 are used for inspection purposes, find the probability of the mean length of a sample exceeding 8.51 cm.

(8) The thickness of washers has to fall within certain limits in order to fit in as assembled unit. If it is planned that the mean thickness is to be 0.8 mm with a standard deviation that does not exceed 0.02 mm, find where the control lines on a quality control chart for means must be drawn if samples of size 20 are to be used.

(9) Electrical resistors having a nominal resistance of 50 ohms are to be produced. Production is to be monitored and the first 10 samples of size 5 are shown in Table 21.10. Find the mean of the sample means and the mean range. What will be the position of the control lines on (a) a means chart (b) a range chart?

Table 21.10

Sample	1	2	3	4	5	6	7	8	9	10
	50.1	50.0	49.9	50.0	49.9	50.0	50.1	50.0	49.8	50.1
	50.3	49.7	50.0	50.2	50.0	50.2	49.9	50.0	50.2	49.8
Resistances	50.1	50.0	50.0	49.8	50.1	50.0	49.9	50.0	50.1	50.1
(ohm)	49.9	49.9	50.0	49.8	50.1	49.8	50.1	50.1	50.0	50.2
	50.2	50.3	50.0	49.9	49.7	49.9	50.1	50.0	50.1	50.0

(10) Samples of size 4 are used in controlling production of radio components. If they are required to have a mean life of 500 h with a standard deviation from this of 15 h, determine the positions of the lines on a control chart for means and on a corresponding chart for ranges.

(11) In taking samples of six breaking strengths, the mean value of sample means is 1584 N and the mean range is 48.4 N. Use these values to set up a control chart for means. Would the 10 further sample means 1598, 1588, 1594, 1562, 1574, 1589, 1593, 1591, 1588 and 1583 indicate the process to be out of control?

Part 21.7

(12) The internal diameters of tubes vary normally with a mean value of 1.84 cm and a standard deviation of 0.026 cm. If, because of machine drift, the mean diameter falls to 1.83 cm, the standard deviation being unchanged, find the probability that a point will fall (a) below the lower inner line, (b) above the upper line, if samples of size 8 are used for inspection purposes.

(13) Use the data in question 12 to find the probabilities of a point falling outside the outer lines.

(14) A canning company grows fruit in its own orchards. For successful canning, the weights of the fruit must be normally distributed with a mean value of 200 g and a standard deviation of no more than 15 g. Using samples of size 4, find the probability that the mean sample weight will fall outside the inner lines of a control chart if the standard deviation reached 20 g.

(15) Given samples with a mean value of 20 and a mean range of 4.5, find the smallest sample size needed to ensure that if the sample mean increased to 20.5, 1 point in 10 will fall outside the upper warning line of a control chart for means.

(16) If units are to be produced with a mean length of 18.5 cm and a standard deviation of 0.3 cm, find the sample size needed so that at least one point in five falls outside an inner line of a control chart if the length changes by 0.05 cm.

Part 21.8

(17) In a quality control test for attributes, 10 samples are taken and the numbers of defective items are respectively 0, 1, 0, 0, 1, 0, 2, 1, 0 and 0. Find the mean value of these figures and use it to determine the positions of the warning and action lines on a quality control chart for attributes.

(18) Find the positions of the 1 in 40 (warning) and the 1 in 1000 (action) lines on a quality control chart for attributes given an average of one defective item per sample. Using the value $m = 1$, obtain Poisson probabilities $P(0)$, $P(1)$, $P(2)$, $P(3)$ . . . and verify the results above by finding integers x_1 and x_2 such that $P(x_1)$ is less than 0.025 and $P(x_2)$ is less than 0.001.

(19) Ten samples of furnishing materials were inspected and the mean number of finish defects per sample was found to be 0.7. Using this value, set up a quality control chart for attributes and use it to determine whether the standard of finishing is unchanged if a further 10 samples have the following numbers of defects within them: 0, 0, 1, 0, 2, 1, 2, 0, 0 and 1.

(20) It is planned to set up production with quality control monitoring. If it is considered that a 2% defect rate is acceptable, where should the inner and outer lines on a quality control chart for attributes be drawn based on the defect rate given above, if samples of size 10 are taken? Suggest why a larger sample size would be more sensible.

(21) If only a limited number of samples are to be inspected, it may be preferable to set up a quality control chart for attributes with just one control line designated 1 in 10. If the mean number of defective items per sample is m, obtain a formula to determine the position of the line.

PART 22: STATISTICAL TESTING

22.1 Biased and Unbiased Estimates

In Part 21 the properties of samples taken from a population with known parameters were discussed. It is more usual, however, to face the reverse problem—deducing the properties of a statistical population from random samples.

If a number of samples are taken, it is possible to extract information from each—its mean, its median, its standard error, its variance (square of the standard error) and so on. From these sampling distributions, it is possible to estimate, in more than one way, the population mean μ and variance σ^2. For example, the mean value of sample means gives a reasonably accurate estimate of μ but the mean value of the sample variances, perhaps surprisingly, tends to be rather smaller than σ^2. The former is said to be an UNBIASED ESTIMATOR of μ but the latter is a BIASED ESTIMATOR of σ^2. If s^2 is the mean sample variance, a more accurate estimate of the population variance, is $(n - 1)s^2/n$ where n is the sample size. Obviously for large values of n, the mean sample variance will be close to σ^2. However, if the variance of the sample is calculated using the formula

$$\frac{\Sigma(x - m)^2}{n - 1}$$

where m is the sample mean, the result is an unbiased estimator of σ^2.

These facts can be illustrated using the figures in Table 22.1. The 50 values are normally distributed integers with a mean value, μ, of 50 and a variance, σ^2, of 188. If we regard these figures as a statistical population and each column as a sample selected from it (which strictly speaking makes the samples nonrandom), it will be found that the mean values of the samples,

Table 22.1

55	38	34	55	36
76	47	70	36	50
56	48	64	40	58
43	50	53	58	57
66	46	59	60	17
47	50	65	52	46
22	69	26	46	73
45	33	62	46	51
36	34	69	64	31
79	40	48	51	43

m, are 52.5, 45.5, 55.0, 50.8 and 46.2. The sums of the squares of the deviations from each sample mean, $\Sigma(x - m)^2$, are 2834.5, 1024.5, 2002, 711.6 and 2209.6 respectively. Estimating the population variance using the formula $\Sigma(x - m)^2/10$ gives 175.6. Using the formula $\Sigma(x - m)^2/9$, the result is 195.2, a value much closer to the true value of σ^2.

22.2 Significance Testing

Suppose that a coin is spun 10 times and lands 'heads' on 8 occasions. Which is more likely on the next spin—that the coin will land 'heads', it will land 'tails' or that there is an evens chance of either?

The sort of arguments that supporters of each view will offer run along one of the following lines:

(a) In the long run, a coin lands 'heads' and 'tails' an equal number of times. Since eight 'heads' have occurred in 10 spins, 'tails' are overdue and are more likely to occur on the next spin.

(b) Each spin of a coin is a separate event, independent of any previous spin. Whether it lands 'heads' or 'tails' is an evens chance.

(c) A coin which has landed 'heads' on 8 occasions out of 10 is biased towards 'heads'. There must therefore be a greater chance of the coin landing 'heads' than 'tails'.

Which of these views is correct?

View (a) can be quickly dismissed. While a coin will land an equal number of times 'heads' and 'tails' in the long run, this applies in the *very* long run. It is possible, and not that unusual, to have sequences of repeated results so that 8 'heads' in 10 spins has no special significance 'in the long run'. In any case, we have no knowledge of how the coin behaved on any previous spins, so the question of which outcome is overdue is a meaningless one.

View (b) is correct *if the coin is unbiased*. So is view (c) *if the coin is biased*. Which is the correct view depends upon whether the evidence that we have suggests that the coin is biased or unbiased.

At this point, it should be stressed that it is impossible to prove absolutely whether or not a coin is biased. It is certainly possible, in either case, to have a coin landing 'heads' 8 times out of 10. However, let us assume for the moment that the coin is unbiased and calculate the probability of obtaining the result in question.

The probabilities of obtaining 0, 1, 2 . . . 10 'heads' in 10 spins are found using a binomial or a normal distribution. Both give the following probabilities correct to three places of decimals

$$P(0) = 0.001 = P(10)$$
$$P(1) = 0.010 = P(9)$$
$$P(2) = 0.044 = P(8)$$

$P(3) = 0.117 = P(7)$

$P(4) = 0.205 = P(6)$

$P(5) = 0.246$

In each case, $P(r)$ is the probability of obtaining r 'heads' in 10 spins.

On the basis of the assumption that the coin is fair, is it a rare occurrence to obtain 8 'heads' in 10 spins? If we are to conclude that this is rare then so is obtaining 9 or 10 'heads' and also 0, 1 or 2 'heads' since this means 10, 9 or 8 'tails'.

The probability of obtaining any of these results is 2(0.001 + 0.010 + 0.044), which equals 0.110. This is about once in every 9 sets of coin spins and would not generally be regarded as especially unusual. The conclusion that most observers would draw from the result is that it does not constitute *strong* evidence that the coin is biased. There is some suggestion that it *might* be biased and further spinning would be helpful. If that took place, the evidence would no longer be based on the outcomes of 10 spins and a new conclusion could be arrived at.

All statistical conclusions are arrived by arguments made along these lines. The process is called SIGNIFICANCE TESTING. An original assumption, often expressed in a negative form (i.e. nothing has changed, there are no differences) is made. This is called the NULL HYPOTHESIS and the probability of the outcome in question is determined. On the basis of null hypothesis, if this probability is less than 0.05, the result is said to be SIGNIFICANT at the 5% level. Significance at this level is generally regarded as reasonable evidence for rejecting the null hypothesis, while significance at the 1% level is taken as strong evidence.

In the example cited, the null hypothesis was that the coin was unbiased. On that basis, it is no more unlikely that 2 (or fewer) 'heads' could occur in 10 spins than 8 (or more) 'heads' and that is why both sets of probabilities were included in the investigation. When probabilities at both ends of the range are included, the testing is said to be TWO-TAILED. In contrast, if a new industrial process is compared with an established one to see if it leads to improved results, the probability of *better* (as opposed to *different*) results is of interest. Only probabilities at one end of the range are considered and the test is ONE-TAILED.

Since all statistical measurements are spread over a range of values significance tests are designed to distinguish between real variations and those arising by chance. The significance level indicates the margin for error because there is always a possibility of rejecting a correct null hypothesis or of accepting a false one. The former possibility is called a TYPE 1 error and the latter a TYPE 2 error.

The majority of significance tests involve a normal distribution. To test whether a sample of size n with a mean value of m differs significantly from a population with a mean value μ and a standard deviation σ, the TEST STATISTIC $z = (m - \mu)/(\sigma\sqrt{n})$ is calculated. This means that using the

Table 22.2 Critical Values of z

Two-tailed test		One-tailed test	
5% sig. level	1% sig. level	5% sig. level	1% sig. level
1.96	2.58	1.65	2.33

areas under a normal distribution curve, the critical values of z are given in Table 22.2, since with the usual notation, we have

$$A(1.96) = 0.975$$
$$A(1.65) = 0.95$$
$$A(2.58) = 0.995$$
$$A(2.33) = 0.99$$

These figures are as accurate as Table 21.1 allows. Critical values of z for other significance levels can be obtained using Table 21.1.

An alternative way of expressing these ideas is to say that 95% of normally distributed sample means lie in the range $\mu \pm 1.96\sigma/\sqrt{n}$. This range of values is called the 95% CONFIDENCE INTERVAL and its extreme values are the 95% CONFIDENCE LIMITS for m. These confidence limits determined the positions of the inner lines on a quality control chart while the outer lines were at positions determined by the 99.8% confidence limits for m.

It is more usual to obtain confidence limits for the population mean μ from the sample mean m. With the usual notation, the 95% confidence limits for μ are $m \pm 1.96\sigma/\sqrt{n}$. This, of course, assumes that σ is known. In practice, it is unlikely that σ is known when μ is not and it must be estimated from the standard deviation of the sample s. When n is large, say greater than 25, s is a reasonably unbiased estimator of σ and the confidence limits can be taken as $m \pm 1.96\ s/\sqrt{n}$. When n is smaller than this, other considerations must be taken into account. This is discussed in Part 22.4.

Example 22.2(a)

A manufacturer claims to supply goods with a 2% defect rate. Would a total of 2 defective items in a sample of 10 be sufficient evidence to indicate that the manufacturer is exaggerating his claim?

The null hypothesis is that the manufacturer is not exaggerating. Using a binomial distribution with $n = 10$, $p = 0.02$ and $q = 0.98$ (or a Poisson distribution as an alternative), the probabilities of no defective and 1 defective item respectively in a sample of 10 are

$$P(0) = 0.817$$
$$P(1) = 0.167$$

and hence

$$P(2 \text{ or more}) = 0.016$$

This result is significant at the 5% level and almost significant at the 1% level and, on that basis, we reject the null hypothesis and conclude that the manufacturer appears to be exaggerating. In fairness, however, this is a very small sample on which to interpret the result and further testing would be sensible.

Example 22.2(b)
A company manufactures support struts which have a mean breaking strength of 1250 kN with a standard deviation from that mean of 185 kN. As a result of trials with more expensive raw material, a batch of 25 struts with a mean breaking strength of 1310 kN is produced. Is this evidence that the new material is producing stronger struts?

The null hypothesis is that the trial has not produced stronger struts. The test statistic

$$z = (1310 - 1250)/(185/\sqrt{25})$$
$$= 1.62$$

Since we are only interested in improved results, a one tailed test is relevant here and z is not significant at the 5% level.

This does not mean that stronger struts are not being produced—merely that such statistical evidence as has been produced is not convincing enough.

Example 22.2(c)
The lives of cutting tools produced by a manufacturer are normally distributed with a mean value of 68 min and a standard deviation from that mean of 9 min. Find the 95% and 99% confidence limits for a sample of six cutting tools.

The 95% confidence limits are $68 \pm 1.96 \times 9/\sqrt{6} = 75.2$ and 60.8 min.

The 99% confidence limits are $68 \pm 2.58 \times 9/\sqrt{6} = 77.5$ and 58.5 min.

Example 22.2(d)
A batch of electrical resistors are sold as having a mean resistance of 500 ohm with a standard deviation from that mean of 25 ohm. A test sample of 50 is found to have a mean resistance of 493 ohm. Is there any reason to suppose that the batch is below specification?

The null hypothesis is that the batch is not below specification. The test statistic $z = (493 - 500)/(25/\sqrt{50}) = -1.98$. The test here is two-tailed and z is significant at the 5% level.

While the result is statistically significant, whether this is important depends upon the use to which the resistors will be put and whether the result will be affected by the drop in resistance.

22.3 Sums and Differences of Normally Distributed Variables

If two independent normal distributions *A* and *B* have means μ_1 and μ_2 and variances σ_1^2 and σ_2^2, then it can be shown that the distribution of both the sum and the difference of the variables from *A* and *B* are normally distributed. It can also be shown that for the distribution of sums, the mean is $\mu_1 + \mu_2$ and the variance is $\sigma_1^2 + \sigma_2^2$. For the distribution of differences the corresponding statistics are $\mu_1 - \mu_2$ and $\sigma_1^2 + \sigma_2^2$, so that there is no differences in variance in the two cases.

The figures in Table 22.3 illustrate this. *A* and *B* are two small independent normally distributed populations each consisting of five members. *C* is the distribution of all possible sums of one member from each of *A* and *B* and *D* is the corresponding distribution of differences. The mean value of *C* is the sum of the means of *A* and *B* while the mean value of *D* is their difference. In both cases, the variance is the sum of the variances of *A* and *B*.

Example 22.3(a)

An assembly is composed of two types of rod linked together. If the first type has lengths which are normally distributed with a mean value of 84 cm and a

Table 22.3

Sample A					
Values:	44	43	30	46	37
Mean of sample:	40			Variance of sample:	34
Sample B					
Value:	27	33	36	30	24
Mean of sample:	30			Variance of sample:	18
Sample C (Sample A + Sample B)					
Values:	71	70	57	73	64
	77	76	63	79	70
	80	79	66	82	73
	74	73	60	76	67
	68	67	54	70	61
Mean of sample:	70			Variance of sample:	52
Sample D (Sample A − Sample B)					
Values:	17	16	3	19	10
	11	10	−3	13	4
	8	7	−6	10	1
	14	13	0	16	7
	20	19	6	22	13
Mean of sample:	10			Variance of sample:	52

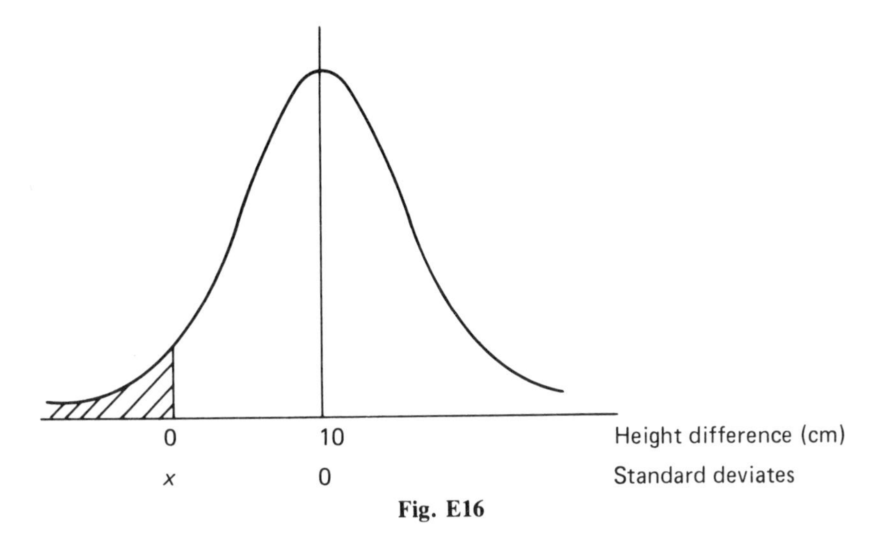

Fig. E16

standard deviation of 0.4 cm and the second type has corresponding statistics of 30 cm and 0.3 cm respectively, find the mean length and the standard deviation from the mean of the assembly.

Mean length of the assembly = 84 + 30 = 114 cm
Variance of assembly lengths = $0.4^2 + 0.3^2 = 0.25$ cm^2
Standard deviation required = 0.5 cm

Example 22.3(b)
The heights of both men and women are normally distributed. If the mean height of men is 175 cm and of women is 165 cm with standard deviations of 5 cm and 3.75 cm respectively, find what percentage of men marry women taller than themselves.

In this example, we are concerned with a difference in heights.

Mean difference = 175 − 165 = 10 cm
Standard deviation of differences = $\sqrt{(5^2 + 3.75^2)} = 6.25$ cm

Since the difference in heights is normally distributed, the percentage required corresponds to the shaded area in Figure E16.

$x = (0 - 10)/6.25 = -1.6.$

$A(x) = 0.055$

The required percentage is 5.5.

Unworked Examples 3

Part 22.1

(1) A sample of 10 items had the following values: 22.3, 19.8, 20.9, 18.8, 21.5, 17.9, 21.8, 19.0, 19.6 and 20.4. Obtain the best estimate of the mean and the standard deviation of the population from which the sample was extracted.

Table 22.4

10.0	10.5	9.2	10.2	11.8
10.4	10.1	9.2	9.7	9.7
9.2	8.0	11.1	9.3	10.0
11.5	9.6	10.4	9.7	9.2
8.4	11.8	9.8	9.1	11.2
11.7	9.5	12.2	10.8	10.2
9.9	11.4	8.5	10.6	9.9
10.9	9.1	10.3	11.0	9.8
10.5	9.2	11.7	11.3	9.3
8.5	8.5	10.5	9.6	8.5

(2) If a sample of eight lengths has a mean value of 17 cm and a standard deviation of 3 cm, estimate the mean and standard deviation of the population from which the sample is drawn.

(3) A random sample is taken from a population and is found to have a standard deviation of 8 units. Find the standard error of a sample mean if the sample size is 25. Determine the sample size needed to give a standard error of 1 unit.

(4) The values in Table 22.4 represent a population of 50 normally distributed variables. By dividing them into five samples of size 10 in any way you wish, estimate the population mean μ and standard deviation σ. For the latter use
   (a) an unbiased estimator of σ,
   (b) the mean sample range with Table 21.7.

(5) Divide Table 22.4 into 10 samples of size 5 and repeat Question 4.

Part 22.2

(6) A manufacturer claims that he supplies 50% of the market of a particular type of machine in a certain area. A large stockist in the area, holding 30 machines carries only 10 made by the manufacturer. Is the claim invalid?

(7) A manufacturer boasts a 1% defect rate. In destructive testing of his goods, there were two failures in 50 tests. Is this result significant?

(8) In comparing products from manufacturers A and B, seven tests were used. In six cases the product from manufacturer A performed better. Is this evidence of the superiority of the products of manufacturer A?

(9) If a coin were spun 100 times, what is the fewest number of heads needed to suggest that the coin is biased towards heads at (a) the 5% level (b) the 1% level?

(10) A manufacturer of electric light bulbs advertises a life of 1200 hours with a standard deviation of 100 h. Would a bulb having a life 1100 h be exceptional? Would a sample of five bulbs with a mean life of 1100 h be exceptional?

(11) A complex manufacturing process takes each operative an average time

of 45 min with a standard deviation of 5 min. After installing special equipment, a group of 12 operatives was able to reduce the process time to an average of 42 min. Is this evidence of the value of installing the equipment?

(12) A sample of 100 units of metal tubing had a mean internal diameter of 2.51 cm with a standard deviation of 0.02 cm. Find the 95% and 99% confidence limits for the tubing diameter.

(13) A manufacturer claims at most a 2% defect rate. Does a total of 4 substandard items in a set of 120 invalidate this claim?

(14) Bottles of cleaning fluid have a nominal content of 1 l. In fact, the bottling plant fills to a mean capacity of 1.004 l. with a standard deviation of 0.002 l. Comment if a sample of six bottles of the fluid has a total capacity of exactly 6 l.

(15) In a standard process for making steel wire, the product has a mean breaking strength of 82 kN with a standard deviation of 4.7 kN. A sample of 20 wires made by a new process had a breaking strength of 84.4 kN. Is this significantly stronger than the previous value?

Part 22.3

(16) My journey to work is partly by bus and partly by walking. The bus journey takes a mean time of 23 min with a standard deviation of 10 min and the walk takes an average of 8 min with a standard deviation of 1 min. Find the mean overall journey time and the standard deviation from the mean.

(17) A machine is kept stable by four weights. These are mass-produced and have a mean value of 4.8 kg. with a standard deviation of 0.1 kg. Find the 95% and 99% confidence limits for the total weight.

(18) A shoe manufacturer produces the right and left feet of a pair on different machines. For a batch of shoes of a particular size, the right shoes had a mean external length of 29.74 cm with a standard deviation of 0.03 cm. The corresponding values for the left shoes were 29.71 and 0.04 cm. Find the mean difference in the lengths of a pair of shoes and the standard deviation from this mean.

The *t* Distribution

The test statistic $z = (m - \mu)/(\sigma/\sqrt{n})$ is normally distributed and this allows us to determine whether values are statistically significant. However, in order to determine z, μ and σ must be known or at least deducible with reasonable accuracy. As we have seen, however, when n is small it is difficult to estimate σ accurately. Since the sample variance s^2 is a biased estimator of σ^2, even the expression $(m - \mu)/(s/\sqrt{(n - 1)})$ could be very different from z in value. The expression is, in fact, generally denoted by t and where n is small, say less than 25, is non-normal. Significant values are given in Table 22.5.

Table 22.5 Critical Values of t

Taken from Table III of Fisher & Yates: *Statistical Tables for Biological, Agricultural and Medical Research*, published by Longman Group Ltd., London (previously published by Oliver & Boyd Ltd., Edinburgh) by permission of the authors and publishers.

ν	Two-tailed test 5% sig. level	Two-tailed test 1% sig. level	One-tailed test 5% sig. level	One-tailed test 1% sig. level
1	12.71	63.66	6.31	31.82
2	4.30	9.92	2.92	6.96
3	3.18	5.84	2.35	4.54
4	2.78	4.60	2.13	3.75
5	2.57	4.03	2.02	3.36
6	2.45	3.71	1.94	3.14
7	2.36	3.50	1.89	3.00
8	2.31	3.36	1.86	2.90
9	2.26	3.25	1.83	2.82
10	2.23	3.17	1.81	2.76
12	2.18	3.05	1.78	2.68
15	2.13	2.95	1.75	2.60
20	2.09	2.85	1.72	2.53
24	2.06	2.80	1.71	2.49
30	2.04	2.75	1.70	2.46
40	2.02	2.70	1.68	2.42
60	2.00	2.66	1.67	2.39
120	1.98	2.62	1.66	2.36
∞	1.96	2.58	1.65	2.33

The value ν, known as the number of DEGREES OF FREEDOM, is the number of *independent* observations in the sample. The value of t is calculated from n observations but, once the first $(n - 1)$ have been taken, the estimated value of μ determines the nth value. Consequently, $\nu = n - 1$. As $\nu \to \infty$, values of t approach corresponding values of z.

Just as the confidence limits for a large sample of size n with a mean value m and a standard deviation s are $m \pm zs/\sqrt{n}$ (the value of z being chosen by the percentage limits required), the corresponding limits for a small sample are $m \pm ts/\sqrt{(n-1)}$.

It can be shown that providing z is not too large and ν is not too small, t is approximately equal to $(z^3 + z(4\nu + 1))/4\nu$. This verifies that $t \to z$ as $\nu \to \infty$ and that Table 22.2 could be replaced by the final line of Table 22.5.

Example 22.4(a)

An order is placed for castings whose weight is to be 8.5 kg. A random selection of 10 castings had the following weights (in kg) 8.41, 8.61, 8.38,

8.52, 8.43, 8.46, 8.39, 8.35, 8.40 and 8.45. Is this evidence that the castings are below the specified weight?

The null hypothesis is that the castings are not below the specified weight.

Mean sample weight, $m = 8.44$ kg

Standard deviation of sample, $s = 0.0725$ kg

$t = (m - \mu)/(s/\sqrt{(n - 1)}) = -2.48$

The sign of t indicates that the mean weight of the castings is below that specified. Its significance requires a one-tailed test. Taking the absolute value of t with $\nu = 9$, it will be seen from Table 22.5 that the result is significant at the 5% level. The statistical evidence that the castings are significantly below the specified weight. This assumes, of course, that the selection of castings in the sample was truly random.

Example 22.4(b)
To test whether a special finish strengthened the lives of cutting tools, one of a pair made by a similar process was subjected to the finish. This was repeated on five further occasions and the six pairs of tools were compared. The differences in lives measured in minutes were 2.1, 8.4, −3.8, 1.6, 7.9 and −2.4, the differences always being taken in the same order. Is there a difference in the lives?

The null hypothesis is that there is no difference, implying that $\mu = 0$.

Mean sample life, $m = 2.3$ min

Standard deviation, $s = 4.63$ min

$t = (m - \mu)/(s/\sqrt{(n - 1)}) = 1.11$

This value based on five degrees of freedom is not significant in a two-tailed test. The null hypothesis is upheld. There is such variation within the differences that a mean figure of 2.3 min cannot be considered significantly different from the assumed zero value.

Example 22.4(c)
Find the 95% confidence limits for the difference in lives of the cutting tools in Example 22.4(b).

Referring to Table 22.5, the appropriate value of t is 2.57. The required limits are $2.3 \pm 2.57 \times 4.63/\sqrt{5}$, i.e. 7.62 and −3.02 min. These are very wide limits because of the high value of s.

Example 22.4(d)
Use the approximation $t = (z^3 + z(4\nu + 1))/4\nu$ to determine the critical values of t for a two-tailed test at the 5% level when (i) $\nu = 20$ (ii) $\nu = 10$ (iii) $\nu = 5$.

In each case $z = 1.96$. Using the formula given, (i) $t = 2.08$ (ii) $t = 2.20$ (iii) $t = 2.43$.

By comparing these values with the values in Table 22.5, it will be seen how the accuracy of the formula drops off as ν decreases.

22.5 Differences Between Means

If two samples were taken from the same population, it is unlikely that their means would be exactly equal, but they would obviously not differ by much. The t distribution can be used to test whether such a difference is significantly different from zero and hence whether the samples are likely to have come from the same population.

If the samples are of sizes n_1 and n_2 with means m_1 and m_2 and variances s_1^2 and s_2^2, the best estimate of the variance of the population from which they are both assumed to be taken is

$$s_b^2 = \frac{(n_1 - 1)s_1^2 + (n_2 - 1)s_2^2}{n_1 + n_2 - 2}$$

If $n_1 = n_2$, s_b^2 simplifies to $\frac{1}{2}(s_1^2 + s_2^2)$.

The standard errors of the sample means can be taken as $s_b/\sqrt{n_1}$ and $s_b/\sqrt{n_2}$ respectively and, using a suitable result from Part 22.3, the standard error of their difference as $s_b\sqrt{(1/n_1 + 1/n_2)}$. The significance of the result can be tested using the formula $t = (m_1 - m_2)/s_b\sqrt{(1/n_1 + 1/n_2)}$ with $n_1 + n_2 - 2$ degrees of freedom.

Example 22.5(a)

A motorist obtains a mean life of 53 300 km from the four tyres on his car with a standard deviation of 350 km. He replaces them with tyres of a different make and obtains a mean life of 56 200 km with a standard deviation of 430 km from the four new tyres. Do these figures suggest that the second set of tyres last longer?

The null hypothesis is that the second set of tyres are not longer lasting, and that the difference arose by chance. The best estimate, s_b, is given by $s_b = \sqrt{\frac{1}{2}(s_1^2 + s_2^2)} = 392$ km.

$$t = (56\,200 - 53\,300)/392\sqrt{(1/4 + 1/4)} = 10.5$$

With $\nu = 6$, this value is significant at well beyond the 1% level in a one-tailed test. The conclusion must be that the second set of tyres give a longer life. It is assumed that both sets of tyres have been used in comparable ways.

Example 22.5(b)

A company produces torch batteries in two factories. The mean life of 100 batteries made in factory A is 67.9 h with a standard deviation of 2.1 h and the corresponding statistics for 80 batteries from factory B are 68.7 and 2.5 h. Is there a significant difference between the means?

The null hypothesis is that there is no difference. The best estimate of the variance is $(99 \times 2.1^2 + 79 \times 2.5^2)/178 = 5.23$. Hence $s_b = 2.29$ and $t = (68.7 - 67.9)/2.29\sqrt{(1/100 + 1/80)} = 2.33$ based on a large number of degrees of freedom.

This requires a two-tailed test and the result is not significant. In this example the normal distribution could have been used as n_1 and n_2 are both large.

22.6 Chi Square Test for Distributions

In the previous sections, tests were carried out to see whether a single representative figure or the difference between a pair of figures differed significantly from an expected result.

If it is suspected that data follow a particular distribution, a test is needed to see if the frequencies in each class fall within expected limits. This is called a CHI SQUARE TEST (χ^2 test) for goodness of fit and the value of χ^2 is calculated using the formula

$$\chi^2 = \sum_{i=1}^{n} (O_i - E_i)^2/E_i = \sum_{i=1}^{n} (O_i^2/E_i) - N$$

Table 22.6 Critical Values of χ^2

Taken from Table IV of Fisher & Yates: *Statistical Tables for Biological, Agricultural and Medical Research*, published by Longman Group Ltd., London (previously published by Oliver & Boyd Ltd., Edinburgh) by permission of the authors and publishers.

ν	5% sig. level	1% sig. level	99% sig. level
1	3.84	6.64	0
2	5.99	9.21	0.02
3	7.82	11.35	0.12
4	9.49	13.28	0.30
5	11.07	15.09	0.55
6	12.59	16.81	0.87
7	14.07	18.48	1.24
8	15.51	20.09	1.65
9	16.92	21.67	2.09
10	18.31	23.21	2.56
12	21.03	26.22	3.57
15	25.00	30.58	5.23
20	31.41	37.57	8.26
24	36.42	42.98	10.86
30	43.77	50.89	14.95

where O_i is the observed frequency of the ith of n classes and E_i is the expected or theoretical frequency. N is the total frequency of all classes. Obviously the size of χ^2 depends upon n, the number of pairs of values of O_i or E_i. Since their totals and possibly other statistics will be equal, the significant value of χ^2 will depend on the number of degrees of freedom with which it is associated. Critical values of χ^2 are given in Table 22.6. This table not only includes values at the 1% and 5% significance levels but also at the other extreme so that data which have been artificially contrived and fit too well can be detected. Values of χ^2 at or below the 99% level should be viewed with suspicion!

Generally, all of the terms of $(O_i - E_i)^2/E_i$ to be summed are of the same order of magnitude unless E_i is small. When this occurs, it may result in one term being large in comparison with the others and so distorting the value of χ^2. A rule of thumb is that $E_i > 5$ and, if a class is smaller than this, it should be combined with a neighbouring group.

It was shown in Part 21.2 that under certain conditions, one distribution may be approximated by another. For this reason, the test may show that it is possible for the data to belong to more than one distribution.

Example 22.6(a)

On a shop floor, five similar machines of different makes are working continuously. The numbers of working hours lost due to breakdowns over a period of six months is as follows

Machine	A	B	C	D	E
Number of hours lost	12.1	14.7	18.9	8.4	5.9

Is there a significant difference between the reliabilities of the machines?

The null hypothesis is that there is no difference. On this basis, the total breakdown time of 60 h would be shared equally between the machines making $E_i = 12.0$ for all five machines.

The calculation of χ^2 is laid out in Table 22.7, working to two places of decimals. There are five classes but as $\Sigma O_i = \Sigma E_i = 60$ only four values are

Table 22.7

	O_i	E_i	$(O_i - E_i)^2$	$(O_i - E_i)^2/E_i$
Machine A	12.1	12.0	0.01	0.00
Machine B	14.7	12.0	7.29	0.61
Machine C	18.9	12.0	47.61	3.97
Machine D	8.4	12.0	12.96	1.08
Machine E	5.9	12.0	37.21	3.10
	60.0	60.0		$\chi^2 = 8.76$

independent, making $\nu = 4$. From Table 22.6, this value is not significant at the 5% level and there are no grounds for rejecting the null hypothesis.

Example 22.6(b)
The distribution of accidents occurring in a factory over a period of 200 working days is as follows

Number of accidents in a day	0	1	2	3
Frequency	138	55	6	1

If accidents occur at random, these figures should follow a Poisson distribution. Do they?

In order to compare the given frequencies with the corresponding Poisson values, it is necessary to obtain the distribution parameter m, its mean. This is readily obtained as 0.35. With the usual notation

$$P(0) = e^{-0.35} = 0.705$$
$$P(1) = 0.35e^{-0.35} = 0.247$$
$$P(2) = 0.35^2e^{-0.35}/2! = 0.043$$
$$P(3) = 0.35^3e^{-0.35}/3! = 0.005$$

and the expected frequencies are these probabilities each multiplied by 200. Taken to the nearest integer these are 141, 49, 9 and 1.

The calculation of χ^2 is laid out in Table 22.8. Two of the classes have been combined because of the size of the final expected frequency. There are three classes used in the calculation. To obtain ν, this figure must be reduced by two because, firstly, the totals were made equal and, secondly, the values of O_i were used to calculate the parameter m. The value of χ^2 is not significant at the 5% level with $\nu = 1$.

Example 22.6(c)
Could the values of Example 22.6(b) be binomially distributed?

Since the mean value of the figures is 0.35, $np = 0.35$ and as $n = 200$, p is equal to 0.00 175, and q to 0.99 825. With these parameters, the probabilities

Table 22.8

Number of accidents	O_i	E_i	$(O_i - E_i)^2$	$(O_i - E_i)^2/E_i$
0	138	141	9	0.06
1	55	49	36	0.73
2 or 3	7	10	9	0.90
	200	200		$\chi^2 = 1.69$

and frequencies would be equal to those obtained in Example 22.6(b) when calculated to the nearest integer. Consequently the same value of χ^2 would be obtained.

Once again, there would be two constraints—totals and the value of p equal—so that ν must be taken as 1 once more.

Example 22.6(d)
The lengths of 100 manufactured components are distributed as follows:

Length (cm)	15.6–15.69	15.7–15.79	15.8–15.89	15.9–15.99	16.0–16.09	16.1–16
	2	14	36	28	15	5

Are these lengths normally distributed?

The theoretical distribution requires a knowledge of the mean and standard deviation of the figures. Together with the total of the expected frequencies being 100, this makes three constraints when deciding on the value of γ.

The figures for the calculation of the mean and standard deviation are laid out in Table 22.9. With the usual notation and formulae

$$m = 15.845 + 5.5/100 = 15.9$$
$$s = \sqrt{(1.55/100 - 0.055^2)} \approx \sqrt{(0.012)} = 0.11$$

The theoretical frequencies corresponding to a normal distribution require the end values of each datum class to be expressed in standard deviates. These are obtained using the formula $z = (x - m)/s$ and the values are shown in Table 22.10 corrected to two places of decimals. It also shows the areas from Table 21.1 corresponding to these values of z and the difference

Table 22.9
Let assumed mean, $A = 15.845$

Length	Frequency f	x	$x - A$	$f(x - A)$	$f(x - A)^2$
15.6–15.69	2	15.645	−0.2	−0.4	0.08
15.7–15.79	14	15.745	−0.1	−1.4	0.14
15.8–15.89	36	15.845	0	−1.8	0
15.9–15.99	28	15.945	0.1	2.8	0.28
16.0–16.09	15	16.045	0.2	3.0	0.60
16.1–16.19	5	16.145	0.3	1.5	0.45
	100			7.3	1.55
				$\Sigma f(x - A) = 5.5$	

Table 22.10

Length (cm) x	Standard deviations z	$A(z)$	Difference
15.595	−2.77	0.0028	
			0.0286
15.695	−1.86	0.0314	
			0.1397
15.795	−0.95	0.1711	
			0.3090
15.895	−0.05	0.4801	
			0.3250
15.995	0.86	0.8051	
			0.1565
16.095	1.77	0.9616	
			0.0347
16.195	2.68	0.9963	

Table 22.11

Length (cm)	O_i	E_i	$(O_i - E_i)^2$	$(O_i - E_i)^2/E_i$
15.6–15.69	2	3 }	1	0.06
15.7–15.79	14	14 }		
15.8–15.89	36	31	25	0.81
15.9–15.99	28	33	25	0.75
16.0–16.09	15	16 }	1	0.05
16.1–16.19	5	3 }		
	100	100		$\chi^2 = 1.67$

between successive values. These differences give the probability of each class and, multiplying by 100 and correcting to the nearest integer, the values of E_i are obtained. These appear in Table 22.11 in which χ^2 is calculated. Once again classes have been combined where values of E_i are small.

As $\chi^2 = 1.67$ when $\nu = 1$ (4 classes − 3 constraints) the value is non-significant. There is no reason to suppose the distribution is other than normal.

22.7 Contingency Tables

Suppose that a large company is deciding on a recruiting policy and is making a decision on whether to employ graduates or school-leavers. It might decide

Table 22.12

	Company service			
	Short	Medium	Long	Totals
School-leaver	46	44	30	120
Graduate	29	16	35	80
Totals	75	60	65	200

to study its employment records and base its policy on the lengths of service that it had received from each type of recruit. The results might be summarised in Table 22.12. A matrix of this kind is called a 2×3 CONTINGENCY TABLE, the 2 indicating two rows and the 3 indicating three columns.

If the null hypothesis that there is no difference in the lengths of service is postulated, the proportion of short stay school-leavers should be the same as the proportion of short stay recruits generally, i.e. $75 \times 120/200 = 45$. Similarly, the expected number of medium stay school-leavers is $60 \times 120/200 = 36$ and of long stay school leavers is $65 \times 120/200 = 39$.

If the calculation is continued in this way, the six observed and corresponding expected figures would be those shown in the first two columns of Table 22.13. The rest of the table is given over to a normal calculation of χ^2. The values in the E_i column are all calculated from values in the totals rows and columns of Table 22.12. Only two internal values need actually be given, as the rest may be determined from the totals and so the table is based on two degrees of freedom. The value of χ^2 is significant at the 1% level and we can almost certainly reject the null hypothesis. An inspection of the table shows that in terms of length of service, graduate recruitment is more profitable.

It can be shown that an $h \times k$ contingency table will be based on $(h - 1)(k - 1)$ degrees of freedom if h and k are both larger than 1.

Table 22.13

O_i	E_i	$(O_i - E_i)^2$	$(O_i - E_i)^2/E_i$
46	45	1	0.02
44	36	64	1.78
30	39	81	2.08
29	30	1	0.03
16	24	64	2.67
35	26	81	3.12
200	200		$\chi^2 = 9.70$

Table 22.14

Class of goods	Costs		
	Raw material	Manufacturing	Total
A	13	49	62
B	22	34	56
C	31	32	63
D	19	49	68
Totals	85	164	249

Table 22.15

O_i	E_i	$(O_i - E_i)^2$	$(O_i - E_i)^2/E_i$
13	21.2	67.24	3.17
49	40.8	67.24	1.65
22	19.1	8.41	0.44
34	36.9	8.41	0.23
31	21.5	90.25	4.20
32	41.5	90.25	2.17
19	23.2	17.64	0.76
49	44.8	17.64	0.39
249	249.0		$\chi^2 = 13.01$

Example 22.7(a)

A company makes four classes of goods. The raw material and manufacturing costs are shown in Table 22.14 as percentages of the selling price. Are there significant differences in these costs between different classes of goods?

The null hypothesis is that there is no difference in costs between classes of goods. The value of χ^2, calculated in Table 22.15, is 13.01. As $\nu = 3$, this value is significant at the 1% level.

Unworked Examples 4

Parts 22.4 and 22.5

(1) Matches are sold in boxes labelled 'average contents 35'. A random selection of twelve boxes had contents 35, 34, 37, 35, 32, 34, 35, 33, 36, 35, 36 and 34. Is the labelling accurate?

(2) Wine is sold having a nominally 11% alcohol content. An analysis of five bottles showed contents of 10.8%, 10.6%, 10.3%, 10.5% and 10.3%.

Do these values suggest that the alcohol content is less than the nominal figure?

(3) A machine should be set to produce units of length 15 cm. The first 20 units produced have the following lengths (in cm): 15.01, 15.07, 15.03, 15.08, 14.99, 15.01, 15.07, 15.09, 15.02, 15.00, 14.98, 15.02, 15.04, 15.08, 15.02, 15.00, 15.01, 15.08, 15.04 and 15.06. Does the machine need re-setting?

(4) Is there a significant difference between the means of the following values:
Set A: 28.0, 26.9, 27.0, 27.6, 27.3 and 27.6
Set B: 27.4, 26.8, 27.1, 27.5, 27.8 and 27.8

(5) In order to test a special finish used to strengthen steel wire, eight samples were taken and cut in halves. One of the halves was treated and the other left untreated. The breaking strengths of the eight pairs are shown in Table 22.16. Do the figures suggest that the finish strengthens the wire?

(6) Ten measurements of the diameter of a ballbearing in millimetres are 12.14, 12.18, 12.17, 12.11, 12.11, 12.15, 12.17, 12.16, 12.14 and 12.17. Find the 99% confidence limits for its diameter.

(7) A company owns two factories both producing steel of the same quality and using the same machinery with a comparable sized workforce. The mean weekly outputs in tonnes over a period of 20 weeks for the factories are 209.8 and 199.2 with standard deviations of 11.5 and 10.9 respectively. Is the difference in output significant?

(8) A research team claims that an extra treatment to the body of a car will allow a better application of paint. To test this, a sample of 40 cars, half treated and half untreated, are selected and the quality of paintwork is carefully inspected. The mean number of surface defects in the treated cars is 13.2 with a standard deviation of 3.8 while the figures for the

Table 22.16

	Breaking strength (kN)	
Sample No.	With a special finish	Without a special finish
1	18.7	19.1
2	18.5	18.8
3	19.1	19.0
4	18.8	18.9
5	18.6	19.1
6	18.6	19.2
7	19.1	18.8
8	18.3	18.4

untreated cars are 15.6 and 4.4 respectively. Do these figures justify the claims?

(9) If, in Question 8, the surface defect figures had been obtained from a sample of 50 treated and 50 untreated cars would the same conclusion have been reached?

Part 22.6

(10) A dairy uses four different types of bottle filling machine. The number of breakages over a period of one week for the machines are 16, 25, 17 and 22. Are these differences significant?

(11) Could the figures in Table 22.17 be distributed (a) binomially (b) Poisson fashion?

Table 22.17

Variable	0	1	2	3	4
Frequency	15	39	30	13	3

(12) In a factory making watches, a random sample of 20 were selected for inspection. It was found that 16 ran slightly fast and 4 slightly slow. Are these figures surprising?

(13) 75 components taken at random from a production line are tested to the limit of their mechanical lives. The results are shown in Table 22.18. Are the lives distributed normally?

Table 22.18

Life (h)	900–919	920–939	940–959	960–979	980–999
Frequency	2	13	38	19	3

(14) Table 22.19 shows the mean weekly value of production and the workforce of five factories belonging to a company. Is factory B exceptionally productive?

Table 22.19

Factory	Size of workforce	Value of production (£)
A	314	285 000
B	208	293 000
C	151	142 000
D	132	129 000
E	117	98 000

Table 22.20

Number of accidents	0	1	2	3	4
Frequency	42	5	2	0	1

(15) The numbers of fatal accidents per week in a particular industry taken over a period of 50 weeks are summarised in Table 22.20. Are these figures distributed in Poisson style?

(16) Could the distribution of Table 22.20 be binomial?

Part 22.7

(17) Table 22.21 shows a 2 × 2 contingency table. Verify that $\chi^2 = T(ad - bc)^2/T_1T_2T_3T_4$ with $\nu = 1$.

Table 22.21

	Class 1	Class 2	Total
Category 1	a	b	T_1
Category 2	c	d	T_2
Total	T_3	T_4	T

(18) If in Question 17, $a = 19$, find all the other variables if $T_1 = T_2 = 50$, $T_3 = 40$ and $T_4 = 60$. Is the resulting value of χ^2 suspiciously small?

(19) The plates on a factory boiler are examined annually. Over a period of three successive years, the boiler was operated at 400, 420 and 440°C and the effect on plate wear is summarised in Table 22.22. Draw a conclusion.

Table 22.22

	Boiler operating temperature			
Effect on boiler plates	400°C	420°C	440°C	Total
Little wear	21	18	17	56
Some wear	18	14	15	47
Much wear	16	23	23	62
Total	55	55	55	165

(20) In a firm, apprentices are trained on the shop floor under the guidance of three instructors Mr. A, Mr. B and Mr. C. The results are shown in

Table 22.23

		Instructor	
	Mr. A	Mr. B	Mr. C
Reaching satisfactory standard	8	11	9
Not reaching satisfactory standard	2	1	3

Table 22.23. Do they suggest that Mr. B is a better instructor than the other two?

PART 23: REGRESSION AND CORRELATION

23.1 Scatter Diagrams

In Parts 21 and 22, we have been concerned with the behaviour of a single distribution. In this part, we will be considering the relationship between two associated sets of distributed variables. These are usually described as BIVARIATE DATA.

If one set of data values are $x_1, x_2, x_3 \ldots$ and the corresponding variables in the second set are $y_1, y_2, y_3 \ldots$ plotting the points (x_1, y_1), (x_2, y_2), $(x_3, y_3) \ldots$ produces a SCATTER DIAGRAM which gives a general picture of the relationship between the data sets. Figure E17 shows three possible forms that the scatter diagram might take. In (a), although there is nothing like a mathematical relationship between the variables, they tend to increase together. A similar situation exists in (b) except that one variable increases as the other decreases. In (c) there is little if any association between the variables. Such association is called CORRELATION and the

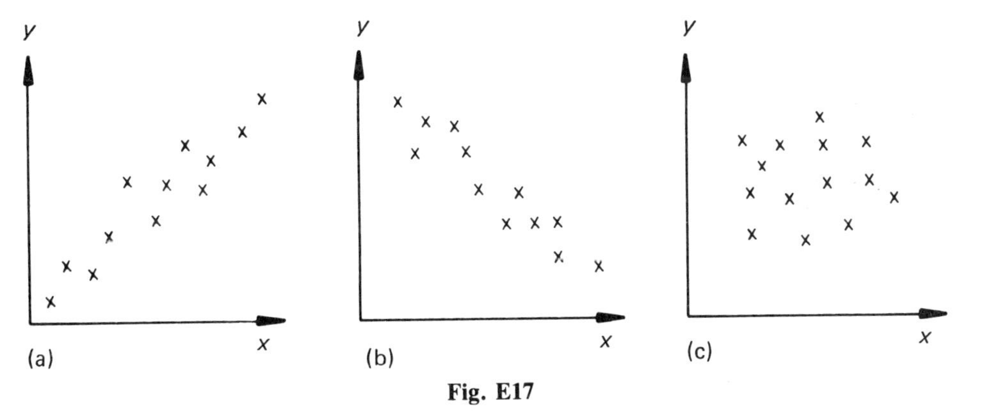

Fig. E17

graph linking the variables is called its REGRESSION CURVE (or REGRESSION LINE). Diagram (a) shows POSITIVE LINEAR REGRESSION, diagram (b) NEGATIVE LINEAR REGRESSION while diagram (c) indicates a lack of correlation.

23.2 Pairs of Regression Lines

With a reasonably accurate regression line, a degree of prediction of one variable for a given value of the other is possible. However, the formula used to determine a value of x from a given value of y is different from the formula used to determine y given x. To see why this is so, consider the heights and weights of 10 men given in Table 23.1.

There will obviously be some correlation, because from our experience, small heights tend to be associated with small weights and large heights with large weights, but there is no precise connection. In addition, it is impossible to decide whether weight is directly associated with height (i.e. a person is heavy because he is tall) or vice versa.

Let us consider both possibilities. The heights in Table 23.1 cover five different values from 160 to 180 cm. The two persons of height 160 cm each weight 55 kg and this can be taken as the mean weight associated with this height. Similarly the men whose height is 170 cm have weights of 55,55, 60 and 70 kg—a mean value of 60 kg. This, in turn, will be taken as the mean weight associated with a height of 170 cm. Continuing in this way, the complete set of figures shown in Table 23.2(a) would be obtained. The values in Table 23.2(b) have been obtained in the reverse manner, calculating the mean height associated with a given weight.

An inspection shows the same general tendency in the sub-sections of Table 23.2 and the complete Table 23.1—a definite but imprecisely defined tendency for height and weight to increase together. What is less clear is which height is associated with which height. Taking the top entries from the sub-sections of Table 23.2, is a weight of 55 kg associated with a height of 160 cm, of 165 cm or of somewhere between the two?

As is often the case when faced with apparently inconsistent figures, a graph or diagram is helpful. No simple curve will fit either set of five pairs of heights and weights, but if it is felt that the relationship between them is approximately linear, a sensible approach is to draw the line that most closely fits the points on a height–weight graph. If this is done with the two sets of points from Table 23.2, two lines are produced. These are shown in

Table 23.1

Height (cm)	160	160	165	170	170	170	170	175	180	180
Weight (kg)	55	55	55	55	55	60	70	65	70	80

Table 23.2

Height (cm)	Associated mean weight (kg)	Weight (kg)	Associated mean height (cm)
160	55	55	165
165	55	60	170
170	60	65	175
175	65	70	175
180	75	80	180
(a)		(b)	

Figure E18, the labelling (a) and (b) corresponding to that in Table 23.2. As line (a) has been drawn using weights associated with given heights, it is called the WEIGHT ON HEIGHT REGRESSION LINE. In the same way, line (b) is called the HEIGHT ON WEIGHT REGRESSION LINE. Unless there is an exact, mathematical relationship linking the variables, this approach will always produce two non-parallel regression lines intersecting in a central data point. Drawing regression lines by eye is, of course, extremely subjective. If, in addition, there are only five points for guidance, the resulting line may be quite inaccurate. An algebraic approach to this problem is considered in

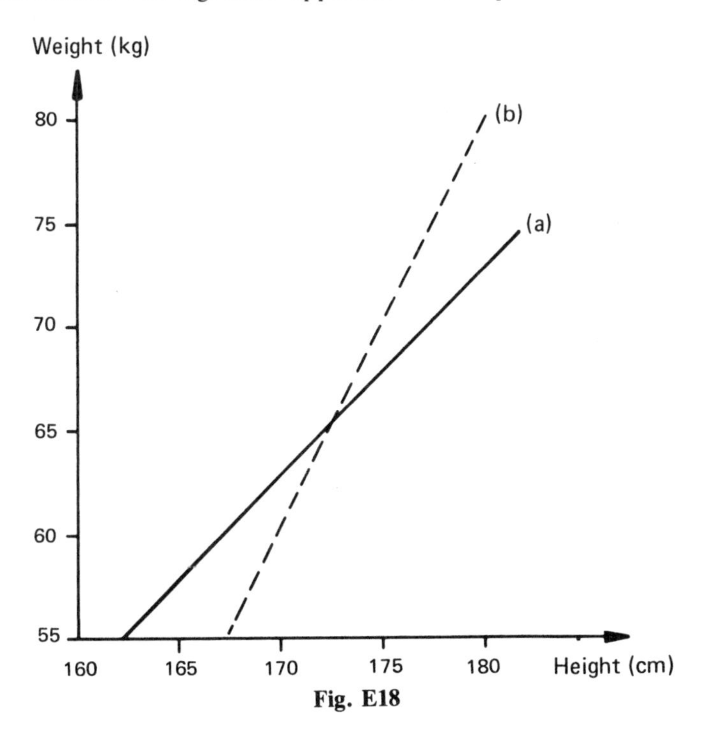

Fig. E18

Part 23.3 and this will give accurate regression lines. It will be shown that if these had been drawn in Figure E18, their point of intersection would be $(\bar{H}, \bar{W})$ where $\bar{H}$ is the mean height and $\bar{W}$ is the mean weight of all the data items (see Example 23.3(a)).

Table 23.3 Examination Marks

M	41	41	48	52	52	52	68	69	69	78
E	47	51	47	61	47	60	61	61	67	68

Example 23.2(a)

The examination marks of 10 students in Mathematics (M) and in Engineering Principles (E) are given in Table 23.3. Draw, by eye, the M on E and the E on M regression lines. Estimate marks for two students who respectively

(a) missed the Engineering Principles examination but scored 75 in Mathematics,

(b) missed the Mathematics examination but scored 55 in Engineering Principles.

Table 23.4 shows the marks for one subject together with the mean associated mark for the other and Figure E19 shows the two regression lines appropriately marked.

(a) From the E on M line, the Engineering Principles mark is estimated at 67.

(b) From the M on E line, the Mathematics mark is estimated at 55.

Example 23.2(b)

Calculate the mean marks in the two subjects using the data of Example 23.2(a). Compare the result with the point of intersection of the regression lines in Figure E19.

By calculation $\bar{M} = \bar{E} = 57$. This is the point of intersection.

Table 23.4

Mathematics marks	Assoc. Engineering Principles marks	Engineering Principles marks	Assoc. Mathematics marks
41	49	47	47
48	47	51	41
52	56	60	52
68	61	61	63
69	64	67	69
78	68	68	78
	(a)		(b)

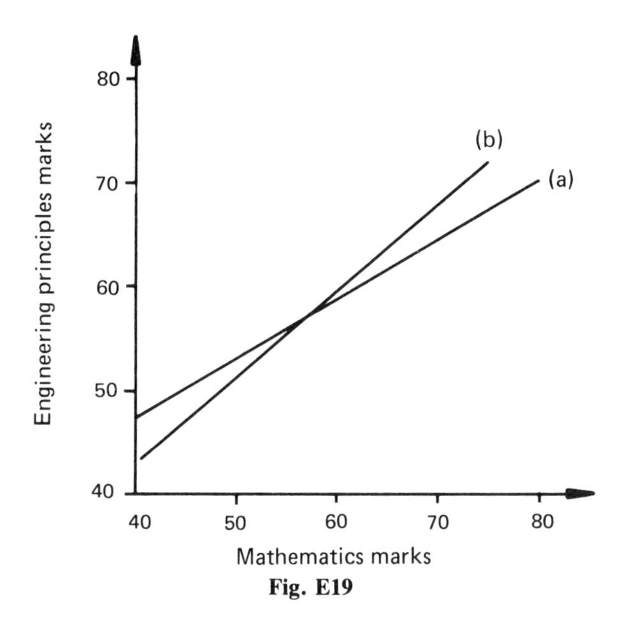

Fig. E19

23.3 Method of Least Squares

The algebraic method normally used to estimate the regression line of best fit involves the use of LEAST SQUARES.

Suppose that we wish to estimate the most accurate y on x regression line to the n points (x_1, y_1), (x_2, y_2), (x_3, y_3) . . . (x_n, y_n). We can take the equation of the line as $y = mx + c$ with the values of m and c to be determined. In Figure E20 it will be seen that the 'error' in the point P, (x_1, y_1) is PQ. Since Q lies on the regression line, its co-ordinates are $(x_1, mx_1 + c)$ and hence $PQ = y_1 - mx_1 - c$. Similarly, the 'error' in (x_2, y_2) is $\pm (y_2 - mx_2 - c)$, in (x_3, y_3) is $\pm (y_3 - mx_3 - c)$ and so on. (The sign depends on whether the point lies above or below the regression line.) To

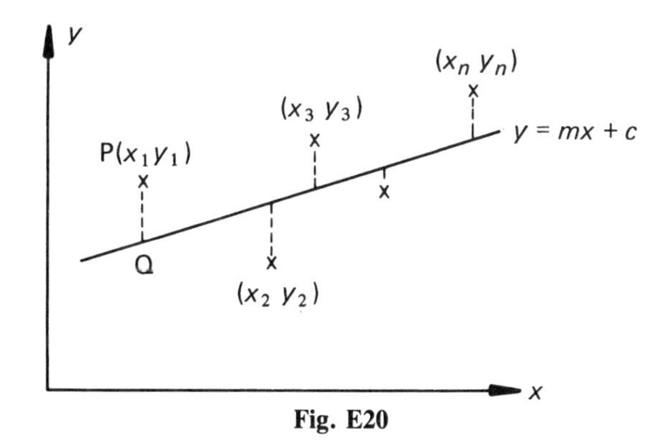

Fig. E20

allow for this ambiguity of sign, the minimum value of

$$S = \sum_{i=1}^{n} (y_i - mx_i - c)^2$$

is required. The size of S depends on the values of m and c and these can be found using the method of part 13.5, which requires the solution of the simultaneous equations $\partial S/\partial m = \partial S/\partial c = 0$

$$\partial S/\partial m = \sum_{i=1}^{n} - x_1(y_i - mx_i - c)$$

$$\partial S/\partial c = \sum_{i=1}^{n} - (y_i - mx_i - c)$$

It is convenient to drop the limits from these expressions and, after simplifying, to write the results in the form

$$\Sigma y = m\Sigma x + nc \quad (1)$$

$$\Sigma xy = m\Sigma x^2 + c\Sigma x \quad (2)$$

(The term nc replaces Σc as there are n pairs of values of x and y.) Equations (1) and (2) are called the NORMAL EQUATIONS for determining m and c. Dividing Equation (1) by n gives $\Sigma y/n = m\Sigma x/n + c$, i.e. $\bar{y} = m\bar{x} + c$ where $\bar{y}$ and $\bar{x}$ are mean values. Thus $(\bar{x}, \bar{y})$ lies on the y on x line, which, if preferred, can be written in the form

$$y - \bar{y} = m(x - \bar{x}) \quad (3)$$

The x on y line can be taken as $x = my' + c'$. Then, by a process comparable with that described above, the normal equations become

$$\Sigma x = m'\Sigma y + nc'$$

$$\Sigma xy = m'\Sigma y^2 + c'\Sigma y$$

These allow m' and c' to be determined and hence the equation of the x on y regression line. It can also be written in the form $(x - \bar{x}) = m'(y - \bar{y})$.

Example 23.3(a)
Find the equations of the height on weight and the weight on height regression lines using the data of Table 23.1. Find the mean height $\bar{H}$ and mean weight $\bar{W}$ and verify that the lines intersect at $(\bar{H}, \bar{W})$.

The calculation of the necessary totals is laid out in Table 23.5. Using these totals, $\bar{H} = 1700/10 = 170$ and $\bar{W} = 620/10 = 62$.

The H on W regression line is given by $H = mW + c$ where

$$\Sigma H = m\Sigma W + 10c$$

$$\Sigma HW = m\Sigma W^2 + c\Sigma W$$

Table 23.5

H	W	HW	W^2	H^2
160	55	8 800	3 025	25 600
160	55	8 800	3 025	25 600
165	55	9 075	3 025	27 225
170	55	9 350	3 025	28 900
170	55	9 350	3 025	28 900
170	60	10 200	3 600	28 900
170	70	11 900	4 900	28 900
175	65	11 375	4 225	30 625
180	70	12 600	4 900	32 400
180	80	14 400	6 400	32 400
1700	620	105 850	39 150	289 450

Substituting the totals in these equations and simplifying gives

$$62m + c = 170$$
$$3915m + 62c = 10\,585$$

Using any suitable method of solution we obtain $m = 0.634$ and $c = 130.7$ so that the H on W regression line is $H = 0.634W + 130.7$.

For the W on H regression line, $W = m'H + c'$, the normal equations simplify to

$$170m' + c' = 62$$
$$5789m' + 34c' = 2117$$

from which we obtain $m' = 1$ and $c' = -108$ so that the required equation is $W = H - 108$. The point (170, 62) satisfies both regression lines.

23.4 Gradients of the Regression Lines

Eliminating c from the normal equations for the y on x line, the gradient m will be obtained in the form

$$m = \frac{\Sigma xy/n - (\Sigma x/n)(\Sigma y/n)}{\Sigma x^2/n - (\Sigma x/n)^2}$$

The denominator may be recognised as the variance of the x values, conveniently denoted by s_x^2. By analogy, the numerator is designated the COVARIANCE of x and y, denoted by s_{xy}. Hence $m = s_{xy}/s_x^2$ and the y on x regression line can also be arranged in the form $y - \bar{y} = s_{xy}(x - \bar{x})/s_x^2$. In the same way, the gradient of the x on y regression line is given by $m' = s_{xy}/s_y^2$, and its equation is $x - \bar{x} = s_{xy}(y - \bar{y})/s_y^2$.

Table 23.6

Fuel costs (£'00)	1.75	1.84	1.82	1.93	2.17	1.95
Production (£'000)	283	297	308	305	325	318

Example 23.4(a)
Use the data of Table 23.5 to obtain the gradients of the H on W and the W on H regression lines.

$$s_{HW} = 105\,850/10 - 170 \times 62 = 45$$
$$s_W^2 = 39\,150/10 - 62^2 = 71$$
$$s_H^2 = 289\,450/10 - 170^2 = 45$$
$$m = s_{HW}/s_W^2 = 0.634$$
$$m' = s_{HW}/s_H^2 = 1$$

Example 23.4(b)
The data in Table 23.6 show fuel costs (F) and production costs (P) for a factory taken over six successive quarters. Find the F on P regression line and estimate the fuel costs for a quarter in which production was valued at £300 000.

The totals are shown in Table 23.7.

$$\bar{F} = 11.46/6 = 1.9 \quad \text{and} \quad \bar{P} = 1836/6 = 306$$
$$s_{FP} = 3516.29/6 - 1.91 \times 306 = 1.59$$
$$s_P^2 = 562\,936/6 - 306^2 = 186.7$$

The F on P line is $(F - 1.91) = 1.59(P - 306)/186.7$ which simplifies to $F = 0.0085P - 0.696$. When $P = 300$, $F = 1.85$.

The regression line could also have been obtained from the normal equations.

Table 23.7

F	P	FP	P^2
1.75	283	495.25	80 089
1.84	297	546.48	88 209
1.82	308	560.56	94 864
1.93	305	588.65	93 025
2.17	325	705.25	105 625
1.95	318	620.10	101 124
11.46	1836	3516.29	562 936

23.5 Correlation Coefficient

The y on x regression line has been used in the form $y - \bar{y} = s_{xy}(x - \bar{x})/s_x^2$. It could equally have been written as

$$\frac{y - \bar{y}}{s_y} = \frac{s_{xy}}{s_x s_y} \cdot \frac{x - \bar{x}}{s_x}$$

and then the expressions

$$\frac{y - \bar{y}}{s_y} \quad \text{and} \quad \frac{x - \bar{x}}{s_x}$$

are the x and y variables expressed in standard deviates. It is convenient to denote these by Y and X respectively, so that

$$Y = \frac{s_{xy}}{s_x s_y} X$$

Replacing $s_{xy}/s_x s_y$ by r, the y on x regression line simplifies to $Y = rX$. In exactly the same way the x on y regression line simplifies to $X = rY$ or $Y = (1/r)X$. These lines are shown in Figure E21. Their point of intersection is $(\bar{X}, \bar{Y})$ which will be the origin. (Why ?) Since the gradient of the Y on X line is r, it follows that in the diagram $\tan \alpha = r$. Similarly $\tan (90° - \beta) = 1/r$. But $\tan (90° - \beta) = \cot \beta = 1/\tan \beta$ and hence $\alpha = \beta$. The lines are equally inclinded to the X and Y axes. If there is perfect correlation between the variables, the lines coincide, $\alpha = \beta = 45°$ (or 135° for perfect negative correlation) and $r = 1$ (or $r = -1$). With no correlation between the variables, $\alpha = \beta = 0°$ and $r = 0$. In r, we therefore have a measure of just how closely the variables are associated and it is known as the

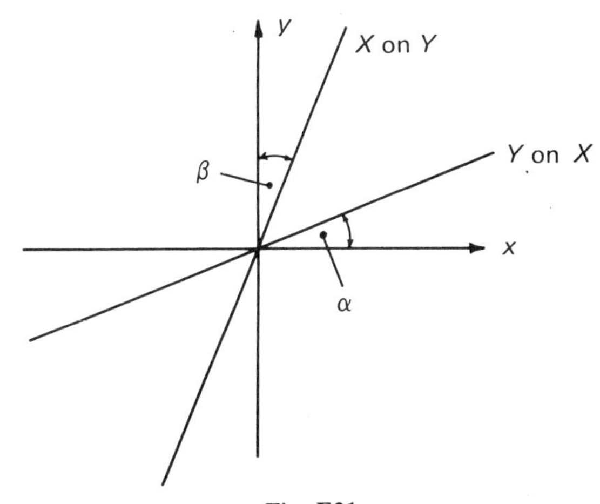

Fig. E21

Table 23.8

Cost per tonne (£)	28.4	30.0	30.5	32.5	33.3	34.7	40.5	46.1
Process time (min)	24.2	25.8	19.3	21.4	22.6	20.8	21.1	19.2

COEFFICIENT OF CORRELATION BY PRODUCT MOMENT. Its complete formula is

$$r = \frac{s_{xy}}{s_x s_y} = \frac{\Sigma xy/n - (\Sigma x/n)(\Sigma x/n)}{\sqrt{(\Sigma x^2/n - [\Sigma x/n]^2)}\sqrt{(\Sigma y^2/n - [\Sigma y/n]^2)}}$$

There are no units for r.

Example 23.5(a)
The operation time required to complete an industrial process depends on the quality of the material supplied. Table 23.8 shows the process time and cost per tonne of eight qualities of material. Find the coefficient of correlation between the variables.

Denoting cost as the x variable and process time as the y variable, the numerical details appear in Table 23.9. Hence

$$r = \frac{5959.44/8-(276/8)(174.4/8)}{\sqrt{(9771.5/8-(276\times276)(8\times8)}\sqrt{(3838.98/8-(174.4\times174.4)/(8\times8)}}$$

$$= -7.17/\sqrt{(31.19\times4.63)}$$

$$= -0.60$$

Example 23.5(b)
Show that the covariance of x and y can be written either in the form $\Sigma xy/n - (\Sigma x/n)(\Sigma y/n)$ or in the form $\Sigma(x - \bar{x})(y - \bar{y})/n$. Hence find an

Table 23.9

x	y	xy	x^2	y^2
28.4	24.2	687.28	806.56	585.64
30.0	25.8	774.00	900.00	665.64
30.5	19.3	588.65	930.25	372.49
32.5	21.4	695.50	1056.25	457.96
33.3	22.6	752.58	1108.89	510.76
34.7	20.8	721.76	1204.09	432.64
40.5	21.1	854.55	1640.25	445.21
46.1	19.2	885.12	2125.21	368.64
276.0	174.4	5959.44	9771.50	3838.98

alternative formula for r based on this. Verify the result of Example 23.5(a) using this formula.

$$\Sigma(x - \bar{x})(y - \bar{y}) = \Sigma xy - \bar{x}\Sigma y - \bar{y}\Sigma x + n\bar{x}\bar{y}$$

Hence

$$\Sigma(x - \bar{x})(y - \bar{y})/n = \Sigma xy/n - \bar{x}\Sigma y/n - \bar{y}\Sigma x/n + \bar{x}\bar{y}.$$

But

$$\bar{x}\Sigma y/n = \bar{y}\Sigma x/n = \bar{x}\bar{y}$$

so that

$$\begin{aligned}\Sigma(x - \bar{x})(y - \bar{y})/n &= \Sigma xy/n - \bar{x}\bar{y} \\ &= \Sigma xy/n - (\Sigma x/n)(\Sigma y/n)\end{aligned}$$

Similarly

$$\Sigma(x - \bar{x})^2/n = \Sigma x^2/n - (\Sigma x/n)^2$$

and

$$\Sigma(y - \bar{y})^2/n - \Sigma y^2/\mathrm{n} - (\Sigma y/n)^2$$

so that

$$r = \frac{\Sigma(x - \bar{x})(y - \bar{y})}{\sqrt{[\Sigma(x - \bar{x})^2\Sigma(y - \bar{y})^2]}}$$

From Table 23.9, $\bar{x} = 276/8 = 34.5$ and $\bar{y} = 174.4/8 = 21.8$. Using these values, the numerical part of the calculation of r is shown in Table 23.10. With these values $r = -57.36/\sqrt{(249.5 \times 37.06)} = -0.60$.

Table 23.10

$x - \bar{x}$	$y - \bar{y}$	$(x - \bar{x})^2$	$(y - \bar{y})^2$	$(x - \bar{x})(y - \bar{y})$
−6.1	2.4	37.21	5.76	−14.64
−4.5	4.0	20.25	16.00	−18.00
−4.0	−2.5	16.00	6.25	10.00
−2.0	−0.4	4.00	0.16	0.80
−1.2	0.8	1.44	0.64	−0.96
0.2	−1.0	0.04	1.00	−0.20
6.0	−0.7	36.00	0.49	−4.20
11.6	−2.6	134.56	6.76	−30.16
		249.50	37.06	−57.36

23.6 The Significance of r

Even if the values of a bivariate distribution are chosen completely at random, they are no more likely to make the coefficient of correlation *exactly* zero than to make it 1 or -1. For such random values, r will be small and either positive or negative and consequently a test of its value is needed to see if it is significantly different from zero. Ignoring the sign of r, it can be shown that $r\sqrt{(n-2)}/\sqrt{(1-r^2)}$ is distributed in exactly the same way as t, described in Part 22.4 and taking γ equal to $n-2$. As a result, its significance can be determined using Table 22.5 in conjunction with a one-tailed test. (The factor $n-2$ enters the calculation because if there are only two points, these will determine the regression line and are therefore perfectly correlated to it.)

Example 23.6(a)
Does the value of r obtained in Example 23.5(a) lie significantly above zero?

Here $|r| = 0.60$ when $n = 8$.

$$t = 0.60\sqrt{6}/\sqrt{(1 - 0.60^2)} = 1.84$$

with $\nu = 6$. This value is not significant at the 5% level.

Example 23.6(b)
Find the numerically smallest value of r significant at the 5% level if six pairs of values are taken.

When $n = 6$, t is given by the formula $t = 2r/\sqrt{(1-r^2)}$ with 4 d.f. The critical value of t at the 5% level is 2.13. Hence $4r^2/(1-r^2) = 2.13^2 = 4.54$. Solving this, $r = \pm 0.73$.

23.7 Correlation by Ranking

The calculation of the coefficient of correlation, r, is a cumbersome process and, even if it yielded only approximate results, a simpler method would be desirable. One such method calculates a correlation coefficient by comparing RANKS or ordered positions. The resulting value, designated R to distinguish it from its product-moment counterpart, is given by the formula

$$R = 1 - \frac{6\Sigma d^2}{n(n^2 - 1)}$$

where n is the number of pairs of observations and d is the difference in rankings between corresponding data values. If, for example, the x values were 12, 15, 18 and 22 and the corresponding y values were 42, 64, 50 and 58

Table 23.11

x values	y values	x rankings	y rankings	d	d^2
28.4	24.2	1	7	−6	36
30.0	25.8	2	8	−6	36
30.5	19.3	3	2	1	1
32.5	21.4	4	5	−1	1
33.3	22.6	5	6	−1	1
34.7	20.8	6	3	3	9
40.5	21.1	7	4	3	9
46.1	19.2	8	1	7	49
		36	36		142

then the x ranks would be 1, 2, 3 and 4 (in ascending order) and the y ranks would be 1, 4, 2 and 3. The differences d would be 0, −2, 1 and 1 so that $\Sigma d^2 = 0 + 4 + 1 + 1 = 6$. The value of R is therefore equal to $1 - (6 \times 6)/(4 \times 15) = 0.4$. If two values had been ranked in equal second position, their rankings would have been taken as 2.5 and so on.

If the largest of the y values were 164 or even 1064, it would not have affected the rankings and therefore would have left R unchanged. This possibility must be borne in mind when deciding how reliable is the value of R as a measure of association.

Rank correlation is useful when dealing with quantitative data because these can often be ranked even though it may be difficult or impossible to assign numerical values. It is particularly useful for comparing data involving such factors as colour range or quality of finish and can be used to compare the ranking judgement of two independent arbitrators.

When n is large R can be simplified to $1 - 6\Sigma d^2/n^3$.

The t test described in Part 23.6 cannot be used to determine the significance of R.

Example 23.7(a)
The coefficient of correlation by product-moment of the data in Table 23.8 has been found to have a value of $r = -0.60$. Find the coefficient of correlation by ranking.

The rankings are shown in Table 23.11. Useful checks are that $\Sigma d = 0$ and that the sums of the x and the y rankings total $\frac{1}{2}n(n + 1)$.
$R = 1 - (6 \times 142)/(8 \times 63) = -0.69$

Example 23.7(b)
Find the coefficient of correlation by ranking of the repeated data in Table 23.12.

Table 23.12

x values	10	10	10	12	12	13
y values	5	8	10	5	10	12

The three items of x data with a value of 10 are each ranked 2, i.e. rankings 2, 2 and 2 replace rankings 1, 2 and 3 while those with a value 12 are ranked 4.5 and so on. The complete layout is shown in Table 23.13.

$$R = 1 - (6 \times 16.50)/(6 \times 35)$$
$$\approx 0.53$$

Example 23.7(c)
Show that when correlating by rank, if the x and y rankings correspond exactly $R = 1$ while if they are in exact reverse order $R = -1$.

If the rankings correspond exactly, all values of d are zero so that $\Sigma d^2 = 0$ and hence $R = 1$.

If the rankings are in exact reverse order

$$\Sigma d^2 = (n-1)^2 + (n-3)^2 + (n-5)^2 + \cdots (5-n)^2 + (3-n)^2 + (1-n)^2$$
$$= 2[(n-1)^2 + (n-3)^2 + (n-5)^2 + \cdots]$$
$$= 2[(n^2 + n^2 + n^2 + \cdots) - 2(n + 3n + 5n + \cdots) + (1^2 + 3^2 + 5^2 + \cdots)]$$

There are $\frac{1}{2}n$ terms in each bracket

$$(n^2 + n^2 + n^2 + \cdots) = \tfrac{1}{2}n^3$$
$$(n + 3n + 5n + \cdots) = \tfrac{1}{4}n^3$$
$$(1^2 + 3^2 + 5^2 + \cdots) = \tfrac{1}{3}n(n^2 - 1)$$
$$\Sigma d^2 = \tfrac{1}{2}n^3 - \tfrac{1}{2}n^3 + \tfrac{1}{3}n(n^2 - 1) = \tfrac{1}{3}n(n^2 - 1)$$

and, substituting this value in the formula for R makes $R = -1$.

Table 23.13

x values	y values	x rankings	y rankings	d	d^2
10	5	2	1.5	0.5	0.25
10	8	2	3	−1	1.00
10	10	2	4.5	−2.5	6.25
12	5	4.5	1.5	3	9.00
12	10	4.5	4.5	0	0
13	12	6	6	0	0
		21.0	21.0		16.50

The expressions representing the sums of bracketed terms may be checked by taking even values for n and substituting.

Unworked Examples 5

Parts 23.1, 23.2 and 23.3

(1) Table 23.14 shows the initial cost (y) of five different makes of machine suitable for an industrial process and the annual production time lost due to breakdown (x).
(a) Draw a scatter diagram.
(b) Calculate the equation of the y on x regression line.

Table 23.14

Initial cost (£'000)	18.4	19.7	25.3	28.0	37.2
Loss of production time (h)	31	27	57	18	2

(2) For the data in Table 23.15 find
(a) the mean value of y for each x value,
(b) the mean value of x for each y value.
Draw the y on x and the x on y regression lines by eye and obtain their equations from the graph.

Table 23.15

x 100	100	120	120	120	130	130
y 50	64	64	70	70	70	70

(3) Use the method of least squares to find the equations of the regression lines of Question 2. Verify that they meet at $(\bar{x}, \bar{y})$.

(4) A manufacturer responsible for aircraft instrumentation is designing a unit able to stand a heavy impact. The relationship between the least impact velocity that damages the instrument and the thickness of casing is shown in Table 23.16. Find the regression lines linking the variables.

Table 23.16

Impact velocity (km h^{-1})	(x)	5	8	9	15	14
Casing thickness (mm)	(y)	1.8	2.0	2.2	2.4	2.6

(5) Find the equation of the straight line joining the points $(x_1\ y_1)$ and $(x_2\ y_2)$. Verify that the formulae of Part 23.3 give this for the y on x and the x on y regression lines.

(6) Show that the points $(1, -1)$, $(1, 1)$, $(-1, 1)$, and $(-1, 1)$ form a square whose centre is the origin. Using the method of least squares, find the x on y and y on x regression lines. What can you say about these lines?

Part 23.4

(7) If, for bivariate data, $\bar{x} = 10$, $\bar{y} = 20$, $s_{xy} = 1200$, $s_x = 80$ and $s_y = 60$ find the equation of the regression lines.
(8) Find the regression coefficients using the data of Table 23.12.
(9) Using the data in Table 23.17, find the regression lines. Estimate the value of x when $y = 5$ and the value of y when $x = 25$.

Table 23.17

x	10	12	18	24	30	32
y	7	4	6	4	2	1

(10) Confirm for the data in Table 23.17 that $\bar{x} = 21$ and $\bar{y} = 4$. Verify that the lines whose equations were determined in Question 9 intersect at (21, 4).

Part 23.5

(11) Find the coefficient of correlation by product-moment of the data in Table 23.14.
(12) Given a set of bivariate data (x_1y_1), (x_2y_2), (x_3y_3), . . . (x_ny_n), what is the effect on the product-moment correlation coefficient of
 (a) increasing all x values by 1 leaving all y values unchanged
 (b) increasing both x and y values by 1
 (c) doubling the x values leaving the y values unchanged.
 (d) doubling both x and y values.
(13) Verify the results of Question 12 using the data in Table 23.17
(14) Table 23.18 shows the relationship between the percentage quality of mined ore and the mining depth. Find the y on x regression line and the coefficient of correlation linking the variables.

Table 23.18

Depth (km)	(y)	0.4	0.5	0.6	0.7	0.8
% ore quality	(x)	2.9	3.1	3.5	3.6	3.4

(15) Table 23.19 shows randomly selected y and x integers. Find their coefficient of correlation using the product-moment method.

Table 23.19

y	3	−2	9	4	−7
x	1	7	8	2	4

Part 23.6

(16) The relationship between production and unit cost is shown in Table 23.20. Find r, the coefficient of correlation linking the variables.

(17) Test the significance of the value of r obtained from (a) Question 11 (b) Question 14 (c) Question 15.

(18) Find the smallest value of a correlation coefficient, r, which is significant at the 5% level if (a) 10 (b) 20 (c) 50 pairs of values are used.

Table 23.20

Daily production ('000)	25	28	35	47	65	73
Unit cost (pence)	71	69	67	63	61	58

Part 23.7

(19) The value of r, the product-moment coefficient of correlation is 0.813 for the data for Table 23.18. Find R, the rank coefficient of correlation.

(20) Find the rank coefficient of correlation for the data in (a) Table 23.15 (b) Table 23.19.

(21) Taking rankings 1 to 6 for one variable and reverse order rankings 6 to 1 for the second, show that $R = -1$. Repeat the process with rankings 1 to 10.

(22) Show that $R = 0$ for the ranked data of Table 23.21.

Table 23.21

Variable 1	1	2	3	4	5	6	7
Variable 2	5	4	6	1	2	3	7

Self-Assessment Questions

This section consists of a series of short questions with answers in many cases consisting of a formula or a few words. They should be used for self-assessment to check that the basic principles of each area of work have been absorbed and to supplement, rather than replace, the worked and unworked examples given in the text.

Section A

Part 1

(1) Express $\cosh x$ in the terms of e^x.
(2) Express $\sinh x$ in terms of e^x.
(3) Express $\tanh x$ in terms of $\sinh x$ and $\cosh x$.
(4) Find the value of $\cosh 0$.
(5) State whether the following functions are odd or even (a) $\sin x$ (b) $\cosh x$ (c) $\tan 2x$.
(6) Complete the formula $\cosh^2 x - \cdots = 1$.
(7) Use Osborn's formula to complete the formula $1 - \tanh^2 x = \cdots$.
(8) Write out the power series for (a) $\cosh x$ (b) $\sinh x$.

Part 2

(1) Add the complex numbers $(3 + j)$ and $(2 - j4)$.
(2) Subtract $2 + j$ from $6 - j5$.
(3) Express j^5 as an imaginary number.
(4) Write down the conjugate of $2 - j^3$.
(5) Express $2(\cos 30° + j \sin 30°)$ in the form $a + jb$ leaving a and b in surd form.

(6) Write down in polar form the product and quotient of $6(\cos 120° + j \sin 120°)$ and $2(\cos 40° + j \sin 40°)$.
(7) Obtain in polar form $\sqrt{(4(\cos 40° + j \sin 40°))}$.
(8) If $x + jy = r(\cos \theta + j \sin \theta)$, obtain formulae expressing r and θ in terms of x and y.
(9) State the modulus and argument of $2 + j2$.

Part 3

(1) Express $\cos \theta + j \sin \theta$ in exponential form.
(2) Complete the formula $e^{-j\theta} = \cdots$.
(3) Complete the theorem $(\cos \theta + j \sin \theta)^n = \cdots$.
(4) Whose name is associated with the formula in Question 3?
(5) Complete the relationships (a) $\cos jx = \cdots$ (b) $\sin jx = \cdots$
(6) Expand $\cos (x + jy)$, where x and y are real.
(7) Express $\sin x$ in terms of e^{jx} and e^{-jx}.
(8) Express $\cos x$ in terms of e^{jx} and e^{-jx}.

Part 4

(1) What is the name given to individual figures in a determinant?
(2) What is the name given to the subdeterminant remaining when the row and column containing a particular figure are removed?
(3) Complete the sentence 'If two rows or columns of a determinant are . . . the value of the determinant is zero'.
(4) If the position of all figures in a determinant are denoted by + and −, that in the top left-hand side will always be . . . ?
(5) In the following equations, (1) + (2) = (3):

$$3x + 4y - 2z = 1 \quad (1)$$
$$x - 3y + 5z = 7 \quad (2)$$
$$4x + y + 3z = 8 \quad (3)$$

(a) Is it possible to obtain unique values of x, y and z?
(b) Is it possible to obtain the ratio $x:y:z$?
(c) What is the name given to such a set of equations?

(6) If Equation (3) of Question 5 is changed to

$$4x + y + 3z = 9$$

(a) Is it possible to obtain unique values of x, y, z?
(b) What is the name given to such a set of equations?

(7) If Equation (3) of Question 5 is changed to $4x + 2y + 3z = 8$ is it

possible to obtain unique values of x, y and z? If so what value does y take?

(8) If the determinant solution of a set of simultaneous equations in unknowns x, y and z can be written

$$\frac{x}{D_1} = \frac{y}{D_2} = \frac{z}{D_3} = \frac{1}{\mathrm{D}_4}$$

state the condition for the equations to be (a) indeterminate (b) inconsistent.

(9) What name is given to simultaneous equations such that a small change in the coefficients produces a much larger change in the solution?

Part 5

(1) Are the following true or false?
 (a) Matrices need not have the same number of rows and columns.
 (b) Square matrices have a value.
 (c) All square matrices have an inverse.
 (d) For conformable matrices **A** and **B**, **AB** is always different from **BA**.

(2) If **A**, **B** and **C** are square matrices of the same order, write true or false to the following
 (a) det **AB** = det **A** det **B**,
 (b) **ABC** = **CBA**.

(3) If **AB** = **O** which of the following statements is correct?
 (a) **A** and **B** are inverses.
 (b) Either **A** or **B** must be null.
 (c) Both **A** and **B** must be null.
 (d) Either det **A** or det **B** or both are equal to zero.

(4) Simplify the matrix expression $\mathbf{ABCC}^{-1}\mathbf{B}^{-1}\mathbf{A}^{-1}$.

(5) If det **A** = 4
 (a) is **A** a singular matrix? If not
 (b) obtain det $\mathbf{A}^{-1}$.

(6) Complete the statement $(\mathbf{AB})^T = \cdots$

(7) Complete the statement $(\mathbf{A}^{-1})^T = \cdots$

(8) What is another name commonly given to a column matrix?

Part 6

(1) If a set of simultaneous equations are expressed in matrix form as **Au** = **b**
 (a) state the known matrices,
 (b) write down the solution in matrix form.

(2) State the solution of the equations

$$\begin{aligned} x \quad\quad\quad &= 5 \\ x + 2y \quad\ &= 7 \\ 3x - \ y - 6z &= 2 \end{aligned}$$

(3) What is the name given to a matrix, the rows of which consist of the coefficients of the unknowns together with the constant from the right-hand side of the equation?
(4) What term is used for two matrices **A** and **B**, such that $\mathbf{A} \sim \mathbf{B}$?
(5) If **L** is a 3×3 lower triangular matrix what is the minimum number of zero elements that it contains?
(6) If $\mathbf{U}_1$ is an $n \times n$ unit upper triangular matrix what is (a) the minimum (b) the maximum number of unity elements that it contains?
(7) With the usual notation if $\mathbf{Au} = \mathbf{b}$ and $\mathbf{A} = \mathbf{L}_1\mathbf{U}$, complete the following matrix equations
(a) $\mathbf{L}_1\mathbf{Uu} = \cdots$,
(b) If $\mathbf{Uu} = \cdots$, then $Ly = \mathbf{b}$.

Part 7

(1) If $\mathbf{a} = a_1\mathbf{i} + b_1\mathbf{j} + c_1\mathbf{k}$ and $\mathbf{b} = c_2\mathbf{i} + b_2\mathbf{j} + c_2\mathbf{k}$, express $\mathbf{a} \cdot \mathbf{b}$
(a) in terms of $|\mathbf{a}|$, $|\mathbf{b}|$ and the angle θ between **a** and **b**,
(b) in terms of the components of **a** and **b**.
(2) If a vector has direction cosines α, β and γ, what is the relationship connecting $\cos \alpha$, $\cos \beta$ and $\cos \gamma$?
(3) Add $3\mathbf{i} + 4\mathbf{j} + 2\mathbf{k}$ and $2\mathbf{i} - 5\mathbf{k}$.
(4) Are the following vector propositions true or false?
(a) $\mathbf{a} \cdot \mathbf{b} = \mathbf{b} \cdot \mathbf{a}$,
(b) $\mathbf{a} \cdot (\mathbf{b} + \mathbf{c}) = \mathbf{a} \cdot \mathbf{b} + \mathbf{a} \cdot \mathbf{c}$.
(5) With vectors **a** and **b** defined as in Question 1 express $\mathbf{a} \times \mathbf{b}$ in terms of $|\mathbf{a}|$, $|\mathbf{b}|$, θ and a vector **n** perpendicular to both **a** and **b**.
(6) What is the scalar or vector value (as appropriate) of
(a) $\mathbf{i} \cdot \mathbf{j}$
(b) $\mathbf{i} \times \mathbf{j}$
(c) $\mathbf{i} \cdot \mathbf{i}$
(d) $\mathbf{i} \times \mathbf{i}$
(7) Does $\mathbf{a} \times \mathbf{b} = \mathbf{b} = \mathbf{a}$?
(8) If $\mathbf{a} \times \mathbf{b}$ is a null vector, what can be said of **a** and **b**?
(9) If $\mathbf{p} = 2\mathbf{i} + \mathbf{j}$ and $\mathbf{q} = \mathbf{i} - 2\mathbf{j}$ find (a) $\mathbf{p} \cdot \mathbf{q}$ (b) $\mathbf{p} \times \mathbf{q}$, assuming they are both three dimensional vectors.
(10) Find the value of the triple scalar product $\mathbf{a} \cdot \mathbf{b} \times \mathbf{c}$ when (a) $\mathbf{a} = \mathbf{b}$ (b) $\mathbf{a} = \mathbf{c}$ (c) $\mathbf{b} = \mathbf{c}$.

(11) What is the relationships between $\mathbf{a} \times (\mathbf{b} \times \mathbf{c})$ and $(\mathbf{b} \times \mathbf{c}) \times \mathbf{a}$?
(12) What can you say about $\mathbf{a} \times (\mathbf{b} \times \mathbf{c}) + \mathbf{b} \times (\mathbf{c} \times \mathbf{a}) + \mathbf{c} \times (\mathbf{a} \times \mathbf{b})$?

SECTION B

Part 8

(1) Are the following explicit or implicit functions
  (a) $y = x^2 + x - 7$
  (b) $xy + 2 = 0$
  (c) $\ln x^2y - \sin xy = e^x$.
(2) Differentiate y^2 w.r.t. x.
(3) If x and y are expressed parametrically in terms of t, obtain an expression for dy/dx in terms of x, y and t and their derivatives.
(4) Differentiate $d/dx\ (\ln f(x))$.
(5) Find $\sin^{-1} 0 \cdot 5$.
(6) Make x the subject of the formula $3 \cos x = 2\theta$.
(7) What is the relationship between dy/dx and dx/dy?
(8) If x and y are expressed parametrically as $x = 2 + t$, $y = t - 5$, what is the equation connecting x and y?

Part 9

(1) If the substitution $x = \sin t$ is used to solve the integral

$$\int \frac{dx}{\sqrt{(1 - x^2)}}$$

what expression replaces (a) dx (b) $\sqrt{(1 - x^2)}$ (c) the substituted integral in its simplest form.
(2) In an integral of the form $\int \sin^m t \cos^n t\, dt$ what substitution is used if
  (a) $m = 2, n = 3$,
  (b) $m = 3, n = 2$,
  (c) $m = 3, n = 3$?
(3) If the substitution $t = \tan x$ is used, state a substitution used for
  (a) $\cos x$ (b) $\sin x$ (c) dx.
(4) What is the value of

$$\int \frac{f^1(x)\, dx}{f(x)}$$

(5) If

$$\frac{5x - 13}{(x + 1)(x - 5)} = \frac{A}{x + 1} + \frac{B}{x - 5}$$

use the cover-up rule to determine values of A and B.
(6) State the formula for integration by parts.
(7) If In represents an integral with a parameter n, what name is attached to a formula expressing In in terms of I_{h-1}.
(8) If $In = \int \sin^n x \, dx$, find I_1.

Part 10

(1) State Maclaurin's theorem.
(2) Write down the power series for e^x.
(3) If $\tan^{-1} x = x - x^3/3 + x^5/5 - \cdots$ find $\tan^{-1} 0.1$ in radians correct to four places of decimals.
(4) State Taylor's theorem.
(5) Use the power series for $\sin x$ to obtain a power series for $\sin 2x$.
(6) State the power series for $\sinh x$ and $\cosh x$.
(7) Evaluate $e^{-0.1}$ correct to three places of decimals.
(8) The power series for $\ln (1 + x)/(1 - x) = 2(x + x^3/3 + x^5/5 + \cdots)$. Is $\ln (1 + x)/(1 - x)$ an odd or an even function?

Part 11

(1) An expression a/b where a and b both tend to infinity is called . . .
(2) Find

$$\lim_{n \to \infty} \frac{n}{n - 2}$$

(3) Find

$$\lim_{x \to 0} \sin ax/x$$

where a is a constant.
(4) Find

$$\lim_{x \to 0} \sin 2x/\sin x$$

using the identity $\sin 2x = 2 \sin x \cos x$.
(5) Use the series form of $\ln (1 + x)$ to determine

$$\lim_{x \to 0} \ln(1 + x)/x.$$

(6) State L'Hospital's rule.
(7) Obtain

$$\lim_{x \to 1} (x^2 - 1) (x - 1).$$

Part 12

(1) The series $1 + \frac{1}{2} + \frac{1}{4} + \frac{1}{8} + \cdots$ has a sum which continually approaches but never reaches 2. This is an example of a $\cdots$ series. The value 2 is called its $\cdots$.
(2) Name the three types of limit of the sum of an infinite series?
(3) A power series $a_0 + a_1 x + a_2x^2 + \cdots$ with all terms positive is convergent if

$$\lim_{n \to \infty} a_{n-1} x/a_n$$

is smaller than what limit?
(4) The series $1 + \frac{1}{2} + \frac{1}{3} + \frac{1}{4} + \cdots$ is called the harmonic series. Is it convergent?
(5) The ratio

$$\lim_{n \to \infty} a_n/a_{n+1}$$

where a_n is the coefficient of x^n in an infinite power series is called rule.
(6) Is the series 1/4, 2/5, 3/6, 4/7 $\cdots$ divergent? Give a reason for your answer.
(7) Is the power series for $e^x - 1 - x$ convergent for all values of x? Justify your answer.
(8) State the type of limit of the series $1 + x + x^2 + x^3 + \cdots$ if (a) $x > 1$ (b) $-1 < x < 1$ (c) $x = -1$ (d) $x < -1$ (e) $x = 1$.

Part 13

(1) If $z = 3x^4y - 2x^2y^2$ obtain z_x and z_y.
(2) Obtain $\delta z/\delta x$ if $z = y^4$.
(3) For a continuous function $z = f(x, y)$ what can you say about zxy and zyz?
(4) Given a function $z = f(x, y)$ state the total differential formula for δz.
(5) If $z = 3x^2y^3\ xy^3$ find the value of (a) z (b) zx (c) zy (d) zxx (e) zyy (f) zxy when $x = y = 1$.
(6) In Example 13.2(b), $u = e^x (\sin y + \cos y)$ satisfied the partial differential equation $\partial^2u/\partial x^2 + \partial^2u/\partial y^2 = 0$. Would you expect $u = e^y (\sin x + \cos x)$ to satisfy it?

(7) If $T = 4l^3/Q$ and l and Q are both in error by 1%, what is the maximum percentage error in T?
(8) If the side of a cube is measured 2% too large what percentage error would be expected in its calculated volume?

SECTION C

Part 14

(1) In the differential equation $\mathrm{d}y/\mathrm{d}x = 12x^2$ which is
(a) the independent variable,
(b) the dependent variable?
(2) Is $y = 4x^3 + A$ an analytical solution to the differential equation in Question 1?
(3) In Question 2 A is referred to as ____.
(4) What is the order and degree of the equation in Question 1?
(5) Line segments which are joined to give a graphical solution of an equation are called ____.
(6) What is the name given to a whole group of curves all of which satisfy the same differential equation?
(7) Which is simpler, a physical system or the mathematical model representing it?
(8) Is the solution in Question 2 a particular solution or a general solution? Given that $y = 1$ when $x = 1$, what is the value of A? Is this an initial or a boundary condition?

Part 15

(1) State the general solution of the differential equation $\mathrm{d}x/\mathrm{d}t = -kx$ if k is a positive constant.
(2) What name is given to a differential equation which can be expressed in the form $\mathrm{d}y/\mathrm{d}x = f(y/x)$?
(3) By means of a standard substitution, any equation of the type described in Question 2 can be changed into the type in which the variables separate. What is the substitution?
(4) By which factor must the left-hand side of the differential equation $\mathrm{d}y/\mathrm{d}x + Py = Q$ be multiplied in order to make it an exact differential?
(5) What is the name given to the factor described in Question 4? If this factor is used to multiply the left-hand side of the differential equation, what does it give when integrated?
(6) Simplify the expression $e^{\int P\,\mathrm{d}x}$ when P is equal to $1/x$.

(7) What is the name given to a family of curves whose members always intersect another family at right angles?

(8) Is $y^2 = x^2 + 9$ the particular solution of the differential equation $\mathrm{d}y/\mathrm{d}x = x/y$, $y = 4$ when $x = 5$?

Part 16

(1) Verify that $y = A\mathrm{e}^{-x}$ and $y = B\mathrm{e}^{-2x}$ both satisfy the differential equation $\mathrm{d}^2y/\mathrm{d}x^2 + 3\,\mathrm{d}y/\mathrm{d}x + 2y = 0$. State a third solution and the principle which allows this to be obtained.

(2) Obtain the auxiliary equation of the differential equation of Question 1.

(3) If the auxiliary equation has repeated roots m, state the general solution of the associated differential equation in terms of its independent variable x.

(4) Repeat Question 3 giving the form of solution if the auxiliary equation has complex roots $m \pm \mathrm{j}n$.

(5) Verify that $y = \mathrm{e}^{-2x}$ satisfies the differential equation

$$\mathrm{d}^2y/\mathrm{d}x^2 + 4\,\mathrm{d}y/\mathrm{d}x + 4y = 0$$

What are the associated initial conditions which would give this particular solution?

(6) Show that if $\mathrm{d}^2x/\mathrm{d}t^2 + \omega^2x = 0$, then $x = A\cos\omega t + B\sin\omega t$. What is the name given to motion, such that its displacement x at time t is described above?

(7) Obtain the general solution of the differential equation $\mathrm{d}^2z/\mathrm{d}t^2 - 4z = 0$.

Part 17

(1) What name is given to a differential equation $\mathrm{d}^2y/\mathrm{d}x^2 + a\,\mathrm{d}y/\mathrm{d}x + by = f(x)$ if $f(x)$ is (a) zero (b) non-zero.

(2) If condition (b) occurs in Question 1, what are the names of the parts which make up the complete solution?

(3) Obtain a solution which satisfies the differential equation $\mathrm{d}^2y/\mathrm{d}x^2 - 16y = 16$. Hence obtain a complete general solution.

(4) Verify that $y = 2x$ satisfies the differential equation $\mathrm{d}^2y/\mathrm{d}x^2 + \mathrm{d}y/\mathrm{d}x = 2$. Is this a complete solution?

(5) Suggest a trial solution to be used when solving the differential equation $\mathrm{d}^2y/\mathrm{d}x^2 + 4y = 8\mathrm{e}^{2x}$.

(6) A suitable trial solution for the particular integral of the equation $\mathrm{d}^2y/\mathrm{d}x^2 + \mathrm{d}y/\mathrm{d}x + 2y = \sin x$ is $V = a\sin x$. Is this correct?

(7) If the equation $\mathrm{d}^2y/\mathrm{d}t^2 - 5\,\mathrm{d}y/\mathrm{d}t + 4y = 2\mathrm{e}^{2x} + 5\sin 2x$ is to be

solved, which trial solution should be used in obtaining the particular integral?

(8) If the complementary function of a differential equation is $Ae^{x} + Be^{-3x}$, the particular integral is $x^2 + 2x + 6$, obtain values of A and B if the associated conditions are $y = dy/dx = 2$ when $x = 0$.

(9) Obtain the complementary function of the differential equation $d^2y/dx^2 + 4\, dy/dx = 8$ and suggest a trial solution that could be used to obtain a particular integral.

(10) Find a if ate^{t} is a particular integral of the differential equation $d^2y/dt^2 - dy/dt = e^{t}$.

Part 18

(1) What name is given to a symbol such as D which can be treated as an algebraic variable or a derivative?

(2) Obtain De^{2x} and $D \sin 3x$.

(3) Calculate $(D^2 + 2D - 4)e^{2x}$.

(4) State the shift rule.

(5) Obtain $(D^4 + D^2 + 1) \sin 2x$.

(6) Calculate

$$\frac{1}{D + 1} e^{2x}$$

(7) Evaluate

$$\frac{1}{D - 2} (\sin x - \cos x)$$

(8) Use the shift rule to obtain

$$\frac{1}{D + 1} 2xe^{-x}$$

(9) Obtain

$$\frac{1}{D + 1} (x^2 + x + 1)$$

(10) Express

$$\frac{1}{D^2 + 4D - 5}$$

in partial fractions. Hence obtain

$$\frac{1}{D^2 + 4D - 5} (5x^2 - 8x - 2)$$

(11) Evaluate

$$\frac{1}{D^2 + 4D + 2} 8$$

(12) Find

$$\frac{1}{D + 2} e^{-2x}$$

(13) Write down the result of

$$\frac{1}{D^2 + 1} (e^x + 6 \sin 2x)$$

(14) Find $\int x e^x \, dx$ using a D operator method.

SECTION D

Part 19

(1) If $x_{n+1} = 2x_n - 3$ and $x_1 = 1.6$ find x_2 and x_3.
(2) Obtain an approximate solution to the equation $x^3 - x^2 - x - 3 = 0$, which is an integer.
(3) What happens to x_{n+1} if $x_n = 1.25$ when using the iterative formula $x_{n+1} = 6x_n - 4x_n{}^2$? Comment.
(4) Can iterates be in any other state than convergent to a root or divergent from it? (Consider the iterative equation $x_{n+1} = -x_n$.)
(5) State the Newton Raphson iterative formula.
(6) What happens to this formula when it is used to solve the equation $f(x) = 0$ if a value is chosen which is a turning point of the curve $y = f(x)$?
(7) Obtain an iterative formula based on the Newton Raphson formula which converges to the cube root of a number N.
(8) How are simultaneous equations which converge rapidly using the Gauss Seidel method to be recognised?
(9) Is the Gaussian elimination technique an iterative method?
(10) Is it possible to obtain the roots of a set of simultaneous equations using Gaussian elimination without including a check?

Part 20

(1) If $f(x)$ is a polynomial of degree 3 which is the first column in its difference table in which every value is zero?
(2) If $f(x)$ is the polynomial $a_1x^n + a_2x^{n-1} + \cdots a_{n+1}$ and a difference

table of step length h is obtained, what is the value of the constant difference? In which column will it be found?

(3) Write out $x^3 - 7x^2 + 15x - 4$ in nested form. Find its value when $x = 1.2$.

(4) Which checks are available when examining the difference tables of an exactly calculated polynomial function $f(x)$?

(5) Name the three parts of a difference table for a rounded function.

(6) What type of difference operator is denoted by (a) Δ (b) δ (c) ∇.

(7) State Newton's forward difference formula expressing f_p in terms of f_0.

(8) State Newton's backward difference formula expressing f_0 in terms of f_p.

(9) Does Simpson's rule require an odd or an even number of tabular values?

(10) Express ∇f_2 in terms of (a) the operator Δ (b) the operator δ.

(11) Obtain $\delta^3 f_{2(1/2)}$ from Table 20.11.

(12) How many tabular values are required when evaluating a definite integral using (a) the three-eighths rule (b) Weddle's rule.

SECTION E

Part 21

(1) Define a probability distribution.

(2) Express the probability of obtaining r successes, $P(r)$, in terms of the binomial parameters n, p and q.

(3) Express the probability of obtaining r successes, $P(r)$, in terms of the Poisson parameter m.

(4) State, in parametric terms, the formula for
 (a) the mean value of a binomial distribution
 (b) the standard deviation of a binomial distribution
 (c) the mean value of a Poisson distribution
 (d) the standard deviation of a Poisson distribution.

(5) What is the area under a standardised normal distribution curve lying between the limits $\pm$ 1.96 ?

(6) State the equation of a standardised normal distribution curve.

(7) State Stirling's formula for $n!$

(8) Give the terms for which AQL and LTPD are abbreviations.

(9) Which are the variables plotted on an operational characteristic curve?

(10) What are the positions of (a) the action lines (b) the warning lines on a quality control chart for means ? Express your result in terms of population parameters μ and σ and the sample size n.

(11) What are the position of (a) the action line (b) the warning line on a quality control chart for attributes ? Express your answer in terms of the sample mean m.

Part 22

(1) With the usual notation state whether
  (a) m is an unbiased estimator of μ
  (b) s^2 is an unbiased estimator of σ^2.

(2) State the best estimate of the population standard deviation from a sample of size n whose mean value is m.

(3) In significance testing, which level is generally taken as being (a) reasonable evidence (b) strong evidence for rejecting a null hypothesis?

(4) In comparing two industrial processes, would a one-tailed test or a two-tailed test be appropriate if we were interested in whether there was a difference rather than an improvement?

(5) State the formula for the 95% confidence limits of a population with a mean value μ and a standard deviation from that mean of σ.

(6) State the formula for the test statistic generally designated z.

(7) How is z distributed?

(8) If normal distributions N_1 and N_2 have parameters μ_1 and σ_1 and μ_2 and σ_2 respectively, state the formula for
  (a) the mean of $N_1 + N_2$,
  (b) the mean of $N_1 - N_2$,
  (c) the standard deviation of $N_1 + N_2$,
  (d) the standard deviation of $N_1 - N_2$.

(9) To which distribution does the t distribution tend as $\gamma \rightarrow \infty$.

(10) State the best estimate of the variance of a population from which two samples of sizes n_1 and n_2 and having parameter (m_1, s_2) have been drawn.

(11) State the standard x^2 formula.

(12) How many degrees of freedom are used in testing whether a distribution follows
  (a) a binomial pattern
  (b) a Poisson pattern
  (c) a normal pattern.

(13) How many degrees of freedom are associated with a 3×3 contingency table?

Part 23

(1) Draw a scatter diagram illustrating positive linear regression.

(2) With the usual notation, state the point of intersection of the y on x and x on y regression lines.

(3) Can the two regression lines (a) coincide (b) lie parallel but non-coincident?

(4) Name the algebraic method normally used to estimate the equations of the regression lines.

(5) Write down the normal equations for determining m and c, the gradient and intercept of the y on x regression line.
(6) Write down the normal equations of determining m' and c', the gradient and intercept of the x on y regression line.
(7) Express the equation of the regression lines in terms of the variance and the co-variance of the variables.
(8) Express the equations of the regression lines in terms of X, Y and r.
(9) State three formulae for r, the coefficient of correlation by product-moment.
(10) In what units is r measured?
(11) State the formula for determining the significance of r using the t distribution.
(12) How many degrees of freedom are associated with t in this case?
(13) Write down the formula for R, the coefficient of rank correlation.

Answers to Self-Assessment Questions

SECTION A

Part 1

(1) $\frac{1}{2}(e^x + e^{-x})$ (2) $\frac{1}{2}(e^x - e^{-x})$ (3) $\sinh x/\cosh x$
(4) 1 (5) (a) Odd (b) Even (c) Odd (6) $\sinh^2 x$
(7) $\text{sech}^2\, x$
(8) (a) $\cosh x = 1 + x^2/2! + x^4/4! + \cdots$
(b) $\sinh x = x + x^3/3! + x^5/5! + \cdots$

Part 2

(1) $5 - j3$ (2) $4 - j6$ (3) j (4) $2 + j3$ (5) $\sqrt{3} + j$
(6) $12(\cos 160° + j \sin 160°), 3(\cos 80° + j \sin 80°)$
(7) $2(\cos 20° + j \sin 20°)$ (8) $r = \sqrt{(x^2 + y^2)}, \theta = \tan^{-1}(y/x)$
(9) $2\sqrt{2}, 45°$

Part 3

(1) $e^{j\theta}$ (2) $\cos\theta - j\sin\theta$ (3) $\cos n\theta + j\sin n\theta$
(4) De Moivre (5) (a) $\cosh x$ (b) $j \sinh x$
(6) $\cos x \cosh y + j \sinh x \sinh y$ (7) $\frac{1}{2}(e^{jx} - e^{-jx})$
(8) $\frac{1}{2}(e^{jx} + e^{-jx})$

Part 4

(1) Elements (2) The minor of that figure (3) Proportional
(4) + (5) (a) No (b) Yes (c) Indeterminate

(6) (a) No (b) Inconsistent (7) Yes, 0
(8) (a) $D_1 = D_2 = D_3 = D_4 = 0$
(b) $D_4 = O, D_1, D_2, D_3$ are all non-zero
(9) Ill-conditioned equations

Part 5
(1) (a) True (b) False (c) False (d) False
(2) (a) True (b) False (3) (a) False (b) False (c) False (d) True
(4) I (5) (a) No (b) $\frac{1}{4}$ (6) $\mathbf{B}^T\mathbf{A}^T$
(7) $(\mathbf{A})^{T-1}$ (8) A vector or a column vector

Part 6

(1) (a) $\mathbf{A}, \mathbf{b}$ (b) $\mathbf{u} = \mathbf{A}^{-1}\mathbf{b}$ (2) $x = 5, y = 1, z = 2$
(3) Augmented matrix (4) Row equivalent matrices (5) 3
(6) (a) n (b) $\frac{1}{2}n(n + 1)$ (7) (a) b (b) y

Part 7

(1) (a) $|\mathbf{a}| \cdot |\mathbf{b}| \cos \theta$ (b) $a_1 a_2 + b_1 b_2 + c_1 c_2$
(2) $\cos^{\alpha} + \cos^2 \beta + \cos^2 \gamma = 1$ (3) $5\mathbf{i} + 4\mathbf{j} - 3\mathbf{k}$
(4) (a) true (b) true (5) $|\mathbf{a}||\mathbf{b}| \sin \theta \, |\mathbf{n}|$
(6) (a) 0 (b) k (c) 1 (d) 0 (7) No, $\mathbf{a} \times \mathbf{b} = -\mathbf{b} \times \mathbf{a}$
(8) If a and/or b are nill or they are parallel (9) (a) 0 (b) $-5\mathbf{k}$
(10) (a), (b) and (c) all zero (11) $\mathbf{a} \times (\mathbf{b} \times \mathbf{c}) = -(\mathbf{b} \times \mathbf{c}) \times \mathbf{a}$
(12) It is a null vector

SECTION B

Part 8

(1) (a) Explicit (b) Implicit (c) Implicit (2) $2y\,dy/dx$
(3) $(dy/dt)/(dx/dt)$ (4) $f^1(x)/f(x)$ (5) $\pi/6$
(6) $x = \cos^{-1}(2\theta/3)$ (7) They are reciprocals (8) $y = x - 7$

Part 9

(1) (a) $\cos t\,dt$ (b) $\cos t$ (c) $\int dt$
(2) (a) $s = \sin t$ (b) $c = \cos t$ (c) Either $s = \sin t$ or $c = \cos t$

(3) (a) $1/(t^2 + 1)$ (b) $t/(t^2 + 1)$ (c) $dt/(t^2 + 1)$ (4) $\ln f(x) + C$
(5) $A = 3, B = 2$ (6) $\int u\, dv = uv - \int v\, du$
(7) A reduction formula (8) $-\cos x + C$

Part 10

(1) $f(x) = f(0) + xf'(0) + xf''(0)/2! + \cdots$
(2) $1 + x + x^2/2! + x^3/3! + \cdots$ (3) 0.0997
(4) $f(x + h) = f(x) + hf'(x) + h^2f''(x)/2! + \cdots$
(5) $2x - \frac{4}{3}x^3 + \frac{4}{15}x^5 - \cdots$
(6) $\sinh x = x + x^3/3! + x^5/5! + \cdots$, $\cosh x = 1 + x^2/! + x^4/4! + \cdots$
(7) 0.905 (8) An odd function

Part 11

(1) An indeterminate form (2) 1 (3) a (4) 2
(5) 1 (6) $\lim_{x\to a} f(x)/g(x) = f'(a)/g'(a)$ (7) 2

Part 12

(1) Convergent; Sum to infinity
(2) Convergent, Divergent, Oscillatory
(3) 1 (4) No (5) Radius of convergence
(6) Yes. Terms are increasing
(7) Yes. The series for e^x is convergent for all values of x and so is any part of it. The series is $e^x - 1 - x$
(8) (a) Divergent (b) Convergent (c) Oscillatory (d) Divergent (e) Divergent

Part 13

(1) $z_x = 12x^3y - 4xy^2$, $z_y = 3x^4 - 4x^2y$ (2) 0
(3) They are equal
(4) $\delta z = z_x\, \delta x + z_y\, \delta y$
(5) (a) 4 (b) 7 (c) 12 (d) 6 (e) 33 (f) 21 (6) Yes
(7) 4% (8) 6%

SECTION C

Part 14

(1) (a) x (b) y (2) Yes (3) An arbitrary constant
(4) Both 1 (5) Isoclines (6) A family of curves

(7) The mathematical model
(8) A general solution; $A = -3$; a boundary condition

Part 15

(1) $x = Ae^{-kt}$ (2) Homogeneous differential equation
(3) $y = vx$ (4) $e^{\int P\,dx}$ (5) Integrating factor, $ye^{\int p\,dx}$
(6) x (7) Its orthogonal trajectories
(8) No, it satisfies the differential equation but not the associated condition.

Part 16

(1) $y = Ae^{-x} + Be^{-2x}$ Principle of superposition
(2) $m^2 + 3m + 2 = 0$ (3) $y = (Ax + B)e^{mx}$
(4) $y = e^{mx}(A \cos nx + B \sin nx)$
(5) $y = 1, dy/dx = -2$ when $x = 0$ (6) Simple harmonic motion
(7) $z = Ae^{4t} + B$

Part 17

(1) (a) Homogeneous (b) Inhomogeneous
(2) Complementary function, particular solution
(3) $-1, Ae^4u2x + Be^{-4x} - 1$ (4) no
(5) ae^{2x} (6) No, it must be $v = a \sin x + b \cos x$
(7) $ae^{2x} + b \sin 2x + c \cos 2x$ (8) $A = -3, B = -1$
(9) $Ae^{-x} + B, v = ax$ (10) 1

Part 18

(1) An operator (2) $2e^{2x}, 3 \cos 3x$ (3) $4e^{2x}$
(4) $f(D)\,Ve^{ax} = e^{ax}f(D + a)V$ (5) $13 \sin 2x$ (6) $\frac{1}{3}e^{2x}$
(7) $(\cos x - 3 \sin x)/5$ (8) $x^2\,e^{-x}$ (9) $x^2 - x + 2$
(10) $\frac{1}{6}[1/(D - 1) - 1/(D + 5)], -x^2$ (11) 4 (12) xe^{-2x}
(13) $\frac{1}{2}e^x - 2 \sin 2x$ (14) $(x - 1)e^x$

SECTION D

Part 19

(1) $x_2 = 0.2, x_3 = -2.6$ (2) 2
(3) $x_{n+1} = x_n = 1.25$ This value is an exact root of the equation

(4) Yes, oscillatory (5) $x_{n+1} = x_n - f(x_n)/f'(x_n)$
(6) At that point $f'(x)$ 0 and the formula cannot be used.
(7) $x_{n+1} = \frac{1}{3}(2x_n - N/x_n^2)$
(8) The variables have dominant coefficients
(9) No (10) Yes, but it is unwise to do so

Part 20

(1) The fourth (2) $n!\, a\, h^n$; column number n.
(3) $-4 + x(15 + x(-7 + x))$, 5.648
(4) The constant difference and the value of $f(x)$ for some simple value of x
(5) The regular columns, the transitional column and the irregular columns
(6) (a) Forward difference (b) Central difference
(c) Backward difference
(7) $f_p = (f_0 + p\Delta f_0 + p(p-1)\Delta^2 f_0/2! + \cdots)$
(8) $f_0 = f_p - p\Delta f_p + p(p-1)\Delta^2 fp/z! - \cdots$
(9) An odd number (10) (a) Δf_1 (b) $\delta f_{1^1/_2}$ (11) 0.0005
(12) (a) $3k + 1$ (b) $6k + 1$ In each case k is an integer.

SECTION E

Part 21

(2) $\binom{n}{r} q^{n-r} p^r$ (3) $e^{-m} m^r/r!$
(4) (a) np (b) $\sqrt{(npq)}$ (c) m (d) $\sqrt{m}$ (5) 0.95
(6) $y = e^{-1/2x^2}/\sqrt{(2\pi)}$ (7) $n! = (n/e)^n \sqrt{(2\pi n)}$
(8) Acceptance quality level; lot tolerance percentage defects
(9) Probability of acceptance against proportion of defectives
(10) (a) $\mu \pm 3.09\sigma/\sqrt{n}$ (b) $\mu \pm 1.96\sigma/\sqrt{n}$
(11) $m \pm 3.09\sqrt{m}$ (b) $m \pm 1.96\sqrt{m}$

Part 22

(1) (a) Yes (b) No (2) $\sqrt{\Sigma(x - m)^2/(n - 1)}$
(3) (a) 5% (b) 1% (4) Two-tailed (5) $\mu \pm 1.96\sigma$
(6) $z = (x - \mu)/(\sigma/n)$ (7) Normally
(8) (a) $\mu_1 + \mu_2$ (b) $\mu_1 - \mu_2$ (c) $\sqrt{(\sigma_1^2 + \sigma_2^2)}$
(d) $\sqrt{(\sigma_1^2 + \sigma_2^2)}$ (9) A normal distribution

(10) $(n_1 s_1{}^2 + n_2 s_2{}^2)/(n_1 + n_2)$ (11) $\chi^2 = \Sigma(O_i - E_i)$
(12) (a) $n - 2$ (b) $n - 2$ (c) $n - 3$ (13) 4

Part 23

(2) $(\bar{x}, \bar{y})$ (3) (a) Yes (b) No (4) Method of least squares
(5) $\Sigma y = m\Sigma x + nc$, $\Sigma xy = m\Sigma x^2 + cx$
(6) $\Sigma x = m'\Sigma y + nc!$, $\Sigma xy = m'\Sigma y^2 + c'\Sigma y$
(7) $(y - \bar{y}) = s_{xy}(x - \bar{x})/s_x{}^2$, $(x - \bar{x}) = x_{xy}(y - \bar{y})/s_y{}^2$
(8) $Y = rX, X = rY$
(9) $s_{xy}/s_x s_y$, $\Sigma XY\surd(\Sigma X^2\ \Sigma Y^2)$,
$(\Sigma xy - \Sigma x \Sigma y/n)/\surd(\Sigma x^2 - (\Sigma x)^2/n\surd(\Sigma y^2 - (\Sigma y)^2/n$
(10) It is dimensionless (11) $t = r\surd(n - 2)/\surd(1 - r^2)$
(12) $n - 2$ (13) $R + 1 - 6\Sigma d^2/n(n^2 - 1)$

Answers to Unworked Examples

SECTION A

Unworked Examples 1

(1) (a) 1, 0 (b) $1 + x^2/2! + x^4/4! + \cdots$ (c) $x + x^3/3! + x^5/5! + \cdots$
(2) $y = \sinh x$ and $y = \cosh x$ (6) (a) 1.54, 1.18 (b) 1.54, −1.18
(12) (a) Odd (b) Odd (c) Odd (d) Odd (e) Odd (13) −2.178
(15) (a) $\sin(\tan^{-1} x) + C$ (b) $\tanh(\sinh^{-1} x)$; $+ C$, Both $\sin(\tan^{-1} x)$ and $\tanh(\sinh^{-1} x)$ are equal to $x/\sqrt{(1 + x^2)}$
(17) $x = 0.39$

Unworked Examples 2

(3) $\sinh(x \pm y) = \sinh x \cosh y \pm \cosh x \sinh y$
$\cosh(x \pm y) = \cosh x \cosh y \pm \sinh x \sinh y$
(4) $\tanh 2x = 2 \tanh x/(1 + \tanh^2 x)$, 0.8 (6) 0.367
(7) $x = 0$ or $x = 1.39$ (8) 0.916 (9) 0.603 (10) $\ln 3 \approx 1.10$
(11) $x = \ln 2 \approx 0.693$; No, only when $x \to \infty$ does $\tanh x \to 1$

Unworked Examples 3

(1) $x = -2 \pm \mathrm{j}$ (2) (a) 2 (b) −4 (c) 2 + j4 (4) 8 + j2
(5) 6 + j3 (6) 2 + j6 (7) 1 + j2 (10) $a = \pm 3$, $b = \mp 1$
(11) 2 (12) 3 − j2 (13) −7 + j24

Unworked Examples 4

(1) (a) 7 + j2 (b) −3 − j4 (c) 3 + j4 (2) 7 − j4
(3) An equilateral triangle

(4) Either $1 + j5$ or $2 - j2$; in both cases AB $= 5$
(5) (a) $\sqrt{2}(\cos 315° + j \sin 315°)$ (b) $10(\cos 36°52' + \sin 36°52')$
(6) $a = 1.597$, $b = 1.204$
(8) 2.236, 5. Yes, we have $\arg(z_1) = 333°25'$ $\arg(z_1^2) = 306°52' = 2 \times 333°26' - 360°$ (to fetch it into the range 0° to 360°)
(10) $\pm(4 + j3)$ (11) $10(\cos 2\pi/3 + j \sin 2\pi/3)$
(12) $(1 + j)/\sqrt{2}$ (13) $-1 + j3$

Unworked Examples 5

(1) $1.551 + j1.551$ (2) $1.5 + j0.87$
(4) $j(\pi + 2\pi n)$ where n is any integer. Here $x = \ln(-1)$
(5) $5(\cos 53°8' + j \sin 53°8')$ (a) $-7 + j24$, $-117 + j44$
(b) $25(\cos 106°16' + j \sin 106°16')$, $125(\cos 159°24' + j \sin 159°24')$
(6) $\cos 2n\pi + j \sin 2n\pi$ ($n = 0, \frac{1}{5}, \frac{2}{5}, \frac{3}{5}$, or $\frac{4}{5}$) (7) $512 - j512\sqrt{3}$
(8) $\cos\theta - j \sin\theta$ (11) 0.9397, −0.1736, −0.7660

Unworked Examples 6

(2) $x = \cos\theta \pm j \sin\theta$
(3) $(e^{j\theta} - e^{-j\theta})/j(e^{j\theta} + e^{-j\theta})$
(5) $\frac{1}{8}(\cos 4\theta - 4\cos 2\theta + 3)$ (6) $0, \pi/6, 5\pi/6, 7\pi/6, 11\pi/6, 2\pi$
(7) -0.368 (8) $\frac{1}{16}(\cos 5\theta + 5\cos 3\theta + 10\cos\theta)$
(11) Real part $\cos x \cosh y$; imaginary part $-\sin x \sinh y$

Unworked Examples 7

(1) −2, 0 (2) 3 (3) $x = 5$, $y = -1$ (4) $a = 1$, $b = 3$
(6) 19 (7) 2 (8) $p = q = r = 1$ (9) 4, 16 (11) 0
(12) −51 (14) 1 (15) 1
(16) (a) $x = 1, y = 1, z = 2$ (b) $l = 3, m = -1, n = 1$ (17) $x = 1$

Unworked Examples 8

(2) $a:b:c = 1:4:1$ (3) $x = y = z = 0$ (4) 5
(5) (a) $x = 1$, $y = 0$, $z = 2$ (6) $x = 17.25$, $y = -0.93$
(7) (a) (8) (a) $x = 100$, $y = 298$ (b) $x = -48.26$, $y = -148.26$
(c) $x = 25.69$, $y = 74.31$
(9) $x = y = 1$. They are well conditioned

Unworked Examples 9

(1) (a) No (b) Yes (c) Yes (d) No (e) Yes (2) $\begin{pmatrix} -2 \\ 8 \end{pmatrix}, \begin{pmatrix} 3 \\ 5 \end{pmatrix}$ (3) 6, 3

(4) $\begin{pmatrix} 1 \\ 17 \end{pmatrix}$ (5) $\begin{pmatrix} 5 & -1 \\ 5 & 2 \end{pmatrix}, \begin{pmatrix} 13 & -2 \\ 11 & 5 \end{pmatrix}$ (7) Yes (8) Yes

(9) $\begin{pmatrix} 6 & 12 & 18 \\ 17 & -2 & -5 \end{pmatrix}$ (13) $\mathbf{A}^2 = \begin{pmatrix} 7 & 10 \\ 15 & 22 \end{pmatrix}$ det $\mathbf{A} = -2$, det $\mathbf{A}^2 = 4$

(14) $\dfrac{1}{ad - bc}\begin{pmatrix} d & -b \\ -c & a \end{pmatrix}$ (16) 10 (17) $\frac{1}{5}\begin{pmatrix} 4 & -17 & 7 \\ -1 & 8 & -3 \\ -6 & 23 & -8 \end{pmatrix}$

(18) $\mathbf{AB} = \begin{pmatrix} 0 & 0 \\ 0 & 0 \end{pmatrix}$, $\mathbf{BA} = \begin{pmatrix} 10 & 20 \\ -5 & -10 \end{pmatrix}$ det $\mathbf{AB}$ = det $\mathbf{BA} = 0$ (19) $\frac{1}{2}\mathbf{B}$

Unworked Examples 10

(1) $l = 3, m = 4, n = -11$ (2) $a = 1.5, b = 7.5, c = 4$
(4) $p = -5, q = 6, r = -7, s = 8$
(5) (a) $x = 4, y = 2, z = 3$ (b) $x = 2, y = 1, z = -3$
(c) $x = 5, y = 2, z = 4$
(8) $a = 2, b = 3, c = 4$ (10) $a = 4, b = 2, c = 3, d = -5, e = -4$

(11) $x = 4, y = -2, z = -5$ (12) $\begin{pmatrix} 2 & 0 & 0 \\ 1 & -4 & 0 \\ -2 & -3 & -11 \end{pmatrix}\begin{pmatrix} 1 & 2 & -3 \\ 0 & 1 & -3 \\ 0 & 0 & 1 \end{pmatrix}$

(13) $p = -2, q = \frac{1}{6}, r = \frac{2}{3}$ (18) $x = 0, y = \frac{1}{2}, z = \frac{1}{2}$

Unworked Examples 11

(2) They are equidirectional (6) $|\mathbf{v}| = 9.54, \alpha = 33°$
(7) $5\mathbf{i} - 9\mathbf{j} - 7\mathbf{k}, \mathbf{i} + \mathbf{j} - 3\mathbf{k}$ (11) (a) $\frac{2}{7}, -\frac{3}{7}, \frac{6}{7}$ (b) $\frac{3}{5}, 0, -\frac{4}{5}$
(12) (a) $\cos\alpha, \cos\beta, \cos\gamma$ (b) $-\cos\alpha, -\cos\beta, -\cos\gamma$
(13) (a) $7\mathbf{i} - \mathbf{j} - 11\mathbf{k}$ (b) $-7\mathbf{i} + 5\mathbf{j} - 4\mathbf{k}$ (c) $\sqrt{29}$ (d) $\sqrt{85}$

Unworked Examples 12

(1) $9, \cos^{-1}\dfrac{9}{\sqrt{231}}$, 53°41′ (2) 0.4 J (7) $5\mathbf{i} + 5\mathbf{j} - 10\mathbf{k}$

(9) $\pm(3\mathbf{i} + 2\mathbf{j} - \mathbf{k})$ (11) $2\sqrt{2}$ cm s^{-1} in a direction $\mathbf{i} - \mathbf{j}$
(14) $\mathbf{a} = \mathbf{i} - \mathbf{j} - 2\mathbf{k}, \mathbf{b} = 2\mathbf{i} + 3\mathbf{j} + 3\mathbf{k}, \mathbf{c} = 3\mathbf{i} + 4\mathbf{j} + 5\mathbf{k}$, 1

(18) 6 (19) When the result is expressed in determinant form, the rows are linearly dependent
(21) (a) $11(-\mathbf{i} + 2\mathbf{j} - \mathbf{k})$ (b) $11(2\mathbf{i} - \mathbf{j} - \mathbf{k})$ (c) $11(-\mathbf{i} - \mathbf{j} + 2\mathbf{k})$

SECTION B

Unworked Examples 1

(1) $-x/4y$ (2) At (0, 1)
(4) $(\pm\sqrt{3}, 1)$ and $(\pm\sqrt{3}, -1)$ Gradients of the tangents are $\mp 1/\sqrt{3}$ and $\pm\sqrt{3}$ respectively
(6) $\mathrm{d}y/\mathrm{d}t = \cos t$, $\mathrm{d}x/\mathrm{d}t = -2\sin t$ (8) $-\frac{1}{2}\sqrt{3}$
(11) $$\frac{e^{2x}}{2\sqrt{x}(x^2 + 2x + 4)} + \frac{2\sqrt{x}\, e^{2x}}{x^2 + 2x + 4} - \frac{2\sqrt{x}(x + 1)e^{2x}}{(x^2 + 2x + 4)^2}$$
(14) 3 (16) $35\sqrt{6}/12 \approx 7.14$ (17) $3/\sqrt{(1 - 9x^2)}$
(19) 0.12 (20) $\mathrm{d}y/\mathrm{d}x = -1/\sqrt{(1 - x^2)}$; $\mathrm{d}x/\mathrm{d}y = -\sqrt{(1 - x^2)}$
(22) $1/2\sqrt{x}(1 + x)$
(23) $\mathrm{d}/\mathrm{d}x(\tan^{-1} x) = 1/(1 + x^2)$; $\mathrm{d}/\mathrm{d}x(\tanh^{-1} x) = 1/(1 - x^2)$
(25) $(\sinh x - \sqrt{(x^2 + 1)} \cosh x \sinh^{-1} x)/(\sqrt{(x^2 + 1)} \sinh^2 x)$
(26) $\mathrm{d}/\mathrm{d}x(\coth^{-1} x) = \mathrm{d}/\mathrm{d}x\,(\tan^{-1} x) = 1/(1 - x^2)$
However $\coth^{-1} x$ exists only if $x > 1$ while $\tan^{-1} x$ exists only if $x < 1$ (27) $\mathrm{d}y/\mathrm{d}x = \sec x$

Unworked Examples 2

(2) $\sin 2(\sin^{-1} \frac{1}{2}x) + 2\sin^{1} \frac{1}{2}x + C +$ (3) 98 (4) 2
(5) $C - \sqrt{(1 - x^2)}$ (7) Both are equal to 0.7616
(8) $\sin^5 t + C$
(11) (a) $2/(x - 3) - 1(x - 1)$ (b) $2/(x + 1) - 3/(x + 2) + 4/(x + 3)$
(15) $$\frac{1}{x + 1} - \frac{3}{(x + 1)^2} + \frac{3}{(x + 1)^3} - \frac{1}{(x + 1)^4}$$
(16) $A = 8$, $B = 4$, $C = -11$, 0.098 (17) $\ln 4.5 = 1.504$

Unworked Examples 3

(2) 0.5 (4) $-2x^2 \cos 2x + 2x \sin 2x + \cos 2x + C$
(6) $x \sin^{-1} x + \sqrt{(1 - x^2)} + C$
(9) $I_n = x^n \sin x + nx^{n-1} \cos x - n(n - 1)I_{n-2}$,
$x^4 \sin x + 4x^3 \cos x - 12x^2 \sin x - 24x \cos x + 24 \sin x + C$

(10) $I_4 = e^x(x^4 - 4x^3 + 12x^2 + 24x + 24)$
(11) $I_m = -1/m(\sin^{m-1} x \cos x - (m-1) I_{m-2})$ (12) 1/60

Unworked Examples 4

(1) $x + x^3/3! + x^5/5! + \cdots$ (2) $1 + \frac{1}{2}x^2 + \frac{5}{24}x^4$
(3) $1 - x + x^2/2! - x^3/3! + \cdots$
(4) $1 + nx + n(n-1)x^2/2! + n(n-1)(n-2)x^3/3! + \cdots$ 0.998
(5) 0.693, 1.099 (6) $\frac{1}{2}x^2 + \frac{1}{12}x^4 + \frac{1}{45}x^6 + \cdots$
(7) $x - \frac{1}{6}x^3 + \frac{3}{40}x^5 - \frac{5}{112}x^7$, 0.481
(8) $a = 1, b = \frac{1}{3}, c = \frac{2}{15}$
(9) $x^n + n\,x^{n-1} + n(n-1)\,x^{n-2}/2! + \cdots$ (10) 47°44′
(11) (a) 0.996 (b) 1.036
(12) $\sin x + y\cos x - y^2 \sin x/2! - y^3 \cos x/3! + \cdots$

Unworked Examples 5

(1) 0 (3) a/c (4) -1 (5) 1.5 (6) 3 (7) 3
(8) 1 (9) 1, p (10) $0, \frac{1}{6}, -\frac{1}{6}$ (11) 0

Unworked Examples 6

(1) (a) $-1 \le x \le 1$ (b) $-2 < x < 2$ (c) All values of x
(2) Convergent (3) 0.0998, 0.9950 (4) Its terms are increasing
(5) $1/(1-x), 1/(1-x)^2$ $-1 < x < 1$
(6) (a) Yes (b) No, the series becomes for (a) $1 - \frac{1}{2} + \frac{1}{3} - \frac{1}{4} + \cdots$ which is convergent and for (b) $-1 - \frac{1}{2} - \frac{1}{3} - \frac{1}{4} +$ which is the harmonic series with all signs changed and therefore divergent.

Unworked Examples 7

(1) $3y - 2xy + 4/y^3, 3x - x^2 - 12x/y^4$
(2) (a) $2/x - ye^x, 1/y - e^x$ (b) $3x^2y + 2xy^2 + 3y^5, x^3 + 2x^2y + 15xy^4$
(5) $x = 1, y = -2, u = 1$
(7) (a) $u_x = 3y^3 - 2y\sin 2xy, u_{xx} = -4y^2\cos 2xy,$
$u_y = 9xy^2 - 2x\sin 2xy, u_{yy} = 18xy - 4x^2\cos 2xy,$
$u_{xy} = u_{yx} = 9y^2 - 2\sin 2xy - 4xy\cos 2xy$
(b) $u_x = 3ye^{3xy} + 2xy^2\cos x^2y^2,$
$u_{xx} = 9y^2e^{3xy} + 2y^2\cos x^2y^2 - 4x^2y^4\sin x^2y^2,$
$u_y = 3xe^{3xy} + 2x^2y\cos x^2y^2,$
$u_{yy} = 9x^2e^{3xy} + 2x^2\cos x^2y^2 - 4x^4y^2\sin x^2y^2$
$u_{xy} = U_{yx} = 3e^{xy} + 9xye^{xy} - 4x^2y^2\sin x^2y^2 - 4xy\cos x^2y^2$

(9) They are both equal to $9x^2 + 2/x^2 + 6y^2e^x$
(10) $z_{xx} = 0$, $z_{xy} = -9$ (13) 340.2 (14) 0.0005 N cm^{-2}
(15) 16.95 cm (16) A 3% underestimate (18) 59°22′ (21) $x = 8$, $\theta = 60°$

SECTION C

Unworked Examples 1

(2) (a) x (b) y (c) 1 (d) 1 (3) $y = 2x^3 + 2x^2 - 3x$
(4) $y = \frac{1}{2}x^2 + A$ (5) $y = \frac{1}{5}x^5 - x + A$, $A = 2$ (6) $dy/dx = y/x$
(7) Order 1, degree 2 $\pm\ y = \sin^{-1} x + A$
(12) Atmospheric pressure is neligible (13) g/k

Unworked Examples 2

(1) $y^2 = 2x^3 + x^2 - 2$ (2) $y = Ae^x$ (3) $y^2 = x^3 + A$
(4) The radius of the circle (5) $\sin t = \ln(\frac{1}{2}y)$ (9) $y = 2x \ln x$
(10) $y = x \ln Ay$ or $y = Be^{y/x}$, A and B being arbitrary constants
(11) $(x - y + 6)^4 = 54(2x + y + 6)$ (13) $y = x \sin^{-1} x$
(14) $(e^x + A)/x^2$ (15) $y = 2e^{3x} - x$ (16) $W = 5 - 3e^{-0.004t}$, 5 kg
(17) $v = (u + g/k)e^{-kt} - g/k$ (18) $y = (x + 1)^2 - 16/(x + 1)^2$

Unworked Examples 3

(1) $e^x(x + 1)$ (3) $y = Ae^{-x} + Be^{-5x}$ (4) $z = e^{3t} + 1$
(5) $y = 4e^{-2x}(2 - e^{-2x})$ (6) $y = Ae^{2x} + Be^{-2x}$
(8) $y = At + B$ (11) $x = te^{-4t}$ (13) $d^2y/dx^2 - 4dy/dx + 4y = 0$
(14) $y = e^x \sin x$ (15) $d^2y/dt^2 - 4dy/dt + 13y = 0$
(16) $x = e^{-kt}(A \cos nt + B \sin nt)$ where $n^2 = \omega^2 - k^2$
(17) (a) $x = Ae^{25t} + B$ (b) $x = Ae^{5t} + Be^{-5t}$
(c) $x = A \cos 5t + B \sin 5t$
(19) $T = 2\pi\sqrt{(\rho l/g)}$

Unworked Examples 4

(1) 2
(2) (a) $z = Ae^t + Be^{3t} + 20$ (b) $z = (At + B)e^{2t} + 15$
(c) $z = e^{2t}(A \cos t + B \sin t) + 12$
(4) $y = A + Be^{-2x} + 2x$ (5) $y = \frac{1}{2}(5e^{2x} + 3)$ (6) $e^{2x} \sin x$
(7) $x = a \cos \sqrt{(st/m)} + mg(1 - \cos \sqrt{(st/m)})/s.$

(8) (a) $-3e^t$ (b) $\cos 2t + 2 \sin 2t$ (c) $-t^2 - t - 1$ (d) $e^t \sin t$
(9) $y = \cosh^{-1} x = \ln (x + \sqrt{(x^2 + 1)})$
(10) $y = 12 - 2e^{5x} + \cos x + \sin x$ (11) $e^x(x^2 + x + 1)$
(12) $y = 3x^2/2 + 3x/2 + 1$ (13) $x = -\frac{1}{4}te^{-4t}$ (14) $\frac{1}{2}t^2e^t$
(15) $a = b = 2$ (16) $y = 2e^{2t} - e^{-t} - \frac{1}{3}te^{-t}$
(17) $y = A\cos x + B \sin x$. Trial solution $V = ax \sin x + bx \cos x$
PI $V = -x \cos x$

Unworked Examples 5

(1) $e^{2x}(3x + 1)$ (3) $14 - 8x - 6e^{2x}$ (6) $16e^x$ (8) $4x^3e^{-x}$
(9) $-\frac{1}{10}e^{2x}(\cos 2x - 2 \sin 2x)$ (10) $4t^3e^t$ (11) $e^x + e^{2x} + e^{-3x}$
(12) $e^{-x}(A \cos x + B \sin x) + \sin x$ (13) $ex + (\cos x - 2 \sin x)/5$
(14) e^x (15) $z = e^{2t} + 2e^{-t} - e^t(\cos t + 3 \sin t)$
(16) $3x - 2$ and $\frac{1}{3}x - \frac{10}{9}$ (18) $3e^{4x} - 10e^x + 2x + 1$
(19) $3t^2 - 6t + 7$ (21) $2x^3 + 3x$ (22) $t \sin 2t - 3t \cos 2t$
(23) (a) $y = (t^2 + 3t + 1)e^{2t}$
(b) $y = A \cos \omega t + B \sin \omega t - t \cos \omega t/2\omega$
(24) $z = Ae^t + Be^{-t} + t^2e^t$ (27) $e^{-x}(4 \sin 4x - \cos 4x) + C$

SECTION D

Unworked Examples 1

(2) Approximately 4 and two roots close to -2
(5) $x_{n+1} = 1/4x_n^4 - 2$, $-1.98\,388$ (6) 1.1194
(7) $f'(x_1) = 0$. The method breaks down (9) 1.027 radians $\approx 58°50'$
(10) 1.55 715 (11) $x_{n+1} = 2x_n - Nx_n^2$, 0.02 083 (12) 1.585
(13) 0.443 (14) $x = 0.143$, $y = 0.429$, $z = 0.714$
(15) Take (4) = (3), (5) = (2) − (3), (6) = (1) − (2), $x = 1.14$,
$y = 0.33$, $z = 0.52$
(16) $l = 0.094$, $m = -0.076$, $n = -0.069$
(17) $a = 0.7$, $b = 0.2$, $c = 1.4$ (18) $x = 4.333$, $y = -7.429$, $z = 6.222$
(19) $x = 1.82$, $y = -3.74$, $z = 1.05$

Unworked Examples 2

(1) 4, 18.4336 (2) $7 + x(-5 + x(0 + x))$, 9.784
(3) (a) Yes (b) Yes, value 0.096
(4) $f(0) = -1.0000$, $f(0.1) = -0.9797$, $f(0.2) = -0.9152$,
$f(0.3) = -0.7957$, $f(0.4) = -0.6032$, $f(0.8) = 1.5088$

(5) $-1 + x(0 + x(2 + x(0 + x(3))))$
(6) Third difference column, 0.096 (7) $-3 + x(-3 + x(-3 + x(3)))$
(8) First four differences are regular, the fifth is transitional—see Table 20.15
(9) $f(0.19) = 0.6554$
(10) All total zero. Errors increase one value in the next difference column and decrease the next value by an equal amount
(12) Three (13) 0.7083 (4 pl. dec.) (14) (a) 0.29 237 (b) 0.59 482
(15) Yes, when interpolating near the foot of the table. (16) 1.7333
(17) $f'(0.4) = 6.68$, $f''(0.4) = 10.4$ Integral has a value 2.083
(18) 0.822 (19) 2.1119 (20) $f'(3) = 0.1852$; Integral = 10.1983
(21) 172/3

SECTION E

Unworked Examples 1

(1) (a) 0.377 (b) 0.265 (2) 0.022 (3) 0.046%
(4) (a) 0.9938 (b) 0.1587 (5) 0.19 (6) 0.034 (7) 0.0156 1.
(8) (a) 8.6% (b) 3.5%, 1.2% (9) 0.31 (10) 0.06
(11) $P(0) = 0.091$, $P(1) = 0.218$, $P(2) = 0.262$, $P(3) = 0.209$
(12) 0.18 (13) 0.026
(14) Binomial probabilities 0.015, 0.066, 0.142 Poisson probabilities 0.018, 0.073, 0.147
(15) 0.29 (16) $1/\sqrt{(72\pi)} = 0.066$, 0.066

Unworked Examples 2

(1) Producer's risk 0.008, consumer's risk 0.028
(2) (a) producer's risk 0.0046, consumer's risk 0.011
(b) producer's risk 0.034, consumer's risk 0.005
(3) (a) Scheme B (b) Scheme A (4) 9
(5) Producer's risk 0.064, consumer's risk 0.22 (6) 9.6, 9.5
(7) 0.39 (8) Outer lines at 0.814, 0.786 mm, inner lines at 0.809, 0.791 mm
(9) Mean of sample means 50.008 ohms, mean range 0.34 ohm. For the means chart, the inner lines lie at 50.136 and 49.880 ohm and the outer lines at 50.210 and 49.806 ohm. For the range chart the lines are at 0.615 and 0.802 ohm
(10) Inner lines: 514.7, 485.3 h outer lines at 523,2, 476.8 h
(11) Inner lines: 1599.1, 1568.9 N; outer lines: 1607.8, 1560.2 N. The value 1562 makes it unlikely that the process is still under control

(12) (a) 0.192 (b) 0.0012 (13) (a) 0.025 (b) Too small to measure
(14) 0.071 (15) 38 (16) 46
(17) 0.5; inner line at 1.89, outer line at 2.68
(18) Inner line: 2.96, outer line: 4.09, $P(0) = 0.368$, $P(1) = 0.368$, $P(2) = 0.184$, $P(3) = 0.061$, $P(4) = 0.015$, $P(5) = 0.003$; $x_1 = 2$, $x_2 = 5$
(19) Inner line: 2.34, outer line: 3.29, outer line; 3.29, unchanged
(20) Inner line: 1.07 outer line: 1.58 (21) $m + 1.28\sqrt{m}$

Unworked Examples 3

(1) 20.2, 1.43 (2) 17 cm, 3.21 cm (3) 1.63, 65
(4) $\mu = 10.05$, $\sigma = 1.02$ (5) See previous answer
(6) Expected number to be in stock is 15. The probability of carrying 10 or fewer is 0.05 making the claim doubtful (or the stockist untypical)
(7) The probability of two or more failures is about 0.09. The result is non-significant
(8) No. In a two-tailed test this corresponds to a probability of 0.125
(9) (a) 59 (b) 63. A one-tailed test has been used
(10) (a) $z = -1$ (non-significant)
(b) $z = -2.24$ (significant at the 5% level).
This only indicates that there is reasonable evidence of improvement of time reduction. It may or may not justify installing new equipment. This depends upon costs and similar factors
(12) 95% limits: 2.51 ± 0.004 cm; 99% limits: 2.51 ± 0.005 cm
(13) No
(14) $z = 4.89$. The result is highly significant and it is extremely unlikely that the plant's claim is valid
(15) $z = 2.29$. This requires a one-tailed test and is almost significant at the 1% level
(16) Mean time 31 min; standard deviation 10.05 min
(17) 95% limits: 18.8, 19.6 kg; 99% limits: 18.7, 19.7 kg
(18) 0.03, 0.05 cm

Unworked Examples 4

(1) $m = 34.67$, $s = 1.31$, $t = 0.84$ with $\nu = 11$. This value is not significant using a two-tailed test and there is no reason to suppose that the labelling is inaccurate
(2) $m = 10.6\%$, $s = 0.19$, $t = 0.56$ with $\nu = 4$. Non-significant.
(3) $m = 15.035$ cm, $s = 0.033$, $t = 4.57$ with $\nu = 19$. Highly significant. The machine needs re-setting
(4) (a) $t = 0$. Non-significant

(5) m =0.2, s = 0.29, t = 1.95 with $\nu = 7$. Non-significant using a two-tailed test

(6) 12.15 ± 0.026 mm

(7) $s_b = 11.2, t = 2.99$ with $\nu = 38$. The difference is highly significant

(8) $s_b = 4.11, t = 1.84$ with $\nu = 38$. A one-tailed test is significant at the 5% level and the figures appear to justify the claim

(9) No. They confirm the result more strongly

(10) $\chi^2 = 2.7$ with $\nu = 3$. The value is non-significant

(11) (a) For a binomial distribution the expected frequencies are 15, 37, 33, 13 and 2. $\chi^2 = 0.45$, combining the final classes with $\nu = 2$. The distribution is close to being binomial

(b) allowing for rounding errors the frequencies are 23, 34, 25, 13 and 5. $\chi^2 = 4.68$ with $\nu = 3$. This is non-significant and the distribution could be Poisson

(12) Assuming an even chance of a watch running fast or slow, $\chi^2 = 7.2$ with $\nu = 1$, a very significant result. The figures are surprising or the assumption above is invalid

(13) $m = 951.6$, $s = 16.5$. The expected frequencies are 2, 16, 34, 20 and 3. Taking 3 classes $\chi^2 = 1.01$. Technically $\nu = 0$ but even taking $\nu = 1$, χ^2 is non-significant

(14) χ^2 is highly significant. Factory B is exceptionally productive

(15) $m = 0.26$. Expected frequencies: 39, 10, 1, 0 and 0. Taking two classes $\chi^2 = 1.15$. If $\nu = 1$ (see solution to Question 13), this is non-significant

(16) $n = 4$, $p = 0.065$, $q = 0.935$. Expected frequencies 38, 11, 1, 0 and 0. Using the result of Question 15, this is non-significant

(18) $\chi^2 = 0.17$. Not suspicious (19) $\chi^2 = 3.13$ with $\nu = 4$. Non-significant

(20) Treating this as a 2 × 3 contingency table $\chi^2 = 1.19$ with $\nu = 2$. This value is non-significant and there is no reason to believe that Mr. B is a better instructor

Unworked Examples 5

(1) $y = -0.22x + 31.54$

(2) (a) (100, 57), (120, 68), (130, 70) (b) (100, 05), (110, 64), (125, 70)

(3) $y = 0.45x + 12.18$, $x = 1.32y + 31.05$

(4) $y = 0.007x + 1.48$, $x = 12.5y - 17.3$

(5) $y = (y_2 - y_1)x/(x_2 - x_1) + (x_2y_1 - x_1y_2)/(x_2 - x_1)$

(6) $y = 0$, $x = 0$ These are the co-ordinate axes

(7) $x - 10 = (y - 20)/3$, $(y - 20) = 3(x - 10)/16$ (8) 0.40

(9) $y = -0.21x + 8.48$, $x = -3.46y + 34.84$ (11) −0.58

(12) All leave the correlation coefficient unchanged

(14) $y = 0.44x - 0.86$, 0.813 (15) 0.137

(16) $r = -0.98$, $t = 9.85$ with $\nu = 4$. This is significant at the 1% level

(17) (a) $t = 1.23$ with $\nu = 3$. Non-significant

(b) $t = 2.42$ with $\nu = 3$. Non-significant

(c) $t = 0.28$ with $\nu = 4$. Non-significant

(18) (a) 0.28 (b) 0.12 (c) 0.035 (19) 0.7

(20) (a) 0.821 (b) 0.2

Index